OXFORD BIOGEOGRAPHY SERIES

Editors: A. Hallam, B. R. Rosen, and T. C. Whitmore

OXFORD BIOGEOGRAPHY SERIES

Editors

A. Hallam, Department of Geological Sciences, University of Birmingham.
B. R. Rosen, Department of Palaeontology, Natural History Museum, London.
T. C. Whitmore, Department of Geography, University of Cambridge.

The aim of the series is to publish a range of titles demonstrating the breadth of biogeography, from the biological to the geological ends of the spectrum. The subject is being revolutionized by plate tectonics, molecular phylogeny, and population models, vicariance, cladistics, and spatial classification analyses. For both specialist and non-specialist, the Oxford Biogeography Series will provide dynamic syntheses of new developments.

1. T. C. Whitmore (ed.): *Wallace's line and plate tectonics*
2. Christopher J. Humphries and Lynne R. Parenti: *Cladistic biogeography*
3. T. C. Whitmore and G. T. Prance (ed.): *Biogeography and Quaternary history in tropical America*
4. T. C. Whitmore (ed.): *Biogeographical evolution of the Malay Archipelago*
5. S. Robert Aiken and Colin H. Leigh: *Vanishing rain forests: the ecological transition in Malaysia*
6. Paul Adam: *Australian rainforests*
7. Wilma George and René Lavocat (ed.): *The Africa–South America connection*
8. Yin Hongfu: *The palaeobiogeography of China*
9. Alan J. Kohn and Frank E. Perron: *Life history and biogeography: patterns in* Conus
10. Anthony Hallam: *An outline of Phanerozoic biogeography*

Vanishing Rain Forests

The Ecological Transition in Malaysia

S. ROBERT AIKEN

Department of Geography,
Concordia University, Montreal

and

COLIN H. LEIGH

Melbourne Parks and Waterways,
Melbourne

CLARENDON PRESS · OXFORD

Oxford University Press, Walton Street, Oxford OX2 6DP
Oxford New York
Athens Auckland Bangkok Bombay
Calcutta Cape Town Dar es Salaam Delhi
Florence Hong Kong Istanbul Karachi
Kuala Lumpur Madras Madrid Melbourne
Mexico City Nairobi Paris Singapore
Taipei Tokyo Toronto
and associated companies in
Berlin Ibadan

Oxford is a trade mark of Oxford University Press

Published in the United States
by Oxford University Press, Inc., New York

A catalogue record for this book is available from the British Library

Library of Congress Cataloging in Publication Data
Aiken, S. Robert.
Vanishing rain forests: the ecological transition in Malaysia/
S. Robert Aiken and Colin H. Leigh
(Oxford biogeography series; 5)
Includes bibliographical references and index.
1. Rain forests—Malaysia.   2. Forests and forestry—Malaysia.
3. Rain forest ecology—Malaysia.   4. Rain forest conservation—
Malaysia.   5. Man—Influence on nature—Malaysia.
6. Deforestation—Malaysia.   7. Sustainable forestry—Malaysia.
I. Leigh, Colin H.   II. Title.   III. Series: Oxford
biogeography series; no. 5.
SD235.M37A35   1992   333.75' 09595–dc20   91–41126
ISBN 0 19 854959 8

Set by Joshua Associates Ltd., Oxford
Printed in Great Britain by
Bookcraft (Bath) Ltd
Midsomer Norton, Avon

For
Mark and Jonathan
and
Alan and David

The conclusion I draw is optimistic: to the degree that we come to understand other organisms, we will place a greater value on them, and on ourselves.

> – Edward O. Wilson (1984, p. 2)

I insist that the means for the control of human destructive actions must be found in ourselves— in culture and behavior.

> – John W. Bennett (1976, p. 10)

Whether we use plants and animals for economic or playful and aesthetic ends, we *use* them; we do not attend to them for their own good, except in fables.

> – Yi-Fu Tuan (1984, p. 176)

There is something Newtonian, not yet Einsteinian, besides something morally naive, about living in a reference frame where one species takes itself as absolute and values everything else relative to its utility.

> – Holmes Rolston III (1985, p. 726)

CONTENTS

PLATES

Plates section appears between pages 54 and 55.

TABLES

FIGURES

PREFACE

Humans are transforming the earth. Never before have the world's economic and ecological systems been so closely interconnected, never before has our vulnerable planetary home been subjected to so many outrageous abuses. Problems like global warming, pollution of the oceans, ozone depletion, and declining biodiversity reveal an unprecedented increase in the pace, scale, and complexity of our multifarious and far-flung interactions with the natural world. We are all affected by such problems, and one of the major challenges facing the international community is how to devise management strategies for the sustainable development of the planet.

Another global issue that has attracted much popular and scientific interest in recent years is tropical deforestation. Few processes provide more dramatic evidence of our ability to transform the face of the earth, and few processes are likely to have more profound biological consequences. Humans have been altering and removing the earth's forest cover for millennia, but nothing in that long history matches the current pace of forest destruction in the humid tropics.

The great blocks of forest that straddle the equator in the Americas, Africa, and the Asiatic tropics are the most complex and exuberant expression of the earth's green mantle. Although they include many different forest formations, it is possible to divide them into two very broad categories: tropical rain forest of the ever-wet tropics and tropical seasonal (or monsoon) forest of the seasonally dry tropics. The sum of the two categories is now generally called tropical moist forest.

Although they clothe only some five per cent of the land surface, tropical moist forests are believed to harbour at least half of the earth's estimated five to ten million (perhaps as many as 30 million) species of organisms. That we possess only a very incomplete inventory of life on the earth reflects in large part how little we know about tropical forests. Much will remain for ever unknown, because each year a vast area of forest is converted to other uses. As the forests vanish—as, quite literally, they go up in smoke—so too do countless species of organisms. Current trends threaten to expunge much of the remarkable cornucopia of tropical life. The process is irreversible; extinction is for ever.

The rate of forest removal is in the order of 100 000 km^2 per annum (though this is a matter of some debate among experts), and perhaps as much again is variously degraded. The reasons for this are complex. Among the readily observable processes of change are slash-and-burn farming, cattle ranching, selective logging, land colonization, road building, and dam construction. But behind these processes lurks a variety of others of generally greater importance, the list including landlessness and poverty, population growth, political expediency, greed and corruption, consumerism in the affluent societies, unequal trade relations between rich and poor countries, and foreign debt. Almost all of the tropical moist forest biome occurs in poor countries, and poverty itself is a major cause of deforestation; this, in turn, suggests that the process is related to the interdependent but lop-sided structure of the international economy.

The two main classes of tropical moist forest present humans with rather different opportunities. The tropical seasonal forest is somewhat more open, partially deciduous, and readily combustible. Regions where it occurs, such as Central America, West Africa, Sri Lanka, Java, and mainland Southeast Asia, have long supported substantial human numbers. Tropical deforestation in these regions has a long history, and in some countries very little of the original forest remains.

On the other hand, the ever-wet tropical rain forest is more difficult to clear and burn, grows on soils that are subject to more rapid leaching, and has long been characterized by generally low population densities. Until recently, most of these forests remained substantially intact. The foregoing distinction no longer holds because both categories of forest are now under attack, and in countries like Brazil, Colombia, Indonesia, and Malaysia vast areas of species-rich tropical rain forest have been cleared or degraded in recent decades. The focus of this book is on the human interaction with the forests of Malaysia,[1] one of the world's most biologically diverse regions.

Formed in 1963, the Federation of Malaysia comprises the eleven states of Peninsular Malaysia (formerly the Federation of Malaya,[2] which became an independent country in 1957) and the two Borneo states of Sabah (formerly North Borneo) and Sarawak.[3] Singapore was part of the federation until 1965, when it became an independent republic. The total area is about 330 000 km^2. Peninsular Malaysia (131 794 km^2) is separated from Sabah and Sarawak (together called East Malaysia and covering some 198 000 km^2) by the South China Sea: the two closest points are 650 km apart in the extreme south, but northwards the distance increases to more than 1500 km. Kuala Lumpur is the federal capital.[4]

Malaysia is a constitutional monarchy. The monarchy is unusual in that the king (*Agong*) is elected for four years by the hereditary state rulers and the appointed governors (of Pulau Pinang, Melaka, Sabah, and Sarawak). Only a Ruler can be elected. The country is governed by an elected federal parliament comprising an upper house, the Dewan Negara or Senate, and a lower house, the Dewan Ra'yat or House of Representatives, and by elected state legislative assemblies. Federal and state responsibilities are set out in the Malaysian constitution in three lists: the Federal List, the State List, and the Concurrent List, the last stipulating subjects on which both the federal government and the state governments may legislate. Land, forestry, agriculture, and mining are on the State List, and this has considerable significance for conservation and environmental management.

The total population of Malaysia in 1985[5] was about 15.8 million: 12.96 million in Peninsular Malaysia (82.1 per cent of the total), 1.27 million in Sabah (8.1 per cent), and 1.54 million in Sarawak (9.8 per cent). The population is growing at an average rate of 2.5 per cent per annum (and an estimate revealed that it had risen to 17.4 million in 1989[6]). About 37 per cent of Malaysians lived in towns and cities in 1985, a level of urbanization that is well above the average for South, South-east, and East Asia, though somewhat short of the 1989 world average of 41 per cent. The overall population density is low by Asian standards, and Sabah and Sarawak are still thinly populated. The majority of Malaysians live in the west coast lowlands of Peninsular Malaysia.

Ethnic diversity is the hallmark of the population. Peninsular Malaysia's ethnic compositon in 1985 was approximately 56 per cent Malay, 33 per cent Chinese, and 10 per cent Indian. Making up less than one per cent of the total were more than 50 000 aborigines, or Orang Asli (literally, 'original people'), most of whom were still deep-forest dwellers. East Malaysia is even more diverse. There, in addition to the Malays, the indigenous peoples include Kadazan, Murut, and Bajau in Sabah and Iban, Bidayuh, and Melanau in Sarawak, among a plethora of other groups. The non-indigenous Chinese comprise a substantial minority in both Sabah and Sarawak: 15 and 29 per cent, respectively, in 1985. Ethnic diversity is a major challenge to national unity.

Malaysia is classified as an upper-middle-income developing country by the World Bank. GNP per capita in 1989 was US $1800, the third highest (after Brunei and Singapore) in South-east Asia. (Other countries with a similar level of GNP per capita include Syria and Mexico.) Although Malaysia's economic performance has been impressive in recent decades, widespread poverty has persisted, income distribution between both regions and economic groups has remained very uneven, and intra-ethnic inequality is growing, especially among the Malays. The main exports are manufactures (e.g. electrical goods and textiles), crude petroleum and liquefied natural gas, timber, palm oil, rubber, tin, and cocoa. The manufacturing sector grew at an average rate of 12 per cent per annum during 1971–85, and in 1984 it emerged as the largest sector of the economy.

There are considerable regional variations in the pattern of economic development. First in importance is the extensive humanized landscape that sprawls over the west coast lowlands of Peninsular Malaysia. Here, in the region facing the Strait of Melaka, is the economic heartland of the nation, the hub of the modern economy, the vital seat of power. No other part of Malaysia has been so radically transformed by human action. Rubber and oil palm estates, tin mines, Malay villages with rice fields, thickly populated rural areas, a well-developed hierarchy of urban places, a plural society composed of Malays, Chinese, and Indians—all these are characteristic features of the western lowlands.

East and south of the mountainous spinal column of the Peninsula is a second lowland region of rather different character. Narrow in places and in others opening out into expansive flatlands and undulating hills, this region has long been less developed, poorer, more rural, less thickly settled, more heavily forested, predominantly Malay, and largely given over to small-scale farming and fishing. Since about 1970, however, much of the region has been transformed by vast land development schemes, logging, road building, and other activities that have greatly reduced the region's long-standing isolation. Here, as in the western lowlands, only remnants of the great forests that formerly clothed the region now remain.

Across the South China Sea are Sabah and Sarawak (East Malaysia). Both are thinly populated and both are largely undeveloped. Here, in comparison with the developed core of Peninsular Malaysia, the level of urbanization is low, manufacturing is less important, greater dependence is placed on primary production, infrastructure and most services are poorly developed, many communities remain largely isolated, and ethno-linguistic and cultural diversity are more marked. This is a region apart. Largely bypassed by many of the processes that transformed the landscape and life of the Peninsula, East Malaysia occupies a peripheral position in the national economy. Yet, it is not without economic significance, because from this frontier-like region are extracted and exported large quantities of valuable oil, natural gas, and timber. Land development and widespread, generally highly destructive timber extraction have made major inroads into the forests of East Malaysia in recent decades.

The characteristic natural vegetation of Malaysia is tropical rain forest. Both Peninsular Malaysia and East Malaysia form part of an extensive rain forest block that is centred on the Malay Archipelago, whose distinctive flora constitutes a botanical region called Malesia.[7] In Malaysia there are many different rain-forest formations, depending on factors such as edaphic conditions, drainage, and altitude. Below an elevation of about 750 m, there are species-rich lowland and hill dipterocarp-dominated forests in dryland areas and mangrove, peat-swamp, and freshwater-swamp forests in wetland areas. Lower montane and upper montane rain-forest formations occur above 750 m and 1500 m respectively. Above about 1200 m the upper dipterocarps are replaced by oak- and laurel-dominated forest.

The remarkably rich flora of Malesia is estimated to include at least 25 000 species of flowering plants, of which at least 8000 are represented in Peninsular Malaysia alone. About 2600 of the latter number are trees, whereas in all of North America there are only some 700 native tree species. The high proportion of trees provides a framework and a range of congenial habitats for many other kinds of plants (as well as animals). This is especially the case in the structurally complex lowland rain forest in dryland areas, where there are usually many big woody climbers, an abundance of strangling figs, a rich palm flora, various parasitic plants (of which the giant *Rafflesia* is the most celebrated), and scavengers (saprophytes) that live on dead or dying organic material in the deep shade; epiphytes, including orchids, are also present, although they tend to be more abundant in peat-swamp and freshwater-swamp forests and at higher elevations.

The animal life of Malaysia is typically forest-dwelling and of great richness and diversity. In Peninsular Malaysia alone there are 203 species of mammals (including ten primate species, as well as the large and spectacular elephant, tiger, tapir, Malayan gaur or seladang, and the rare two-horned Sumatran rhinoceros), some 450 species of birds, many lizards, amphibians, and snakes, about 260 species of primary freshwater fishes, and a great diversity of invertebrates. Most species are specialized forest-dwellers and many are arboreal, including flying and gliding lizards, snakes, geckos, frogs, squirrels, bats, and lemurs. The birds include plant-eating hornbills and barbets in the top of the canopy and pheasants in the undergrowth. The structurally complex lowland rain forest is especially rich in species, and its importance is underscored by the fact that 52 per cent of the mammals (in the Peninsula) live below 300 m and 81 per cent below 600 m.

But it is, unfortunately, precisely these forests that have borne the brunt of human activities, and current trends threaten to eliminate much of Malaysia's natural wealth. Almost all of the Peninsula's lowland rain forest has gone, and many species of plants and animals are now endangered. There are still extensive tracts of lowland forest in Sabah and Sarawak, but there too the pace of change is rapid—and generally destructive. The expansion of the area devoted to plantation crops, selective logging of hardwoods, shifting cultivation, and population growth are some of the factors that account for forest loss in Malaysia.

Although anthropogenic forest change in Malaysia spans several millennia, it was not until the nineteenth century that there was any appreciable increase in its pace or scale. The prime mover behind this new departure was expanding colonialism, which resulted in the closer incorporation of the region's agricultural and mineral production into the European-dominated global economy. This was especially the case after about 1874, when British rule or control was rapidly extended over the territories that now comprise the modern state of Malaysia. Throughout the nineteenth century, however, there was unanimous agreement among contemporary observers that the forests were virtually illimitable.

The first major impact on the rain forest resulted from the rapid adoption of rubber cultivation, and in the first two decades of this century extensive tracts of forest along the western lowlands of the Peninsula were replaced by the new boom crop. But elsewhere in that region, and to an even greater extent in North Borneo and Sarawak, most of the forest estate remained substantially intact until a few decades ago. This is no longer the case, because in the relatively short post-colonial period (since 1957 in the Peninsula and since 1963 in Sabah and Sarawak) the pace and scale of forest destruction have increased dramatically. Thus, for example, the proportion of the total land area under forest in the Peninsula declined from 74 per cent in 1958 to about 40 per cent[8] in 1990—a drop of 34 per cent in 32 years. It is the many unfortunate human and environmental consequences of this quite unprecedented rate of forest removal that strike a plangent chord in the imagination of many friends and admirers of Malaysia.

This book is about the vanishing rain forests of Malaysia. It focuses on four interrelated themes. Firstly, we describe the rain forest formations of the study area and their remarkable diversity of flora and fauna, and we provide a sketch of how the rain forest functions. The coverage is selective and condensed because this is not a book on the ecology of tropical nature in Malaysia. Secondly, we outline the processes and policies that have resulted in anthropogenic forest change in the region over the past two centuries, focusing mainly on this century. Thirdly, we look at certain of the environmental, biological, and human consequences of change, both past and present. And fourthly, we describe what has been done to conserve the region's forests, natural systems, and wildlife, and we outline the kinds of changes that will be required to direct Malaysia towards a more sustainable future. An introductory chapter provides a context for the case study of Malaysia. It outlines and documents four topics: tropical rain forest, resource utilization, human impact, and conservation and sustainable development. These are the themes of the four main chapters, as described above.

Malaysia is in the throes of a profound 'ecological transition'. We have borrowed this concept from the anthropologist John W. Bennett, and the idea is reflected in the subtitle of the book.[9] As originally formulated, the ecological transition referred to the progressive incorporation of nature into culture and to the several expressions of the transition—technological, sociological, ecological, and philosophical. We use the concept in the rather restricted sense of 'the human impact on natural systems'; although often implicit, this is the thread, or leitmotiv, that pervades the book.

The spelling of Malaysian (and Indonesian) place-names poses certain problems because a good many familiar names have recently been replaced by new romanized Malay spellings. We have taken as our guide the spellings given in the *Fifth Malaysia Plan 1986–1990*. Only a few Indonesian place-names are mentioned in the text and these are rendered after an official government publication.[10]

ACKNOWLEDGEMENT

Various drafts of the manuscript were typed by Tina Skalkogiannis, and we warmly acknowledge her skill and abiding patience. Our thanks also to Jacqueline Grekin, Anne Pollock-McKenna, Jane Barr, and Carolyn Slack for their help with the manuscript. To Lourdes Meana, who drew the maps with such loving care, we extend our sincere gratitude. Finally, we thank Dr. T. C. Whitmore for his kind encouragement and good counsel.

1 WHITHER TROPICAL RAIN FOREST?

Environmental change is an enduring facet of human history. Vegetation, soils, fauna, water, climate, landforms—in short, all the major components of the biosphere—have been profoundly altered by human agency. In the mid-latitudes, for example, cultivated landscapes and urban systems sprawl over regions where once great forests held sway. Little of the natural vegetation remains. Now the most complex and luxuriant of all forest types is under attack, both from within and from without. Viewed from a historical perspective, the onslaught is only the most recent phase in a long and concerted effort by humans to extend the horizons of the forest frontier.

Tropical rain forests occur in poor countries (Australia is the exception that proves the rule), where somehow the goals of development must be achieved. The forests are viewed as a resource, and it would be idle to assume that most of what remains can be saved. Sustainable agriculture and forestry are required to meet basic human needs and to remove pressure from the still largely undisturbed forests, and reforestation is already sorely needed in many places. There are indigenous resource management strategies that, if adopted more widely, would do much to protect the forests while alleviating poverty and human suffering. Even if given a chance, however, their potential may well be negated by pressures and linkages emanating from the structure of the international economy. Meanwhile a rich part of our biospheric heritage hangs in the balance.

This chapter establishes a broad framework within which the detailed case study of Malaysia can be set. It introduces four broad themes, each of which is the focus of a subsequent chapter on Malaysia: tropical rain forest (Chapter 2), resource utilization and forest conversion (Chapter 3), human impact (Chapter 4), and conservation and management (Chapter 5).

1.1 TROPICAL RAIN FOREST

The forests of the humid tropics vary considerably from place to place and various kinds of forest (or formations) that differ from one another in structure, physiognomy, and floristic composition have been recognized and variously (often confusingly) named. A common procedure is to group the formations into two broad classes: tropical rain forest and tropical seasonal (or monsoon) forest. The now familiar name for the sum of the two classes is tropical moist forest (Sommer 1976, pp. 6–7; cf. Myers 1980, pp. 15–18).

What is tropical rain forest, or rainforest, as some scholars (e.g. Baur 1968) have preferred to call it? Beard (1944, 1955) defined it narrowly, Schimper (1903) less rigidly, others very loosely, while in the popular mind it is 'jungle'. Humboldt called it the *hylaea*, but the term *tropische Regenwald*, or tropical rain forest, was coined by Schimper (1903), who noted that in 'tropical districts with precipitations at all seasons the forest is ... developed as rain-forest' (p. 264), which he described as 'evergreen, hygrophilous in character, at least thirty meters high, but usually much taller, rich in thick-stemmed lianes, and in woody as well as herbaceous epiphytes' (p. 260). This familiar (though curiously much misquoted) definition has been widely accepted, albeit with further elaboration (e.g. P. W. Richards 1952; Whitmore 1984a). In short, tropical rain forest is the characteristic natural vegetation of the perhumid low latitudes

Some oft-mentioned features of tropical rain forest (other than those cited by Schimper) include great species richness and diversity (although some forests are species-poor); the apparent paradox of luxuriant vegetation growing on soils of generally low fertility (although tropical soils vary considerably); general poverty of leaf litter reflecting rapid decomposition (although some litter is relatively thick and decomposition rates are variable); supposedly 'tight' nutrient cycles, with little or no net loss from the forest ecosystem (although direct evidence of such cycles is sparse); involvement of many animal species in herbivory, dispersal, and pollination, hence much variety in the presentation of flowers and fruits (although, as in temperate regions, the majority of the pollinators are insects); many examples of complex coevolutionary relationships between plants and animals; and a wide range of tree architecture (e.g. Hallé *et al.* 1978; Mabberley 1983, pp. 4–9; Proctor 1983; Greenwood 1987). There is growing awareness that forests everywhere share many fundamental features (e.g. Brünig 1987).

Tropical seasonal (or monsoon) forest replaces tropical rain forest in regions with a pronounced dry season. Schimper (1903, p. 260) contrasted tropical rain forest with monsoon forest, which he described as 'more or less leafless during the dry season, especially towards its termination, trophilous in character, usually less lofty ... rich in woody lianes, rich in herbaceous but poor in woody epiphytes'. Other contrasts include lower biomass and different floristic composition (P. W. Richards 1952, chapter 15; Whitmore 1984*a*, chapter 14).

The focus of this book is on tropical rain forest *à la* Schimper, although in places we have found it instructive to draw on studies of forests (and the

human impact on them) that are not *sensu stricto* tropical rain forest.

1.1.1 Distribution and change

Tropical rain forest, which comprises about 40 per cent of the approximately 9 million km² tropical moist forest biome, occurs in three main blocks, with the Neotropics having more than half of the total (see Ayensu 1980, pp. 14–19; also Myers 1979, chapters 8–10, 1984, chapter 2). Figure 1.1 stands in lieu of a detailed description of the distribution.[1]

Climate is the main determinant of world patterns of vegetation. Within the humid tropics, two principal climatic types can be distinguished on the basis of seasonal rainfall contrasts: a tropical wet or perhumid type and a tropical wet-and-dry type. Tropical rain forest occurs in the perhumid tropics (Fig. 1.1; e.g. in the Amazon Basin), and tropical seasonal (or monsoon) forest in areas of more prolonged water stress (e.g. in mainland South-east Asia). Even within the perhumid or so-called ever-wet tropics, however, there are periods of water stress of varying duration, and it is now recognized that such conditions play an important role in plant and animal life and in the distribution of forest types (Whitmore 1984*a*, pp. 53–9).

The current configuration of tropical rain forest is only the most recent of many distributions

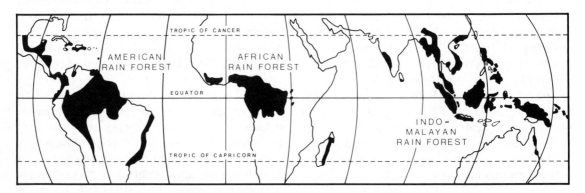

Fig. 1.1 Distribution of tropical rain forest. (From Whitmore 1984*a*, p. 3.)

(Whitmore 1990, chapter 6). We perhaps can imagine Fig. 1.1 as the last, somewhat unsteady, frame in a motion picture that was set running some 2.5 million years ago, at the beginning of the Quaternary. (Earlier geological events are discussed in Chapter 2.) During the Pleistocene glacial and interglacial phases, when there were, according to some scholars, fluctuations between semi-arid and pluvial conditions, the rain forests are thought to have contracted and expanded several times. There is some palynological, geomorphological, and other evidence of climatic change in the humid tropics (e.g. Kershaw 1978; Absy 1985; Bigarella and Ferreira 1985; Morley and Flenley 1987), and obervations of species disjunctions in the Amazon Basin provide indirect evidence that during dry periods the forests contracted into so-called 'refugia' (most of them on higher ground), where genetic differentiation and speciation are thought to have occurred (Prance 1982, 1985).

However, several scholars have rejected or disputed the 'refuge theory' (Endler 1982; Beven et al. 1984; Connor 1986; Gentry 1986), and Colinvaux (1989), who found no convincing evidence of aridity in the Amazon lowlands, has presented a quite different argument: during glacial periods 'some part of the lowlands remains as a relatively warm, wet preserve for rain-forest trees and the uplands—formerly thought to be refuges—become inhospitable' (p. 106).

Although much remains unknown about the vegetational history of the humid tropics, it is becoming increasingly clear that most, if not all, tropical forests have long been subject to continuing disturbance (Unesco 1978, chapter 3; Flenley 1979; Kershaw 1981; Hamilton 1982; Sowunmi 1986). Contrary to previous thinking, tropical rain forests have not been stable for millions of years, consequently the hypothesis that stability accounts for diversity appears untenable (e.g. Federov 1966; P. W. Richards 1969).

Evidence suggests that vegetational change occurs on many spatial and temporal scales, that it has been constant during the Holocene, and that it is related to many causes of disturbance, including volcanic activity, earthquakes, landslides, hurricanes, cyclones, flooding, erosion, drought, fire, and, of course, human activity (Garwood et al. 1979; Lugo et al. 1983; Sanford et al. 1985; R. J. Johns 1986; Salo et al. 1986). Forests are always in a state of flux, and tropical rain forests are no exception; the result is a dynamic patchwork of gaps, successions, and big mature trees (see Chapter 2).

1.1.2 Some attributes of species

The goals and priorities of land-use planning, resource management, and conservation require attention to the following vitally important characteristics of tropical rain-forest species:

Diversity

Most, though not all, plant and animal communities are extremely diverse and many writers have cited examples of great species richness in a hectare or two of forest (e.g. P. W. Richards 1973, pp. 60–2; Whitmore et al. 1985). Several regions—for example, Colombia's Choco, Madagascar, and Peninsular Malaysia—possess remarkable concentrations of species. Some such 'hot spots' are believed to coincide with Pleistocene *refugia* (see above), which have served to identify locations for parks and reserves in tropical Latin America (Wetterberg et al. 1981; Pádua and Quintão 1982; Myers 1983a). Diversity is increasingly threatened by human activities.

Low density

A consequence of diversity is that many species occur at very low densities (e.g. Whitmore 1980, pp. 314–15). But in the case of trees, for example, little is known about the minimum number of individuals that is required to maintain genetic diversity, and it is often difficult to estimate other minimum viable populations (see Franklin 1980). Low density means that many species are probably subject to inadvertent elimination. Certain characteristics of trees—for example, low density, dioecy (that is, the occurrence of male and female

flowers on separate plants), and endemism—
'dictate that conserved forests must be large,
undisturbed, and numerous, to be fully effective
in the conservation of tropical species and communities' (Ng 1983, p. 359).

Patchiness

Many habitats in tropical rain forest occur in
patches, and the patches themselves are dynamic.
Patchiness results from edaphic, topographic,
and climatic heterogeneity and from disturbance
(Foster 1980). Birds, for example, are patchily
distributed, especially in species-rich tropical
areas (Diamond 1980), and most rain-forest tree
species are much more patchily distributed
(clumped) than previously thought (Hubbell
1979; Hubbell and Foster 1983, 1986).
Although invariably ignored by developers,
patchiness is familiar to forest dwellers. Large
reserves are needed to encompass diverse habitats, thus conserving as many habitat specialists
as possible.

Rarity

Usually, but not always, few species are common
and many are rare. Causal factors include habitat
specialization (see above), regeneration niche
specialization, and migration from population
centres elsewhere (Hubbell and Foster 1986).
Again, reserves should be large to encompass
diverse habitats. Attention should be given to
natural disturbance regimes because 'much of the
niche differentiation of tropical trees is expressed
in terms of gap regeneration requirements'
(Hubbell and Foster 1986, p. 229), and centres
of dispersal require protection. Rarity, according
to Terborgh and Winter (1980, p. 132), 'proves
to be the best index of vulnerability'.

Endemism

Many species are localized or parochial and
occur nowhere else. In part this is because many
species are rare and many do not spread very
easily (Denslow 1988, p. 29). A large proportion
of species in the Neotropics are endemics, but in
some countries they are not protected in parks

and reserves (Terborgh and Winter 1983). High
levels of local endemism have been equated with
Pleistocene *refugia* (see above), but Gentry
(1986) has argued that habitat specialization
determines many localized distributions of both
plants and animals. The implication is that conservation efforts should focus on protecting areas
of high habitat diversity.

Mutualism

Many species are products of coevolutionary
processes and complex mutualistic relationships
often occur. Numerous species depend on particular hosts, the elimination of which will cause other
extinctions; an eliminated plant may take 10–30
other species with it (Raven 1976, p. 155). Species
whose elimination would cause cascades of linked
extinctions have been called 'keystone mutualists'
(Gilbert 1980; also Terborgh 1986). High priority
should be given to conserving such species.

Disturbance

Disturbance and succession (patch dynamics)
reduce the threat of extinction for some species
(those that tolerate habitat change) while exacer-
bating it for others (like those with small populations). It is therefore necessary to establish precise
goals for development and conservation. Foster
(1980) has discussed the management options in
relation to different taxa and the size of reserves.
He noted that 'large reserves provide the best
insurance against the deleterious effects of disturbance as well as being the most economical way
to provide a healthy mix of disturbance-induced
habitats' (p. 91).

   The elimination of a vast number of species is
probably inevitable because most development in
the humid tropics ignores the distributional and
other characteristics of species. The emerging
consensus is that parks and reserves should be
large—the larger the better. The goals of conservation, however, must be widened to embrace
species everywhere. There is no substitute for
careful land-use planning and management.

1.2 RESOURCE UTILIZATION AND FOREST CONVERSION

The major part of this section outlines some of the main features of traditional and modern natural resource-use systems. What constitutes a natural resource is largely a matter of cultural appraisal (Spoehr 1956). A natural resource is defined by a group's attitudes, values, and perceptions, by its technological skills and social organization, as well as by its needs and wants (Zimmermann 1933; Fernie and Pitkethly 1985, pp. 2–8). We can expect cross-cultural variation in the utilization and meaning of tropical rain forest. The term 'traditional' means pre-industrial; technologies are palaeotechnic. It is not implied that the resource-use systems are unchanging, and no evolutionary sequence is assumed. The term 'modern' means fossil-fuel-using, neotechnic, or industrial. The section closes with a brief review of the literature on the extent of tropical moist forest and the rate at which it is being destroyed or altered.

1.2.1 Traditional resource-use strategies

Three broad ypes of subsistence strategies are briefly described: hunting and gathering, shifting cultivation, and permanent-field agriculture. Several problems with this familiar typology are underscored. The strategies are viewed as adaptive. As a plan of action in response to either internal or external stimuli, an adaptive strategy 'links together the social institutions and technology of the population and defines what in the environment is culturally meaningful as a resource' (Hardesty 1986, p. 14).

Hunting and gathering

Few hunter–gatherers now remain and those that do are not necessarily representative of the full array of former societies. Contact with Europeans over several centuries has devastated many groups (Crosby 1986, chapter 9). Hunter–gatherers in tropical rain forests include, for example, the Mbuti (Pygmies) in the Ituri forest of Zaïre and the Semang (Negritos) of Peninsular Malaysia. Although they are still often regarded by non-forest peoples as 'inferior savages', anthropologists are more likely to describe the hunter–gatherers as 'the most leisured peoples in the world' (Service 1966, p. 13), to view them as 'the original affluent society' (Sahlins 1974, chapter 1), and to praise the effectiveness and efficiency of their ways of life.

Some widely accepted characteristics of hunter–gatherers include the following: small, relatively stable overall populations and low densities; generally small, fluid social groupings that are marked by fission and fusion in relation to subsistence opportunities and interpersonal friction; culturally sanctioned adjustments of population size to available resources; mobility that is attuned to the seasonality, periodicity, and dispersed nature of rain-forest resources; flexible and variable ways of exploiting local habitats and environmental gradients; exceptionally detailed environmental knowledge; communal control over resources; considerable leisure time and security; and generally adequate nutrition (see, for example, R. B. Lee and DeVore 1968; Unesco 1978, pp. 436–51; Hames and Vickers 1983). On the other hand, research since the 1960s has shown that 'very few "typical" hunter–gatherer patterns emerge' (Hill and Hurtado 1989, p. 437); rather, the life-styles of such peoples are characterized by great variability.

Shifting cultivation

Also called swidden or slash-and-burn farming and known by a variety of regional names, shifting cultivation is a type of agriculture in which fields rather than crops are rotated (Spencer 1966). There are some 140 million shifting cultivators in tropical countries (Unesco 1978, p. 469; Myers 1980, chapter 3). Although many colonial officials viewed shifting cultivators as *mangeurs de bois*, studies by anthropologists and others since the 1950s have revealed that shifting cultivation is usually well adapted to tropical conditions, including generally low soil fertility and competition from weeds and pests (see Whitmore 1990, pp. 133–6, 155–6).[2]

Shifting cultivation is a rational form of inter-cropping involving detailed knowledge and use of many cultigens in a wide range of habitats (Pelzer 1978, pp. 271–4). Geertz (1963, pp. 16–25) argued that the floristic diversity and stratified form of the cultivator's garden (or swidden) reflect part of a design to mimic the nature of the rain forest, although it can be shown that these structural characteristics vary greatly in relation to different cropping patterns: swiddens with high-diversity intercropping bear some resembl-ance to the forest, whereas those with low-diversity intercropping and monocropping generally do not. An alternative to Geertz's hypo-thesis has been offered by Vickers (1983), who argued that garden structure is best understood in terms of the economic utilization of different cul-tigens 'rather than by simple analogy to tropical forest physiognomy' (p. 36).

Clearings are 'abandoned' after a year or two owing to declining soil fertility and/or the invasion of weeds, although fruits and other products are often harvested for many more years and old plots are favoured hunting grounds. Food gathering also contributes to most diets. Generally speaking, total inputs and outputs are low, settlements are of modest size, and population densities are below 40 per km[2] (although the reported range is very wide). Formulating generalizations is difficult because shifting cultivation is a 'variable set of adjustments to specific environmental, agrono-mic, demographic, and cultural constraints' (Moran 1986, p. 108).

Permanent-field farmers

Farmers with permanent fields create humanized landscapes: generalized rain-forest ecosystems yield to specialized agro-ecosystems, species diversity gives way to (greater) uniformity, and closed-over settings are replaced by (more) open ones, although there are many variations. Perma-nent-field cultivators generally work the land more intensively than do shifting cultivators; that is, labour, skills, or capital (inputs) are substi-tuted for land so as to achieve greater production (output) from a constant unit of land over a specified period of time.

The conditions that give rise to intensification have been much disputed since Boserup (1965) advanced the thesis that there is a positive corre-lation between population pressure (density) and agricultural intensification. Some researchers have found evidence to support the thesis; others have modified it by adding certain production and environmental factors; while yet others have argued for greater attention to the relations between labour expenditure and such matters as social rewards and prestige, colonial exploitation, or economic opportunity (e.g. Brookfield 1972; Netting 1974, pp. 36–42; Turner *et al.* 1977).

Intensive, high-yielding farming systems are usually associated with population concentrations in the order of several hundred or more persons per square kilometre, the origins and growth of urban places, and more complex trade linkages. Examples of intensive systems in the humid trop-ics include wet-rice cultivation in parts of South-east Asia and the variety of techniques, including terracing and raised fields, that were practised in parts of pre-Hispanic Central and South Amer-ica (Harrison and Turner 1978; Denevan 1980, pp. 219–34).

Critique

Such a simple threefold division of traditional subsistence strategies disguises much that is revealing about such systems. The following points take their cue from an admirable essay by Padoch and Vayda (1983).

1. Farmers obviously manipulate their bio-physical surroundings, but so too do hunter–gatherers whose activities include, for example, use of fire to promote regeneration of desired species and protection of certain favoured spe-cies—of fruits, for instance. Consequently, the environmental relations of hunter–gatherers and farmers differ in degree rather than in kind.

2. Forest hunter–gatherers and non-forest farmers often maintain social and economic

relationships that are disguised by simple typo-logies. For example, symbiotic relations link Semang foragers and Malay farmers in Peninsular Malaysia, while external trade via Chinese and Malay middlemen has linked the region to the outside world since ancient times (Dunn 1975; Rambo 1982). It is wrong to assume that tradi-tional peoples have always been self-sufficient.

3. It is not uncommon for rain-forest dwellers to 'rely simultaneously or serially on several different ways of exploiting resources' (Padoch and Vayda 1983, p. 306). Various combinations of food-obtaining technologies have been reported from the humid tropics (Netting 1971, pp. 20–1; Bronson 1978), including examples of complex, variable adaptations to micro-environ-mental gradients (e.g. Posey 1985). Many agricul-tural peoples continue to collect forest products and to hunt and fish (Allen and Prescott-Allen 1982; Padoch 1988, pp. 132–6).

4. Pre-industrial foragers and farmers are seldom custom-bound conservatives. There is evidence, for example, of both disintensification and intensification of agriculture, of eager accept-ance of new plant species, and of openness to change. The notion of unchanging traditional societies is a myth.

An important conclusion is that traditional resource-use systems are often characterized by flexibility and variability (Harrison 1978, pp. 4–8). Farming systems, in particular, have been classified in a variety of ways (e.g. Boserup 1965; Ruthenberg 1976; Turner and Brush 1987), but it is doubtful whether any classification can do justice to the diversity of such systems, either current or dormant, that are known, if only to individuals (Denevan 1983). The known options constitute a kind of cultural reservoir of strategies from which potentially adaptive ones can be drawn when new constraints or opportunities are encountered.

1.2.2 Modern resource regimes

Modern, neotechnic, or industrial regimes are associated, among many other things, with:

(1) the emergence of industrial societies, new technological innovations that have changed the availability of resources, further expansion of the capitalist world-economy, and mounting pressure for surplus production;

(2) substitution of human and animal labour with energy from fossil fuels, which is used to run machines like bulldozers and chain-saws and to manufacture fertilizers and other agrochemicals;

(3) processes that increasingly operate at national and global scales (rather than at local and regional levels), coupled with greater organiza-tional and institutional complexity (but also greater inflexibility).

Space precludes more than a brief discussion of each of these points in relation to tropical rain forest.

Capitalism is an inherently expanding system driven by market competition. While its organiza-tional forms vary, and it is by no means the only kind of socio-political formation that has resulted in environmental disruption, the expanding capi-talist mode of production has caused profound social and environmental changes in traditional societies and peripheral regions in recent centuries (Blaikie 1986; Peet 1986). Industrialization, improvements in transportation, and the steeply rising demand for products like tropical hard-woods and palm oil in the metropolitan or core regions have all contributed to widespread tropi-cal deforestation since the nineteenth century (Tucker and Richards 1983a; J. F. Richards and Tucker 1988).

Energy and matter flows differ between sub-sistence and industrial agro-ecosystems, the latter being heavily dependent on energy subsidies supplied by fossil fuels. Studies show that whereas productivity is enhanced by increasing the energy subsidy, energy-use efficiency—that is, the ratio of energy output to input—declines (Jordan 1987*a*, pp. 106–13; also Deshmukh 1986, p. 273). Experiments at Yurimaguas, Peru, have demonstrated the potential for 'technologically sustainable' agricultural production on acid Amazonian soils, provided that there are no restraints on energy subsidies and other inputs (Sanchez *et al.* 1982; Nicholaides *et al.* 1985). However, the intensive fertilization that is called for is probably limited to areas with a well-developed infrastructure, and Fearnside (1987*a*) has cast doubt not only on the economic viability of the 'Yurimaguas technology' but also on its potential for reducing deforestation.

Although it is often assumed that the energy efficiency of traditional agro-ecosystems is greater than that of their industrial successors, this is generally not the case when the energetic values of nature's services (e.g. fire) are included (Westman 1977; Rambo 1980*a*). The important point to emphasize is that when the structure of the tropical rain forest is maintained 'annual yield may be lower, but yield is sustainable'; total output in the long term is likely to be higher, and 'so the efficiency of resource use becomes higher' (Jordan 1987*a*, p. 113).

A development goal of many Third World governments is to link hitherto remote or peripheral areas with national and international markets. This has been a major justification for new roads, hydroelectric power schemes, new settlements for landless peasants, and other kinds of infrastructure. As a result, vast areas of forest have been opened up for cash crops, beef, timber, minerals, and other commodities that are destined primarily for export. Important related processes include consumer demand in the rich nations (usually fuelled by advertising), tax holidays, and other fiscal incentives that encourage private investment (as well as the concentration of wealth and land-ownership), greed and corruption, loans from development banks, and debt repayment. Increasingly, failure to anticipate changes in prices, markets, or other conditions can lead to severe social, economic, and environmental disruptions.

1.2.3 Tropical moist forest: extent and rate of conversion

Several important reports and summary statements dealing either in whole or in part with the extent of tropical moist forest and/or its rate of conversion have appeared since the mid-1970s (Persson 1974; Sommer 1976; Lanly and Clement 1979; Myers 1980; FAO/UNEP 1981*a,b,c*; Lanley 1982, 1983). These sources have formed the bases of:

(1) retrospective comparisons, additional estimates, and assessments of the reliability of the data (Grainger 1983, 1984; Allen and Barnes 1985);

(2) projections of future trends (Barney 1980, pp. 117–20);

(3) a good deal of dispute over the rate of conversion, the meaning of 'deforestation'[3] and, among other things, the implications for the extinction of species (e.g. Sedjo and Clawson 1983; Myers 1984, pp. 178–9, 1985*a*).

In order to avoid a lot of complicated details, we cite below, in addition to the area of tropical moist forest estimated by Sommer (1976), *three calculations* by Grainger (1984) *based on* Persson (1974), Myers (1980), and the FAO/UNEP (1981*a,b,c*) studies co-ordinated by Lanly (1982). Generally speaking, all four estimates refer to tropical moist forest (the sum of tropical rain forest and tropical seasonal forest), although the descriptive designations vary (Table 1.1). As Grainger (1983, p. 388) has noted, the reason for using this broad category 'is that data are not available on the areas of the distinct forest types which this term comprises'. There is considerable agreement over the extent of tropical moist forest (Table 1.1).

On the other hand, estimates of the rate of conversion vary considerably. Referring to the mid-1970s, Sommer (1976, p. 20) stated that 'a regression rate of approximately 11 million ha

Table 1.1 Estimates of the extent of tropical moist forest

	Tropical moist forest	All closed forest	All closed moist broad-leaved forest	All closed broad-leaved forest
Reference	After Sommer (1976)	Based on Persson (1974)	Based on Myers (1980)	Based on FAO/UNEP (1981 a,b,c) and Lanly (1982)
Area (million km²)	9.35	9.81	9.73	10.81

Sources: Sommer (1976, p. 24) and Grainger (1984, table 1 and pp. 4–16).

per year may be assumed'. Myers (1980, pp. 25–6), in a study conducted in the late 1970s, noted that when the activities of forest farmers are considered 'in conjunction with other factors—timber harvesting, planned agriculture, cattle raising, etc.—it becomes possible to credit that something approaching 200,000 km² of TMF [tropical moist forest], and possibly even more, are being converted each year'. On the basis of the most thorough study to date (FAO/UNEP 1981a,b,c), Lanly (1982, p. 77) reported that 'closed broadleaved forests' were being 'cleared and converted to other uses' at an annual rate of 6.9 million ha during 1976–80 and 7.1 million ha during 1981–5, or at about one-third of the rate estimated by Myers (1980; see also Grainger 1984, pp. 20–3). An additional significant finding is that 'primary', undisturbed closed forest is being cleared much less rapidly than is logged-over closed forest: 0.27 per cent as against 2.06 per cent annually (Lanly 1982, p. 78).

Myers (1984, p. 179) has explained the apparent discrepancy between his figures and those of the FAO/UNEP report by noting that his study 'looked at *significant conversion* of primary forests, that is, destruction plus degradation. The FAO/UNEP study focused instead on *outright elimination* of forest, that is, destruction alone. Hence the difference in the two sets of figures . . .' (emphasis in original; see also Sedjo and Clawson 1983, 1984, pp. 155–66; Melillo *et al.* 1985; Myers 1985a).

We make three points by way of conclusion.

Firstly, estimates of the extent of the tropical moist forests and their rates of conversion are just that—they are estimates, and they ought to be accepted with great caution. The fact that there are inaccuracies in the data is suggested, for example, by disagreements over the magnitude of forest conversion in the Amazon Basin (cf. Fearnside 1982, 1986; Lugo and Brown 1982; Moeller 1984). Secondly, there is an urgent need for continuous, preferably centralized, satellite-based monitoring of tropical moist forests using remote-sensing techniques (Green 1983; Grainger 1984; Malingreau and Tucker 1988). Only when periodic surveys are replaced by monitoring will accurate, up-to-date data be readily available. And thirdly, the lack of such data is a major impediment to rational land-use planning and forest management.

1.3 THE HUMAN IMPACT

This section reviews some of the consequences of the human impact on tropical moist forest, a subject that is taken up in detail in Chapter 4. At this juncture, however, we consider it appropriate to begin with a short discussion of the processes that result in deforestation and forest alteration.

1.3.1 Anthropogenic impact: what are the causes?

Much of the literature on tropical rain-forest conversion has emphasized a rather limited set of more or less direct and immediate processes of change, including rapid population growth, depredations caused by slash-and-burn farmers, expansion of ranchland, selective logging of hardwoods, and Third World government-sponsored schemes such as land development, transmigration, road building, and dam construction. A closer look suggests that these processes are largely 'symptoms' of an underlying set of 'causes' and that explanation must be sought in the context of the interdependent but highly distorted world-economy. There is space for only a few comments.

Rapid population growth in the Third World is both a contributor to and a consequence of underdevelopment. There is no denying that it is a major problem (McNamara 1984; Ehrlich and Ehrlich 1986). Contrary to certain views (e.g. Poore 1983a), however, by itself it is usually not a major factor in resource depletion (Repetto and Holmes 1983; Repetto 1986). Population growth is often linked to food shortages and hence to forest destruction by small farmers. Generally more important, however, is the distribution of land. Blaming deforestation on the poor is vicious.

The activities of growing numbers of poor slash-and-burn farmers are wasteful of forest resources. Many of the poor are landless peasants who have been dispossessed by industrial agriculture. Barred from existing agricultural land, which is often very unevenly distributed, they exploit the only resources available to them. Vested interests often oppose land reform, hence it is politically expedient for governments to either ignore the forest farmers or to make them the scapegoat for forest destruction, thus deflecting attention from the ravages of other processes.

Expansion of ranchland for beef production has led to rapid forest conversion in Central America (Myers 1981; Nations and Komer 1983; Myers and Tucker 1987) and parts of the Amazon Basin (Hecht 1981, 1983, 1985). Key explanatory factors include tax incentives to promote corporate investment, increasing land values and land speculation, revised land laws that have turned communal lands into private property, funding of cattle-ranching interests by multilateral develop-ment banks, dispossession of small farmers by investors in beef production, and the social prestige that adheres to the mystique of the cattle baron.

Selective logging is especially important in insular South-east Asia, which dominates world trade in tropical hardwoods. Operating behind the obvious signs of forest alteration and degradation are factors such as greed and political corruption (Plumwood and Routley 1982); new legislation to encourage investment by multinational corpora-tions; tax and credit subsidies for wood-processing industries; and logging concession terms that capture only part of potential rents, thereby providing exceptional profits for concession hol-ders and private contractors (see Repetto 1987, 1988a; Repetto and Gillis 1988).

Third World government policies have resulted in excessive forest depletion (Repetto 1988a, pp. 13–17). Most policies are based on a Western-style development model that emulates the mass-consumption economies of industrialized countries, enhances the privileged positions of governing élites, emphasizes 'growth' rather than social justice and basic human needs, promotes large-scale, export-oriented commodity produc-tion for foreign exchange (which is needed to service huge debt burdens), encourages invest-ment by multinational corporations but places few controls on their exploitation of natural resources, and relies heavily on loans from development banks, whose frequently short-term economic and security interests and lax environmental standards tend to promote environmental degradation (Plumwood and Routley 1982; Rich 1985; Fitz-gerald 1986; Fearnside 1987b). The policies that implement the model call for, among other things, new roads, dams, mines, integrated land develop-ment and settlement schemes, and monocultures of cash crops, all of which have resulted in massive forest depletion.

But pressures on tropical forests also result from consumer demand in the industrialized countries for timber, beef, minerals, and other commodities; from trade monopolies, protectionist policies, and tariff structures that favour developed countries;

from development aid that is disbursed for political or strategic reasons rather than for the eradication of poverty; from the activities of multinational corporations, including logging, mineral extraction, and agribusinesses that emphasize use of chemical fertilizers and specialization in a few crops—in short, from the nature of the international economy and its multifarious linkages (Redclift 1987, chapter 4).

To sum up: behind the direct and obvious processes of change lurks a variety of others of generally greater importance, including poverty and landlessness, political expediency and the class biases of governments, fiscal incentives, consumerism in the affluent societies, unequal trade relations between rich and poor nations, lending policies, unmanageable debt, and the increasing role of transnational companies. Variations in the pace of forest conversion (Table 1.2) are related not only to national policies and socio-economic conditions but also—indeed, fundamentally—to the structure of the international economy; additional factors include initial extent of forest, accessibility, and presence or absence of other readily exploitable natural resources (such as minerals). A synthesis of the complex story of tropical rain-forest conversion remains to be written.

1.3.2 Some consequences of forest change

While much about the human impact remains unknown, speculative, or controversial, it is widely believed that 'tropical habitats are more sensitive, less resilient, and in greater danger of complete destruction than [are] temperate or boreal habitats' (Soulé and Wilcox 1980, p. xi). Although the consequences of human activities are usually interrelated and cumulative, it is convenient to divide them into three arbitrary categories: environmental, biological, and human.

Potential environmental impacts include changes in climate, hydrological conditions, soils, and nutrients. Although there is no firm evidence that 'regional rainfall is either significantly increased by afforestation or decreased by de-

Table 1.2 Annual rates of tropical deforestation in selected countries

Country	Annual rate (10^2 km²)[§]	
	1981–5[†]	1989[‡]
Bolivia	8.7	15
Brazil	136	500
Burma	10.15	80
Cameroon	8	20
Colombia	82	65
Congo	2.2	7
Ecuador	34	30
Gabon	1.5	6
Ivory Coast	29	25
India	13.2	40
Indonesia	60	120
Kampuchea	2.5	5
Laos	10	10
Madagascar	15	20
Malaysia	*25.5*	*48*
Mexico	47	70
Nigeria	30	40
Papua New Guinea	2.2	35
Peru	26	35
Philippines	9.1	27
Thailand	24.4	60
Venezuela	12.5	15
Vietnam	6	35
Zaïre	18	40

† Sayer and Whitmore (1991, table 3, p. 206, citing FAO 1988).

‡ Myers (1989, table 1, p. 7).

§ Note that the considerable discrepancy between the two sets of figures (especially for Asian countries) is at least in part a reflection of different definitions of *deforestation*. Myers (1980, 1989) employed a broad definition (clearance combined with alteration, including logging), whereas FAO (see Lanly 1982, pp. 74–7) has used a more precise definition (essentially clearance followed by other land uses). But regardless of the definitions employed, it would appear that there has been a *real* increase in the rate of deforestation in most countries.

forestation' (Goudie 1986, p. 260), it has been intimated, for example, that forest clearance may result in altered rainfall patterns and intensities (Brünig 1977, p. 190; 1985, pp. 20–1). Whether extensive deforestation will result in regional climatic change is unclear because models predict contradictory results (Dickinson 1981, pp. 425–7; Salati and Vose 1984, p. 136). The burning of tropical forests has been variously implicated in the rising CO_2 content of the atmosphere (e.g. C. S. Wong 1978; Woodwell 1978; Goreau and Mello 1988), but just what impact the 'greenhouse effect' will have on tropical forests and their species remains unknown (Bolin *et al.* 1986, p. xxxi).

Studies suggest that removal of the rain forest results, for example, in higher albedo, lower evapotranspiration, reduced rainfall interception, lower infiltration rates, increased runoff, more rapid soil erosion, river sedimentation, downstream flooding, and higher peak storm flows and reduced low flows (e.g. Brünig 1977; Gentry and Lopez-Parodi 1980; Salati and Vose 1983; Myers 1986*a*). On the other hand, watershed research has revealed many misconceptions about the protective role of tropical forests and about the impact of development activities on the hydrological behaviour of tropical forest lands. For example—and contrary to conventional wisdom—the presence of forests *lowers* the water yield; changing the forest cover in headwater catchments usually produces only a minor impact on downstream flooding; and other vegetation types—grassland for instance—can provide protection that is equal to or better than that of forests (for details, see L. S. Hamilton 1983, 1985*a,b*; Cassells *et al.* 1987). Given current management practices, however, and considering the full range of services they perform, forests are the most *practical* use of watershed areas (L. S. Hamilton 1987*b*; Smiet 1987).

The diversity of life on earth is being depleted by habitat destruction and overexploitation of species (Inskipp and Wells 1979; Fitter 1986, chapters 5 and 6). Some scholars believe that the rate of human-induced extinction is now several hundred times higher than the average 'background' rate of extinction; in addition, the amount of genetic variation within individual species is being eroded. Although it is a world-wide phenomenon, the depletion of biological diversity is especially rapid in tropical moist forests, home of

at least half (perhaps 90 per cent or more) of the earth's estimated 5–30 million species of organisms. There is mounting concern that the human impact on tropical forests could result in the biggest 'extinction spasm' since the 'great dying' of the dinosaurs some 65 million years ago (Myers 1979, 1985*a,b*, 1986*b*; Ehrlich and Ehrlich 1983; Myers and Ayensu 1983; Wilson 1985; Simberloff 1986; WCED 1987, chapter 6).

How many species will be driven to extinction in coming decades is not known. Estimates range from 15–50 per cent of all species by the year 2000, and it has been suggested that a staggering 60 000 plant species could disappear for ever by 2050 (Davis *et al.* 1986, p. xxxiv). Several critics of the assumptions on which such estimates are based have claimed that the rate of extinction has been greatly exaggerated (cf. Raven 1984 and Myers 1985*a* with critics Simon 1984, 1986; Simon and Kahn 1984; Simon and Wildavsky 1984). The dispute highlights the need for a complete inventory of life on earth (Wilson 1985), more comprehensive monitoring of extinctions, greater funding for biodiversity research (Office of Technology Assessment 1987; Shen 1987), and more serious attention to the conservation status of invertebrates, which comprise the majority of species (Collins and Wells 1983; Greenwood 1987).

Wild species are of inestimable real and potential value to humans (Myers 1983*b*; Huxley 1984; Fitter 1986). In addition to performing such vital ecological services as pollination, natural disease control, and nutrient cycling, they contribute in a myriad of ways to our material and non-material welfare. Tropical species are no exception, although hardly one in 20 is known to science (although many more are known to forest-dwellers). Extinction is probably expunging genetic resources that might otherwise have contributed to advances in agriculture, industry, and medicine (Oldfield 1981; National Research Council 1982).

The impact of forest conversion on indigenous forest-dwellers is a matter of growing concern to feeling and caring persons everywhere. In that there is indeed much to be concerned about, it is all the more to be regretted that certain conventional, paternalistic, or otherwise outworn views about such peoples continues to persist. Policies based on such views, no matter how well intentioned, are likely to be ineffective.

Just as natural systems are dynamic, so too are human–environment relations: change rather than stasis, flexibility rather than rigidity, and diversity rather than uniformity have generally characterized such relations. Contrary to widespread opinion, isolation from outside influences since time immemorial has not been typical of the experience of many forest peoples, including hunter–gatherers; change, whose direction is variable, emanates not only from outside but also from within; and far from being unchanging hidebound conservatives, so-called traditional peoples often eagerly embrace new opportunities that are perceived to be beneficial.

The spectrum of human consequences of forest loss and associated changes range from extinction through forced acculturation to creative and successful adaptation to new opportunities. According to Ribeiro (as cited in Smith 1985, p. 358), 87 of Brazil's 230 tribes became extinct between 1900 and 1957. Various extant indigenous peoples in the same region face obliteration of their cultures as coherent ways of life. A similar fate probably faces the Penan of Sarawak, where logging is polluting rivers, depleting game, and rapidly encroaching on remaining ancestral lands. Logging has destroyed or severely disrupted local economies in other parts of South-east Asia (Plumwood and Routley 1982; Hurst 1987).

On the other hand, several studies have shown that people like Bugis migrants in East Kalimantan, Iban shifting cultivators in Sarawak, swidden farmers elsewhere in South-east Asia, and the river-bank people of the Amazon (ribereños) are very responsive to new opportunities for increasing their income; for example, flexibility and resourcefulness have been revealed in such matters as adjusting to price changes and market conditions, accepting new plant species or varieties, adopting irrigation or other techniques where conditions are favourable, and finding new of employment (Kunstadter 1978; Vayda 1980; Padoch 1982, 1988; Padoch and V 1983, pp. 308–10; Vayda and Ahmad Sahur 1985). The creativity, decision-making capabilities, rationality, and knowledge of diverse environments that such peoples possess are an important resource that decision-makers would do well to keep in mind (Vayda et al. 1980).

1.4 CONSERVATION AND SUSTAINABLE DEVELOPMENT

1.4.1 Concepts

Conservation of living, renewable natural resources such as forests is that aspect of management 'which ensures that utilization is sustainable and which safeguards the ecological processes and genetic diversity essential for the maintenance of the resources concerned', according to the *World Conservation Strategy* (IUCN/UNEP/WWF 1980, n.p.). From the perspective of experience, conservation is a point of view, a state of mind, perhaps an emotive call to action; it implies a particular view of land and life. Human well-being depends on conserving living resources and the diversity of life on earth.

Development is a process of change and growth that concerns all nations. Its central goal is to enhance the human condition, both economic and non-material. Economic development is a historical process that in recent centuries has favoured the North, which has created a one-world economy in which the South is largely a dependent part.

Most development strategies, whether international or national, capitalist or socialist, have failed to meet the basic needs of the global underclass (Eckholm 1982; Bartelmus 1986, chapter 1). It is now generally recognized that environmental problems result from both development and the lack of it; that an integrated approach to environment and development is required, because ecological and socio-economic factors are inseparable; that conservation and development are mutually reinforcing; and that many development trends, in both the North and the South, are unsustainable. These considerations are behind the concept of 'sustainable development', which has been defined as 'development that meets the needs of the present without compromising the ability of future generations to meet their own needs' (WCED 1987, p. 43). In short, sustainable development is a process of change that generates ongoing social and economic benefits while maintaining the integrity and productivity of natural systems. This is a key concept in the *World Conservation Strategy* (IUCN/UNEP/WWF 1980) and *Our Common Future*, the report of the World Commission on Environment and Development, which emphasized the basic needs of the world's poor and the limitations of the environment to meet present and future needs (WCED 1987; see also Clark and Munn 1986; Global Tomorrow Coalition 1986; Redclift 1987; SGSNRM 1988).

Everywhere, but especially in the poor countries, such basic requirements as food, shelter, clean water, health care, education, and employment are essential to sustainable development. Indeed, the productivity of natural systems cannot be maintained unless these needs are met because, for example,

The farmer clawing out a living in a fragile tropical forest environment cannot put maintenance of soil productivity or protection of biological diversity ahead of survival. The natural resource base will be managed sustainably only if the basic needs of billions of people are satisfied (Global Tomorrow Coalition 1986, p. 5).

In most cases the results of transferring industrial production techniques from the temperate, developed North to the quite different conditions of the tropical, developing South have been less than impressive (Janzen 1973), and too little attention has been given to the relevance of various indigenous resource management systems (Klee 1980*a,b*; McNeely and Pitt 1985). Many development projects in the humid tropics are probably neither ecologically nor economically sustainable.

1.4.2 Conservation and sustainable land use

According to the *World Conservation Strategy* (IUCN/UNEP/WWF 1980; also R. Allen 1981; Fitter 1986), conservation of living resources like tropical rain forest means doing three things: maintaining essential ecological processes and life support systems, preserving genetic diversity, and utilizing species and ecosystems sustainably. Sustainable development requires the practice of conservation in all places at all times. The two are mutually reinforcing. Conservation that is confined to parks and reserves, which increasingly resemble islands in a sea of humanized landscapes, will not be enough to protect biological diversity, genetic resources, and the integrity of ecosystems; in addition, there is an urgent need to devise and adopt sustainable systems of land use for the vast areas of forest and cleared lands that lie outside specially protected natural areas (Dasmann 1984).

Land-use planning requires particular attention to the ability of natural systems to recover from disturbance. That ability, as Jordan (1986, 1987*a*, *b*) has shown, is related to the intensity, size, and duration of the disturbance. Planners need to recognize that large, intensively used clearings may never be reclaimed by climax (or primary) rain-forest species because:

1. Some or all of the nutrient-conserving mechanisms that operate in undisturbed forest, such as large root mat, concentration of roots at or near the soil surface, and associations between plant roots and mycorrhizal fungi, are likely to be destroyed (Jordan 1982, 1985, chapter 2).

2. The seeds of primary tree species cannot tolerate the light and humidity conditions that occur in large clearings (Ng 1983).

3. Whereas small clearings favour regeneration because birds and mammals that disperse seeds and mycorrhizal propagules will enter and cross them, such animals 'are less likely to carry seeds to the middle of a clearing several kilometers in diameter' (Jordan 1985, p. 154).

4. Keeping in mind that pollination and dispersal of most primary rain-forest species probably only occurs over short distances, 'gene flow between distinct ecotypes is precluded where patches of forest are isolated' (Whitmore 1980, p. 315).

5. Constraints on reproduction of many tree species include sparse distributions, limited size of populations, and dioecy, all of which are likely to be aggravated by large clearings.

6. According to Ng (1983, p. 369), 'the soil seed bank is virtually useless as a source of seeds for the establishment of the original composition of the forest after loss of the mature vegetation'.

7. Fire-resistant grasses and shrubs, which are now common in many degraded areas, can effectively block forest growth (Uhl 1983).

Frequently repeated activities that never permit the forest to progress beyond the building phase of the growth cycle (see section 2.4.3) are likely to cause depletion of biological diversity and loss of nature's services.

Towards sustainable forestry

'Sustainability' usually refers to the ability of a system to maintain productivity in spite of minor stresses or more severe perturbations (see Jordan 1987a, pp. 102–4). A basic requirement of sustainable land-use systems is the allocation of the most appropriate land to uses that will maintain the productive capacity of the natural resource base, with 'the end result being potentially more valuable than any known alternative' (Poore 1976, p. 5). Most constraints are primarily political, economic, and social rather than technological. A major problem is the general absence of integrated national land-use policies and plans that include a forest policy (Wyatt-Smith 1987).

Trees and forests perform several roles or functions, each of which can contribute to basic human needs. These include rehabilitating damaged or degraded forestlands; protecting soils and water catchments; ameliorating site conditions; producing industrial timber, domestic wood needs, and minor forest products; and preserving natural ecosystems and genetic resources for scientific, recreational, and other purposes (this role is discussed below in relation to parks and reserves).

Rehabilitation Damaged forests—there are some 16 million ha in Indonesia alone—include overlogged forests, degraded agricultural lands, eroded hillsides, and overgrazed or repeatedly burned cattle pastures (Grainger 1988; Lugo 1988). Returning such areas to productive use means that succession must be manipulated and managed. Fundamental requirements are 'conservation and restoration of soil organic matter and soil fertility. An ample supply of genetic material and favorable land form are also critical elements' (Lugo 1988, p. 43).

Protection The trend is for residual forests to be confined to upland areas, often on steep slopes and over rugged terrain. However, even these forests are coming under increasing pressure from slash-and-burn farming and other activities. Reduction of ground cover and poor land management in catchment areas can result in increased runoff, soil erosion, river sedimentation, and landslides, all of which may affect downstream agricultural regions (although as already noted above, there are many myths and misconceptions regarding the hydrological consequences of changing land uses). Emphasis should be placed on maintaining effective ground cover, retaining trees on erosion-prone slopes, and protecting riparian buffer zones so as to minimize the impact of upstream developments on downstream areas (see Cassells *et al.* 1987).

Amelioration Trees can reduce soil erosion, improve the nutrient status and physical properties of soils, reduce leaching losses, help to control pests and diseases (especially if communities are diverse), and in various other ways help to improve site conditions (Cruz and Vergara 1987). As a vital part of agroforestry systems, for example, the rehabilitative and ameliorative roles of trees may lead to long-term benefits such as improved and sustained crop productivity, more employment and income opportunities, better health and nutrition, and conservation of natural resources (e.g. Myers 1986c).

Production   Three themes are outlined: supplying industrial timber; farm-forestry planting; and harvesting minor forest products.

Exports of tropical hardwoods have increased dramatically in the post-war period (Laarman 1988). Most supply has been based on reckless 'mining' of primary forests, and timber reserves have been severely depleted in some countries. Meanwhile demand for industrial wood products is rising sharply in many developing nations. Strategies that would remove pressure from remaining primary forests while helping to meet mounting domestic demand include:

(1) creating compensatory plantations of fast-growing species, preferably in areas that have already been damaged or degraded;

(2) bringing more production forests under silvicultural management and encouraging better silvicultural practices;

(3) obtaining more revenue per unit of forest area by promoting local processing, removing trade restrictions, and developing markets for lesser-known species;

(4) changing government policies that promote rapid exploitation while yielding excessive profits for concessionaires and logging companies;

(5) promoting policies in the importing countries that would reduce, or even eliminate, the demand for tropical hardwoods (see e.g. Spears 1979, 1983; Poore 1983b; Repetto 1987; Postel and Heise 1988, pp. 23–35).[4]

One of the consequences of deforestation is depletion or outright elimination of essential sources of products like fuelwood, fodder, and building poles. Replenishing such products requires compensatory replanting in the form of low-cost farm forestry (also called 'social forestry', 'community forestry', or 'agroforestry'), although government-managed plantations also have a role to play in ensuring adequate supplies of fuelwood, charcoal (for urban dwellers), and building poles, as well as providing employment for the landless (Spears 1983). Studies suggest that agroforestry practices (see below) can also contribute to higher agricultural productivity, more employment and income, and better environmental conditions (e.g. Winterbottom and Hazlewood 1987).

So-called minor forest products like fibres and canes, essential oils, exudates, dyes, waxes, and tannins have been harvested and traded for centuries. (The pre-eminence of one product, namely timber, is a very recent phenomenon.) Some non-timber products can be grown on plantations (rubber is an obvious example), while others require harvesting dispersed resources from undisturbed or secondary forest. The dispersed products offer many opportunities for decentralized, small-scale, forest-based industries that would leave the forest intact. In addition to being labour-intensive and potentially conserving of resources, most such industries can be adapted to a wide range of socio-economic and environmental conditions (see e.g. Myers 1984, chapter 12, 1986c).

Towards sustainable agriculture

Sustainable *and* more productive agriculture is required to meet basic human needs and to relieve pressure from the remaining primary forests. Food production has been greatly enhanced by high-yielding crop varieties (mainly rice in the humid tropics), but most small farmers cannot afford the costly inputs that such crops require and many occupy marginal or 'fragile' lands that are unsuited to Green Revolution technologies (Greenland 1975; King 1979). The number of such farmers is growing rapidly and this combined with several other factors—for example, increasing land concentration, distorted patterns of food distribution, unemployment and inability to purchase food, and agricultural yields that have not increased markedly in recent decades—places mounting pressure on the remaining forests.[5]

The basic challenge is 'to balance the need to exploit forests against the need to preserve them' (WCED 1987, p. 136). A critical first step is to assign lands and forests to the most appropriate uses (although the necessary inventories are lacking in many developing countries). For example: intensive food production should be confined to the best land (now largely pre-empted by commercial crops for export, with the result that poor farmers are often confined to marginal areas) or to kitchen gardens or other plots where it can benefit from organic inputs; agroforestry is the best use of many areas—especially rolling or sloping land—and can be utilized to rehabilitate degraded lands; and large areas should remain under managed forests or strict reserves (e.g. Fearnside 1979, 1983). The prospects for continuous cultivation of annual crops on acid soils is still a matter of debate (see above). Generally speaking, polycultures are preferable to monocultures, perennial crops to annual crops. Especially for the millions of small farmers who have been bypassed by Green Revolution technologies, productive and sustainable agriculture calls for 'a new view of agricultural development that builds upon the risk-reducing, resource-conserving aspects of traditional farming, and draws on the advances of modern biology and technology' (Dover and Talbot 1987, p. 7).

Many indigenous peoples of the tropical rainforest biome have devised agro-ecosystems that conserve genetic resources (Oldfield and Alcorn 1987), suppress pests and weeds, yield balanced diets, provide employment throughout the year, and sustain production on a long-term basis; some other attributes of such systems, most of which are polycultures, include detailed environmental knowledge, adaptation to local conditions, structurally complex and diverse plant communities, 'maintenance of soil organic matter, minimization of soil disturbance, and control of the size and shape of the disturbed area'

(Jordan 1985, p. 149; also Dickinson 1972; Dover and Talbot 1987, pp. 32–42; Redclift 1987, pp. 150–7; Gradwohl and Greenberg 1988, chapter 2). Studies reveal that many indigenous resource management practices—for example, intensive cultivation of kitchen gardens, agroforestry systems, and hillside terracing—can be usefully retained, revived, or improved in schemes to enhance agricultural productivity (e.g. Gliessman et al. 1981; Nations and Komer 1983; Posey 1985; Gómez-Pompa 1987) and that there is great potential for combining these practices with new biotechnologies that would boost the yields of many Third World staples (Wolf 1986; Tangley 1987).

Over much of the humid tropics, but especially in upland areas, there is great scope for agroforestry—that is, the deliberate association of trees with crops and/or animals in the same area—to increase the sustainable productivity of the land, thereby helping to meet basic needs while slowing the pace of forest destruction (e.g. Vergara 1985; Rocheleau and Raintree 1986; Vergara and Briones 1987).

Achieving sustainable agriculture in the humid tropics, or elsewhere for that matter, will require many socio-economic and political changes—local, national, and international—including changes that conflict with the prevailing Western capitalist view of agriculture. The basic requirement 'is an attitude toward nature of coexistence, not of exploitation' (Altieri et al. 1983, p. 48; see especially Redclift 1987).

1.4.3 Parks and reserves

Protected natural areas serve many of the goals of conservation. These include preserving diversity; allowing evolutionary processes to continue undisturbed; safeguarding ecological and environmental services; supplying representative ecosystems for scientific research, monitoring, and bench-mark studies; saving wildlife, genetic resources, and gene pools; protecting watersheds; providing forests for recreation, tourism, and education; and affording secure homelands for indigenous forest-dwellers. Adequate protection of representative tropical rain-forest ecosystems and their species may require parks and reserves that cover some 10–20 per cent of the entire tropical moist forest biome, whereas recent coverage was only about two per cent (Myers 1983*a*).

Issues concerning the location, size, number, shape, and disposition of reserves have generated considerable debate, most of it cast in the context of island biogeographical theory (MacArthur and Wilson 1967). Simberloff and Abele (1982) and Soulé and Simberloff (1986) have written admirable reviews, and there is no space here to go over the ground again. Instead, we offer only a few general points:

(1) forest fragmentation is proceeding apace and virtually all nature reserves will eventually occur as islands in a sea of development;

(2) most researchers agree that reserves should be large;

(3) studies of edge- and area-related effects suggest that extensive buffer zones are needed to protect core areas (Lovejoy *et al.* 1983, 1986);

(4) the real and potential contributions of protected species and ecosystems to sustainable development and the welfare of indigenous forest-dwellers should be explicitly recognized in land-use planning and reserve management (Miller 1982);

(5) successful reserves require the support and co-operation of local people, a rational land-use plan, maintenance, long-term management, adequate funding, and local political support (Blower 1984; Gradwohl and Greenberg 1988, pp. 55–101);

(6) sustainable land-use systems must be adopted in surrounding areas (Dasmann 1984, p. 671).

1.5 CONCLUSION

Tropical rain forest is the most complex and exuberant expression of the earth's green mantle. Now much fragmented and tattered round the edges, the tropical mantle has begun to unravel, revealing a thin, worn-out undergarment that provides little protection from the elements. So rapid is the pace of change that much of great scientific, biological, and cultural value is being lost for ever. Very few large, intact blocks of forest are likely to survive much longer. In the landscape of the future there will be less primary forest, large remaining areas of secondary forest, extensive degraded areas requiring rehabilitation, perhaps more plantations and a greater area devoted to agroforestry, more roads, settlements, and people, and various island-like parks and reserves in an otherwise largely humanized scene. Sustainable agriculture and forestry will probably be more widely adopted, but not soon enough to avoid a spasm of plant and animal extinctions. Deforestation will continue for as long as mass poverty persists. Can basic human needs be met, thereby removing pressure from the remaining forests? The now fashionable concept of sustainable development, which focuses on such needs, leaves many questions unanswered: How is the basic-needs approach to be implemented? What does such an approach imply in terms of political change? How is the international co-operation that is required to promote and manage sustainable development to be achieved? How can basic needs be met at the local level while structural linkages at the international level continue to result in marginalization?

2 THE RAIN FORESTS OF MALAYSIA

The Asian rain-forest block is centred on the Malay Archipelago, the name given to the great chain of islands linking mainland South-east Asia to Australia. In the western part of the region, where Malaysia is situated, are dipterocarp-dominated forests of unsurpassed grandeur: in no other rain forest are there loftier trees or a greater number of different species of trees growing together in discrete areas; there are 'no other forests anywhere in the world', according to Whitmore (1988, p. 21), 'which have so many genera and species of a single tree family [namely Dipterocarpaceae] growing together in the same place'; and few other forests can boast a greater wealth of both plant and animal life.

Recent decades have witnessed a frontal attack on the lowland forests of the Archipelago, where vast areas have been cleared for plantation and subsistence agriculture or felled to supply the international trade in tropical hardwoods. The original, near-continuous vegetation cover is now much fragmented, and in some places—for example, in Peninsular Malaysia—only isolated patches of lowland primary forest remain.

This chapter provides an introduction to the rain forests of Malaysia. Coverage is selective, not comprehensive. There are two reasons for this: firstly, there is a rich literature on the subject already;[1] and secondly, it is not our objective to write an ecology of tropical nature in Malaysia, but rather to assess the implications of continuing forest removal and alteration and to evaluate the adequacy of past and current land-use planning and management. We set the Malaysian forests in their regional context, provide a brief description of each of the major forest formations, outline some aspects of the richness and diversity of both the flora and fauna, and present a synopsis of the main functional and dynamic features of the forest system.

The great forests of Malaysia are of value to people everywhere. We hope that this chapter conveys some impression of their richness and diversity and that it conjures up some idea of the value of what is now in danger of being lost for ever.

2.1 THE FORESTS IN THEIR REGIONAL SETTING

The rain forests of Malaysia belong to the Malesian floristic region of archipelagic South-east Asia (Fig. 2.1). This section places the forests in their larger regional setting, thus helping to explain some of their characteristic features.

2.1.1 The Malesian floristic region

Except to the east, the boundaries of the Malesian floristic region are clearly defined (Fig. 2.1). The western boundary lies across the Kra Isthmus, just north of the border between Malaysia and Thailand: here some 200 Asian genera and 379 Malesian genera reach their southern and northern limits, respectively (Steenis 1979). A similarly abrupt boundary between the Asian and Malesian flora occurs immediately north of the Philippines: at least 265 genera that are found on the Chinese mainland (together with Taiwan) do not occur in the Philippines, while some 421 Malesian genera do not extent northwards beyond the Philippines. To the south-east, Torres Strait forms another well-defined boundary, in this case between Malesian and Australian flora: some 340 Australian genera do not extend beyond Cape York and 641 Malesian genera have not migrated into Australia; according to Steenis (1950, p. LXXII), this is 'one of the main demarcations of the Palaeotropic plant world'.

Although Malesia is a well-defined floristic region, the vegetation is far from uniform. Three distinct floristic provinces have been recognized: West Malesia, comprising the Malay Peninsula,

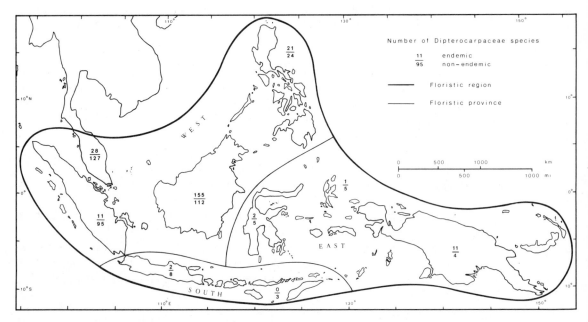

Fig. 2.1 The Malesian floristic region and numbers of dipterocarp species. (From Steenis 1979, p. 102; Ashton 1982, p. 240.)

Sumatra, Borneo, and the Philippines; South Malesia, comprising Java and Nusatenggara; and East Malesia, comprising Sulawesi, Maluku, and New Guinea (Fig. 2.1; Steenis 1950, 1979). Although there are many floristic differences between the three provinces, one of the most striking and significant is the great abundance of the Dipterocarpaceae in West Malesia (Fig. 2.1), where in the lowlands they may account for up to 80 per cent of all emergents and for up to 40 per cent of understorey trees (Whitmore 1981, p. 72; Ashton 1982, pp. 239–40). A staggering 267 species, 59 per cent of which are endemic, are found in Borneo, in contrast to the 15 species found in New Guinea, seven in Sulawesi, and six in Maluku. The deep Makasar Strait between Borneo and Sulawesi has clearly acted as a formidable barrier to dipterocarp migration. Other plants of Asian origin, such as the primitive family Magnoliaceae, are also concentrated in West Malesia, whereas plants of southern (Gondwanic) origin are concentrated or exclusively

found in East Malesia: for example, the southern Pine (*Araucaria*) is restricted to the island of New Guinea.

The flora of South Malesia is depauperate compared with that of the other two provinces (Steenis 1950, 1979), and it boasts only 14 endemic species of flowering plants compared with 132 in East Malesia and 150 in West Malesia. Most species are wide ranging, and most are found in the other two provinces. Dipterocarps are poorly represented (Fig. 2.1). The forests of Java and Nusatenggara are generally more open than are those of the wetter parts of Malesia, and many species have a deciduous habit.

Differences in vegetation between and within the three provinces are related to physical factors and to the migrational history of the Malesian flora. Elevation varies considerably, with peaks in some areas exceeding 4000 metres: for example, Mts. Jaya (4884 m), Mandala (4680 m), and Wilhelm (4510 m) in New Guinea; and Mt. Kinabalu (4101 m) in Sabah. An altitudinal zonation of vegetation occurs, ranging from lowland rain forest through hill and montane rain forest to tropialpine shrubs and grasses (e.g. R. J. Johns 1982).

Variations in vegetation are also related to precipitation totals and to seasonality. Much of Malesia receives an annual rainfall total in excess of 2000 mm that is fairly evenly distributed throughout the year, so that in the lowlands the natural vegetation is evergreen rain forest. Parts of Malesia, including Java, Nusatenggara, southern Papua New Guinea, and parts of Sulawesi, are distinctly drier, receiving between 1000 and 2000 mm of rain per annum (Koteswaram 1974, fig. 27). Here monsoon-type forests occur that are characterized by a more open structure and by deciduous species. Elsewhere in Malesia, localized areas with similarly drier climates are found in the lee of high ground, particularly where it lies across the tracks of moisture-laden airstreams: for example, in the central part of the Malay Peninsula and along the median valley of Sumatra (Dale 1959).

Vegetation changes are often related to lithology and soils and to drainage conditions. Whitmore (1984a) has described distinctive forest types that have developed on soils derived from siliceous sandstones, ultrabasic rocks, limestone, and beach sands. Wetland forests occur in many areas, particularly in low-lying coastal regions where mangrove and peat-swamp forests are widespread.

But the distributions of many genera and species cannot be related to present environmental factors; rather, some distributions can only be explained by reference to the origin and evolution of the Malesian flora, which is a composite of species of Asian and southern continental origin together with endemics, the relative proportions of each varying considerably throughout the region. The explanation for this is related to the region's geological history and to changes in climate and sea-level.

2.1.2 The origin and evolution of the Malesian flora

From land bridges to plate tectonics

Formerly, it was generally assumed that the Malesian flora was essentially of Asian origin and that it had migrated from island to island across now foundered land bridges (Steenis 1962, 1979). It also was assumed that the migration extended to Australia (Burbridge 1960; Barlow 1981).[2] A very different hypothesis began to emerge in the late 1960s, namely that many present-day distributions could be explained by continental drift—that is, by continental masses having 'collided' or 'drifted' apart at various times in the past.[3] It is generally accepted that western Malesia (often called the Sunda Shelf) and New Guinea–northern Australia (often called the Sahul Shelf) 'collided' some 15 million years ago, in mid-Miocene times, allowing interchanges of flora and fauna to occur, although evidence now suggests that some interchange of flora occurred previously during Cretaceous and early Tertiary times when fragments that had broken away from the Australian plate formed 'stepping-stones' between Australia and Asia (Audley-Charles 1987). Fossil evidence, although meagre, suggests that three taxa that were abundant in the Cretaceous and early Tertiary Australian forests— Casuarinaceae, Proteaceae, and the conifer *Dacrydium*—migrated to Sundaland before the mid-Miocene collision (Truswell *et al.* 1987, p. 44).

However, the main interchange of flora undoubtedly took place after the collision of the two plates in the mid-Miocene. Initial contact was submarine and islands gradually emerged, eventually providing stepping-stones for plant and animal migrations. A migration route had probably been established between Sulawesi and eastern Australia by the late Miocene or early Pliocene, and, according to Audley-Charles (1981) this route was probably almost as well established by the late Pliocene as it is at the present day.

Plant migrations took place in both directions,

but not equally: far more species invaded New Guinea and Australia from Sundaland (the name given to the land-masses of the Sunda Shelf) than invaded Sundaland from Australia or New Guinea. The main reason for the virtual one-way traffic would appear to be essentially ecological (Barlow 1981). The emerging islands of the New Guinea Archipelago provided a succession of new environments that were invaded by Malesian elements. The existing cool-adapted flora could not compete and retreated to higher elevations, reinforcing a trend already in progress from the time that New Guinea began to move into tropical latitudes. The New Guinea forest flora is today characterized by species such as *Nothofagus* and families such as Cunoniaceae and Monimiaceae, all of which have southern distributions indicating Gondwanic origins (Balgooy 1976, p. 18). In contrast, relatively few species of Australian origin have migrated westwards into South and West Malesia. Australian species have found it difficult to displace Malesian species, which are environmentally well adapted and occupy virtually all available niches (D. Walker 1982).

Trees of southern origin that have migrated into western Malesia and beyond are conifers. Four genera extend north of the equator into Malaysia: *Agathis*, *Dacrydium*, *Phyllocladus*, and *Podocarpus* (Whitmore 1984*a*, chapter 16). *Agathis* and *Podocarpus* are the most common conifers in the lowland evergreen forests of both Peninsular and East Malaysia, while *Agathis* and *Dacrydium* are locally abundant in Sarawak. Some conifers have considerable economic potential because their softwood timber is of high quality and their high annual volumetric growth rates makes them ideal plantation trees.

The impact of Quaternary perturbations

World sea-levels fell during the Quarternary glacial periods, with a maximum lowering of some 180 m during the last glacial period (D. Walker 1982). In the Malesian region, large expanses of the Sunda and Sahul shelves were exposed and many islands were joined or almost joined (Biswas 1973, fig. 2); for example, northern Australia was

joined to New Guinea, the Malay Peninsula was connected to Sumatra and Borneo, and the distance between Borneo and Sulawesi was reduced to only a few kilometres. The presence of land bridges facilitated the migration of animals and plants eastwards and westwards, while the isolation of many land areas during periods of higher sea-levels was conducive to the evolution of endemics.

A considerable number of cool-demanding plants migrated into the Malesian region during the colder phases of the Pleistocene. Three main migration routes have been postulated: a Sumatran track utilized by taxa from the Himalayan region (e.g. *Primula imperialis*); a Taiwan–Luzon track for taxa from the north (e.g. *Lilium*); and a New Guinea track for taxa of southern origin (e.g. *Drypetes*, *Astelia*) (Flenley 1979, pp. 79–81).[4]

The impact of Quaternary climatic changes on the extent of the lowland rain forests of Malesia is uncertain, although mounting evidence now suggests that the forests retreated in some areas during the glacial periods (Morley and Flenley 1987). Verstappen (1975) postulated that rainfall totals during such periods would have been considerably lower throughout Malesia than they are at present, owing to changes in the location of the Intertropical Convergence Zone and to the emergence of some 3 million km^2 of land as the sea-level eustatically fell. The replacement of extensive areas of warm shallow seas by land, together with changes in the pattern of ocean current circulation and generally lower temperatures, would have combined to reduce rates of evaporation and hence rainfall totals. It would seem likely that the Malesian lowland rain forests contracted in some areas, particularly in marginal environments (Whitmore 1981). The occurrence in parts of Java and Nusatenggara of patches of rain forest in the midst of depauperate deciduous forests was considered by Steenis (1979) to indicate that the evergreen rain forests were formerly more widely distributed, and Morley and Flenley (1987, p. 54) have suggested that savannah-like conditions existed in parts of the Malay Peninsula, thereby allowing plants and animals from seasonal areas of

Thailand to migrate to Java and Sulawesi. That there was a previously drier climate in the Malay Peninsula is suggested by the occurrence of alluvial spreads and by fossil conifer and grass pollen.

The Malaysian flora was significantly altered by the profound physical changes that occurred during the Pleistocene. It would appear that the lowland forests contracted in some areas, while fluctuating sea-levels and river courses (together with tectonic movements) influenced the migration of certain flora (particularly the dipterocarps) and the evolution of endemics (Ashton 1969).

2.2 THE RAIN FORESTS: FORMATIONS AND VARIATIONS

Although both Peninsular and East Malaysia are located within the West Malesian floristic province (Fig. 2.1), the Malaysian forests are far from uniform. There is a well-defined altitudinal zonation of vegetation, and there are distinctive variations within the lowland and montane forests. The observed distributions have been attributed to several factors, including the following: migrational history of the flora, particularly with respect to physical barriers; isolation and the emergence of endemics; variations in lithology, soils, topography, and water availability; and disturbance resulting from landslides, cyclones, and other natural events. Some species appear to be randomly distributed.

This section presents a brief description of the various forest formations and examines the topic of intraformation variations; it also includes some information on species numbers and a discussion of why so many species occur.

2.2.1 The forest formations

A note on classification

Various criteria have been used to classify vegetation. Those most commonly employed relate to habitat, physiognomy, and floristics (Poore 1963). The more sophisticated classifications are based on floristic criteria, which require not only lists of species present but also quantitative

information on distribution and densities. Webb and Tracey (1984) have prepared a floristic classification of the Australian rain forests but this is not yet possible for the much richer, far more extensive, and often less accessible Malesian rain forests, although some attempts have been made to include subdivisions based on dominant species (Wyatt-Smith 1964; Brünig 1969a; Fox 1978, 1983).

What has emerged over the years for the Malesian region is a variety of classifications based mainly on habitat and structure and generally embracing a single country or part thereof (see Whitmore 1984b; also Paijmans 1975). Terminology is far from uniform because local names are often used and more than one name is sometimes given to the same formation, all of which has led Whitmore (1984a, p. 156) to comment that there are almost as many names as there are authors!

Recently, two Malesia-wide classifications have appeared, one proposed by FAO (FAO/UNEP 1981c), the other by Whitmore (1984a). Both are based on structure and physiognomy and both have been developed from earlier schemes (Champion 1936; Burtt-Davy 1938; Steenis 1950). This book follows the classification and nomenclature proposed by Whitmore.

Figure 2.2 shows the forest formations that occur in Malaysia. Although they are differentiated on the basis of structure and physiognomy, the formations can be grouped according to habitat—wetland, dryland, and montane. The montane forests appear around 750 m above sea-level (Fig. 2.3), although this may vary with aspect and with the mass of hill or mountain (Steenis 1972). It should be kept in mind that most of the lowland forests in Peninsular Malaysia and extensive forest tracts in Sabah and Sarawak have been logged, hence their structure and species composition differ from those of virgin stands (Whitmore 1988).

The lowland evergreen formation

The lowland evergreen rain forest is the most extensive forest formation in Malaysia (Fig. 2.2

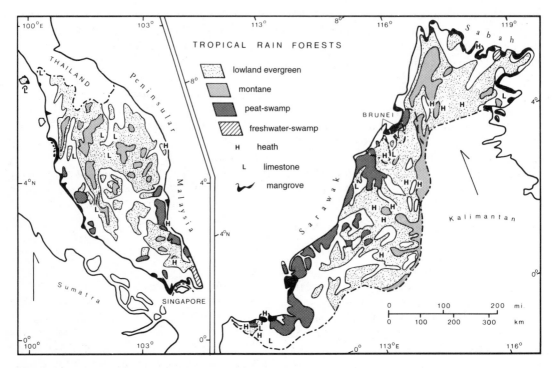

Fig. 2.2 Malaysia: forest formations. (From Whitmore 1984*b*.)

and Plate 1). This is the 'true' rain forest (or jungle, although many popular images misrepresent reality). It is found on dry land below approximately 750 m and comprises both the lowland and hill dipterocarp forests (Symington 1943/1974; Wyatt-Smith 1964; Fox 1978; Whitmore 1984*a*). The climate of such forested areas is uniformly hot throughout the year and annual rainfall totals are high (1500–3000 mm), with a fairly even distribution throughout the year (FAO/UNEP 1981*c*, p. 26).

The trees in this formation are tall, the canopy being some 21–30.5 m above the ground (Wyatt-Smith 1964). Emergents rise to 46 m and some giants exceed 70 m: for example, along the ridges of the Andalau Forest Reserve in Brunei (Ashton 1964, fig. 25). A number of workers have recognized three distinct tree layers in the lowland forests: a main canopy layer, an emergent layer, and a lower canopy of suppressed trees and saplings (Wyatt-Smith 1964; Robbins and Wyatt-Smith 1964; FAO/UNEP 1981*c*). However, the reality of such stratification in tropical lowland forests has been questioned (e.g. Hallé *et al.* 1978) and the debate remains unresolved, although observation clearly shows that emergents do rise above the main canopy layer. Most trees have straight trunks that are unbranched to considerable heights, although a variety of architectural forms have been identified and named (Hallé *et al.* 1978). Lianas and epiphytes add to the architectural complexity and visual experience. Contrary to popular believe, the forest floor is generally open and not a tangled mass of vegetation, while the litter layer is generally thin and sometimes patchy, the result of rapid decomposition and surface wash (Leigh 1982*a*).

Foresters and ecologists distinguish between the lowland and the hill dipterocarp forests (e.g. Symington 1943/1974; Wyatt-Smith 1964; Fox

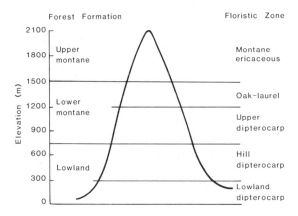

Fig. 2.3 Altitudinal zonation of vegetation in Peninsular Malaysia. (From Whitmore 1984a, p. 243.)

1978). In the latter the canopy is generally lower and less continuous (Robbins and Wyatt-Smith 1964), species distributions change, and new species appear. Many areas of hill forest in Peninsular Malaysia are characterized by stands of seraya (*Shorea curtisii*) growing along the ridges, and other species of Shorea such as *S. laevis* and *S. glauca* are frequently encountered (Symington 1943/1974; Wyatt-Smith 1964; Burgess 1968); a feature of the shrub layer in some areas is the presence of dense thickets of bertam palm (*Eugeissona tristis*).

Neither lowland nor hill dipterocarp forest is floristically uniform. Wyatt-Smith (1964) identified nine types of lowland and six types of hill forest in Peninsular Malaysia and Fox (1978, 1983) identified five types of lowland and two types of hill forest in Sabah. Both suggested that the distribution of these forest types can in part be related to topography and lithology.

A feature that distinguishes the forests of Sabah from those of other parts of Malaysia is the abundance in many areas of species of *Parashorea* (Fox 1983). In the north-east and south-east coastal lowlands, *Parashorea tomentella* and *Parashorea malaanonan* are dominant species, and almost pure stands of the latter in the hinterland of Darvel Bay were among the first forests to be commercially exploited in North

Borneo (now Sabah; see John 1974). Today, few such stands survive near the coast or along the lower reaches of the major rivers.

The heath forest formation

Heath forests replace the evergreen lowland rain forest in parts of Malesia. They are usually found on free-draining podzolic soils (often with a hardplan layer) derived from siliceous parent materials (Browne 1952; Brünig 1970, 1971, 1973; Ashton 1971; Whitmore 1984a). Along coastal lowlands they are characteristically found on raised beaches and on soils derived from sandy marine and fluviatile sediments, both legacies of fluctuating Pleistocene sea-levels. Further inland they are found on sandstone plateaux and cuestas (P. W. Richards 1936a,b). The most extensive heath forests in Malaysia are in Sarawak (Fig. 2.2, Plate 2), where they are generally referred to as *kerangas*.[5] Other areas where heath forests occur include the east coast of Peninsular Malaysia, the east coast of Sumatra, and the Indonesian islands of Bangka and Belitung (Whitmore 1984a).

Even to the casual observer, the heath forests are strikingly different from other forests. Although none of the tree species is unique to the heath forest, there is a much smaller range of species than in adjoining evergreen forest: 860 species v. 1800–2300 species in the case of Sarawak, according to Brünig's (1973) estimates. Forest structure is also very different. The canopy of the heath forest is relatively low and there are few emergents. Small saplings and poles predominate, often forming an almost impenetrable barrier, in contrast to the generally open floor of the evergreen forest. Colours also differ, the greens of the evergreen forest being replaced by reddish brown hues in the heath forest (Ashton 1971).

Species of *Hopea*, *Shorea*, and *Cotylelobium* are often prominent in more favoured sites. A distinctive and interesting feature of many heath forests is the presence of members of the Myrtaceae family, particularly *Tristania*, and the conifers *Podocarpus* and *Dacrydium*, which are of Australian origin (Ashton 1971, p. 158; Whitmore 1984a, p. 167). The survival of the conifers may be due to the heath forest environment being un-favourable for flowering plants.

Much speculation has focused on the factors responsible for the distinctive structure and xero-morphic physiognomy of the heath forest vegeta-tion. Two hypotheses have been proposed: periodic water deficiency and nutrient deficiency. Brünig (1969b, 1970, 1971) favoured periodic drought as the causal factor, noting that the following structural and physiognomic charac-teristics of the heath forest are likely to minimize water loss or reduce the heat load on the leaves: relatively low roughness of the canopy surface; small leaf size; steeply inclined leaves and twigs; and pale tone of the foliage. Droughts certainly do occur from time to time in Borneo and are occasionally severe, as they were in 1914 and in 1982–3 (Cockburn 1974; Malingreau et al. 1985). The exceptional droughts are undoub-tedly associated with major variations in global-air and ocean-current circulations, which are believed to be triggered by changes in the location of El Niño—the ocean current that originates off South America.

The nutrient deficiency hypothesis was first suggested by P. W. Richards (1952) and sub-sequently examined in detail by Proctor et al. (1983a,b) in conjunction with the Royal Geo-graphical Society's Gunung Mulu Expedition. Studies at Mulu showed that soils under virgin heath forest are not systematically poorer in plant mineral nutrients than those under other, adjacent forest formations, and it was concluded that the heath forest is not an adaptation to drought conditions but rather that it is an adaptation designed to limit the flow of toxic hydrogen ions and phenols to root systems from extremely acid soils.

Current thinking would seem to favour the drought hypothesis, although Whitmore (1984a, p. 172) has noted that both drought and nutrient deficiency appear to have similar effects on forest structure and physiognomy. The relative impor-tance of the two factors still needs to be deter-mined on a site-to-site basis (see especially Whitmore 1990, pp. 144–6).

The montane forest formations

The lowland rain-forest formations give way to the montane forest formations above elevations of about 750 m. In Peninsular Malaysia, extensive areas of montane forest occur along the Main Range from the border with Thailand to the vicinity of Kuala Lumpur, and along the East Coast Range between the Kelantan and Pahang rivers. In Sarawak, the montane forests are found in the interior along the border with Kalimantan, while in neighbouring Sabah they are most exten-sive in the Crocker Range, which extends from the Sarawak border northwards to the Kinabalu massif (Fig. 2.2).

Structure and floristics change with elevation, giving rise to separate lower and upper montane formations and to two floristic zones within the lower montane formation (Fig. 2.3) whose eleva-tion extends from about 750–1500 m. The main structural and physiognomic changes that occur with elevation are a decrease in structural com-plexity and in biomass, a decrease in leaf size, and an increase in the numbers of epiphytes, bryo-phytes, and ferns; in terms of floristics, families with a predominantly tropical distribution, including the Dipterocarpaceae, are replaced by families with temperate and subtropical affinities (Whitmore 1984a, p. 250).

Two distinct zones can be recognized within the lower montane formation in Peninsular Malaysia: upper dipterocarp forest from about 750–1200 m and oak–laurel forest between about 1200–1500 m (Symington 1943/1974; Robbins and Wyatt-Smith 1964; Whitmore 1984a, p. 243). The height of the main canopy of the upper dip-terocarp forest ranges from 23–30.5 m above ground level and that of the oak–laurel forest from 18–21 m; emergents are few in the former and generally absent in the latter.

Only a few species of Dipterocarpaceae are generally present in the upper dipterocarp for-ests, but those that do occur are often abundant (Symington 1943/1974; Wyatt-Smith 1964). *Shorea platyclados* and *Shorea ovata* are the dominant species in many areas, while the shrub layer is often composed of dwarf palms and rattans. Dipterocarps generally do not extend into the oak–laurel forests, where species of Fagaceae, Lauraceae, and Myrtaceae dominate except on drier sites where conifers (*Agathis alba*, *Dacrydium* spp., and *Podocarpus* spp.) thrive. Above elevations of between about 1500–1830 m the oak–laurel forest is replaced by the upper montane formation, a low single-layered forest some 1.5–18 m high. The trees are often gnarled and twisted and, because of the frequent cloud cover, there is a proliferation of mosses, liverworts, and filmy ferns that cling to the trees and carpet the forest floor (Robbins and Wyatt-Smith 1964, p. 211). Numerous peaty mounds, exposed roots, and fallen debris hamper overland travel.

Species such as rhododendron are common in the upper montane forest (see Plate 3), which is often referred to as moss forest or ericaceous forest, although conifers and species of Myrtaceae are also present. There are many beautiful orchids and pitcher plants (*Nepenthes*), although these also occur at lower elevations. On Mount Kinabalu in Sabah, for example, 800 species of orchids in 72 genera have been recorded (Meijer 1971) and 16 species of pitcher plants, seven of which are almost exclusively found in the upper montane forest zone, have been described (Shigeo Kurata 1976).

Although no part of Malaysia extends beyond the climatic tree-line, the summits of at least two of the major peaks are devoid of tree cover: on the summit area of Gunung Tahan (2190 m), the highest peak in Peninsular Malaysia, only stunted woody plants (0.3–0.6 m high) grow among the sandstone outcrops (Soepadmo 1971); and on Mt. Kinabalu in Sabah (4101 m), the highest mountain in South-east Asia, trees do not extend beyond 3505 m. At higher elevations grasses and sedges are scattered among the wide expanses of bare rock, a legacy of glaciation during the late Pleistocene (Koopmans and Stauffer 1966; Meijer 1971).

The wetland forest formations

Extensive stretches of the Malaysian coastline and adjoining lowlands are covered by wetland or swamp-forest formations. Forests of this type occur along the western and southern coastal areas of Peninsular Malaysia, behind the coastal dune ridge complexes of the south-eastern part of the Peninsula, and along the coastlines of Sabah, Sarawak, and Brunei (Fig. 2.2). They are generally absent from the high-energy east coast of Peninsular Malaysia and from the more rocky coastline of north-western Sabah. Relatively small patches of swamp forest occur inland in some areas, as is the case, for example, at Tasek Bera in central Pahang.

There are three main wetland-forest formations: peat-swamp forest, freshwater-swamp forest, and mangrove forest; only the last occupies a saltwater environment. Peat-swamp and freshwater-swamp forests have different sources of water supply: for the former it comes exclusively from rainfall, whereas the latter are periodically inundated by stream and river water (Wyatt-Smith 1961*a*, 1964; Whitmore 1984*a*, p. 178), which of course is a richer source of nutrients.

The mangrove forests grow in saline water and are generally best developed along shelving coastlines, particularly in bays and estuaries sheltered from wave attack and strong currents (Paijmans 1976), although they also extend upstream for considerable distances along some of the larger rivers. They are extensive along the shelving and sheltered coastline of western Peninsular Malaysia and along parts of the Sabah and Sarawak coasts, whereas they are virtually absent from the east coast of Peninsular Malaysia, which is exposed to the full force of the north-east monsoon.

Mangroves conjure up images of impenetrable swamps and densely packed trees with stilt roots. Such trees, which are species of *Rhizophora*, are certainly common, but many other species may be present. Mangrove vegetation is extremely sensitive to changes in tidal regimes and water salinity, with the result that there is often a distinct zonation of vegetation. In Peninsular Malaysia, Watson (1928) identified five subtypes related to frequency of inundation, to which Berry (1972) added a sixth, a 'stream' zone, and P. K. K. Chai (1975) identified nine major types and seven minor subtypes in Sarawak. There is generally a transition zone that merges into the adjoining dryland forest on the landward side. Structurally, the mangrove forests vary from scrub-like to dense. Most are relatively low and of even height (6–24 m) with an unbroken canopy (Wyatt-Smith 1964). The understorey and shrub layer are either poorly developed or absent. Most of the tree species are not found in other formations.

Peat-swamp forests extend along virtually the length of the Sarawak and Brunei coasts and at some localities stretch inland for tens of kilometres. In Peninsular Malaysia, they are generally restricted to the coastal areas of Selangor, Pahang, and Johor (Fig. 2.2). Depth of peat varies from 0.5–20 m. The peat swamps are probably of relatively recent origin. According to Whitmore (1984a, p. 183), they have developed since the end of the last glacial period; no endemic tree species has been found, again probably reflecting their relatively recent age. The peat developed in depressions behind mangrove swamps, with the mangrove vegetation gradually being replaced as conditions became less saline (Anderson 1964a).

The peat-swamp forests of Peninsular Malaysia reach a height of 30.5 m and have an even, often open, canopy; emergents are rare (Wyatt-Smith 1959). Brunei and Sarawak boast structurally and floristically more complex forests. Anderson (1963) identified six concentrically zoned forest types in Sarawak, although not all of these are ubiquitous. Only one species, *Dactylocladus stenostachys*, is common to all six types. Structurally, the sequence of types varies from a close similarity to the adjoining dryland forests to an even-canopied but stunted xeromorphic forest to an open savannah-like woodland. A unique feature of two of the forest types is the dominance of *Shorea albida*, which often reach massive proportions.

Much less extensive are the freshwater-swamp forests that are scattered throughout Malaysia (mainly because they have been widely cleared for padi). Those of Peninsular Malaysia vary in structure and species composition, and they have been described in detail by Wyatt-Smith (1961a, 1964). Structurally, they range from scrub forest to dense pole forest to a forest similar to that developed in the peat-swamp areas. Floristically, they are generally similar to adjoining dryland forests.

2.2.2 Intraformation variations

Floristic and structural variations within the forest formations contribute to the complexity of the Malaysian forests and have implications for forest management. Some variations are regional in nature, others local; some can be related to causal factors, but others appear to be random. Discussion in this section focuses on the more intensively studied lowland evergreen forests.

Regional variations

Some regional variations can be related to the migrational history of the flora. Because Malaysia is located within the West Malesian floristic province, species of Asian origin predominate and dipterocarps are widespread. However, the province should not be viewed as a closed or static system. In recent geological times, a number of species have migrated into West Malesia from the east, and there have been internal migrations aided by lower sea-levels during the Pleistocene. Migrations are undoubtedly still occurring, but the continuing removal of vast tracts of forest makes their elucidation increasingly unlikely (Whitmore 1984a, p. 222). Regional variations in species composition are of sufficient magnitude to enable forest types to be distinguished in some areas. A particularly good example is the *Parashorea* forests of Sabah (see above), whose close affinities with forests in the Philippines suggest an invasion route through the Sulu Archipelago.

Some migrations clearly have been affected by physical barriers. Mountain ranges are obvious barriers but in Sarawak rivers have also proved to be effective. Of the 136 dipterocarp taxa with limited distributions in Sarawak, the limits of all but 36 are determined by one of four river systems: Lupar, Rajang, Kemena, and Suai-Sibuti (Ashton 1969, fig. 12). The Kemena and the Suai-Sibuti are relatively small and insignificant but their valleys are thought to have been formerly occupied by the much larger proto-Rajang and proto-Baram rivers. It is striking that the small Kemena River is a much more effective barrier than is the present-day Rajang, Sarawak's largest river. But the distributions of many species cannot be so readily explained. *Neobalanocarpus heimii*, for example, is widely distributed throughout Peninsular Malaysia but is absent from some areas for no apparent reason, while *Dryobalanops aromatica* is, except for one possibly introduced outlier, restricted to the eastern side of the Peninsula, again without any obvious limiting factors (Symington 1943/1974; Whitmore 1984*a*, p. 221).

Some regional variations in Malesia are probably related to disturbance resulting from earthquakes, cyclones, volcanic activity, and other natural processes (e.g. R. J. Johns 1986). Forests that are occasionally affected by such events tend to differ from neighbouring unaffected forests. Fortunately for its inhabitants, Malaysia has no active volcanic areas, experiences few and then relatively minor earthquakes, and is too close to the equator to be affected by cyclones.[6]

Although fires are not generally associated with tropical rain forests, there is increasing evidence that they occasionally have devastating effects.[7] Fire has destroyed extensive areas of forest in Borneo on at least three occasions in this century: some 81 000 ha of forest was destroyed on the Sook Plain in south-western Sabah in 1915; there were widespread fires near Tenom in Sabah in 1973; and there were conflagrations in Sabah and Kalimantan in 1982–3 that have been described as 'ecological events of unprecedented size and intensity' (Malingreau *et al.* 1985, p. 320; also Cockburn 1974; Mackie 1984; Beaman *et al.* 1985). The three fire events all followed periods of abnormal drought. The 1982–3 drought has been related to an unusual configuration of the El Niño Southern Oscillation and it has been suggested that such events have an average recurrence interval of once or twice in a century. The long-term impact of fire on the Borneo forests is unknown, although investigations by Woods (1989) indicated that

some tracts of forest affected by the 1982–3 fires may never revert to their original state (see Chapter 4 this volume). More frequent fires could occur owing to increasing anthropogenic modification of the forest. Evidence indicates that many of the 1982–3 fires were started by humans and that they burned more severely in areas affected by human activities (Mackie 1984).

Local variations

Distinctive floristic associations ranging in size from many square kilometres to a few hundred square metres or less can be recognized within many lowland forests. Variations in species composition have been related to lithology, topography, soil type, soil characteristics, drainage, landslides, and flooding.

In both Peninsular Malaysia and East Malaysia distinctive floras have been related to certain types of rock. For example, Fox (1978) noted that the *Dipterocarpus*/*Shorea* forests in the east and north of Sabah are generally found on sandstone escarpments. Discernable changes can be observed in some localities where there are ultrabasic rocks: for example, on Mt. Kinabalu in Sabah (Whitmore 1984*a*, p. 176). Scattered throughout Malaysia are limestone hills whose form varies from isolated, spectacular karst towers to more extensive, less dissected hills. These hills have a most distinctive vegetation, with noticeable variations on summits, crags, slopes, and basal areas (Whitmore 1984*a*, pp. 172–3), and 11 per cent (143) of the species on the limestone hills of Peninsular Malaysia are endemics (Chin 1977, 1979).

Species distributions at the micro-level are related to edaphic factors and to topography. Some species show a marked preference for certain topographic sites. At the Pasoh Forest Reserve in Negeri Sembilan, densities of light and red meranti species and of cengal (*Neobalanocarpus heimii*) vary with slope (Ashton 1976*a*), the former being concentrated on hill and undulating areas, the latter on flat land. As already noted, *Shorea curtisii*, which is often the dominant species in the hill dipterocarp forests of Peninsular Malaysia (Burgess 1968), is characteristically found in stands on ridge tops, where it has probably adapted to the relatively dry edaphic environments of such localities.

Variations within the mixed dipterocarp forest of Brunei and Sarawak have been associated with soil type and with soil properties. At three sites in Brunei, Ashton (1964) identified two distinct forest types: a *Shorea parvifolia*-dominant forest on clay soils and a *Dryobalanops aromatica*-dominant type on sandy soils. More detailed analyses of soil properties indicated complex interrelationships. A re-examination of the Brunei data (Austin *et al.* 1972) suggested relationships with soil texture and with leaching gradient, and the work of Baillie *et al.* (1987) in central Sarawak demonstrated that *Dryobalanops lanceolata* is often associated with shallow soils, *Shorea ferruginea* with fine-textured soils, and *Shorea quadrinervis* with deep soils.

Variations also have been correlated with the chemical properties of soils. At the Gunung Mulu National Park in Sarawak, variations within the heath forest appear to be related to soil organic carbon and to cation exchange capacity and within the alluvial forest to pH and calcium (Newberry and Proctor 1984). Ashton (1977) and Baillie *et al.* (1987) suggested relationships between species distributions and phosphorus and exchangeable magnesium, with the former being critical. Once concentrations of these two chemicals drop below threshold levels, species that can tolerate such conditions have a competitive edge. The elucidation of species–soil interrelationships could result in improved logging operations and in better silvicultural practices (Baillie *et al.* 1987).

Intraformation variations also are related to disturbances such as those associated with landslides and with flooding. Landslides are relatively frequent in areas of hill forest in Peninsular Malaysia, particularly over granitic terrain and where slopes are greater than 40° (Burgess and Tang 1972; Burgess 1975). Few really large trees occur on steep slopes because the recurrence

interval of landslides is generally less than the hundreds of years required for rain-forest trees to grow into forest giants. Flooding may also affect the composition of the vegetation. It has been suggested that floristic variations in the now cleared Jengka Forest Reserve in central Pahang were related to a combination of soil properties and a severe flood event (Ho *et al.* 1987).[8]

2.2.3 On the richness of the flora

Tropical rain forests harbour a great wealth of species. Why this should be so has long challenged scientific explanation. This section presents some Malaysian examples of the richness of the Malesian flora and outlines some of the arguments that have been advanced to account for diversity. (The richness of the fauna is discussed in a subsequent section.)

Estimates suggest that there are at least 25 000 species of flowering plants in Malesia, or about 10 per cent of all identified plant species on Earth. There are at least 8000 such species in Peninsular Malaysia alone and possibly as many as 10 000 in Borneo. On Mt. Kinabalu in Sabah, according to Myers (1984, p. 53), 'there are five times as many oaks as in the whole of Europe, together with 400 species of ferns and 800 species of orchids'. The herbaceous flowering plants of Peninsular Malaysia comprise some 2580 species in 551 genera and 94 families, but this is probably an underestimate because several families await detailed study (B. H. Kiew 1988). There are some 846 species of orchids in the Peninsula, which also boasts many endemic herbs, each of eight families having more than 100 endemic species.

Research on tree species indicates that there are 99 families in Peninsular Malaysia and that within 82 of these there are 2398 species, 654 (27.3 per cent) of which are endemic (Ng and Low 1982). Surveys show that many species of trees occur within relatively small areas. While the recorded numbers are certainly impressive (Table 2.1), they would have been even more so if both trees smaller than 0.3 m in diameter and woody climbers had been enumerated (e.g. Whit-

Table 2.1 Species richness among trees in Malaysia†

Location	Vegetation type	Area (ha)	Number of species	Minimum diameter (cm)	Source
South Pangkor, Selangor	Lowland rain forest	62.7	173	120	Cousens (1958)
Sungai Menyala, Negeri Sembilan	Lowland rain forest	2.0	240	10	Wyatt-Smith (1949)
Bukit Lagong, Selangor	Hill dipterocarp forest	2.0	495	10	
		1.6	444	10	
Gunung Mulu Sarawak	Alluvial forest	1.0	223	10	Proctor *et al.* (1983*a*)
	Dipterocarp forest	1.0	214	10	
	Heath forest	1.0	123	10	
	Forest over limestone	1.0	73	10	
Wanariset, East Kalimantan	Lowland rain forest	1.6	239	10	Kartawinata *et al.* (1981*a*)
Lempake, East Kalimantan	Lowland rain forest	1.6	205	10	
Pasoh, Negeri Sembilan	Lowland rain forest	11.0	460	—	Hadley and Lanly (1983)
Pasoh, Negeri Sembilan	Lowland rain forest	4.0	328	30	Wong and Whitmore (1970)
Jengka, Pahang	Lowland rain forest	20.0	375	91	Poore (1968)
Andalau and Belalong, Brunei	Lowland rain forest	40.5	760	10	Ashton (1964)
Lambir, Sarawak	Lowland rain forest	0.2	79 (lianas)	1.0	Putz and Chai (1987)

† Whitmore (1990, p. 151) has noted that the 'most species-rich plot so far enumerated is in Peru with 283 species of trees 0.1 m in diameter or over amongst 580 stems on one hectare' (see his fig. 2.27, p. 31).
Source: Compiled by authors.

more *et al.* 1985). As Putz and Chai (1987) have shown, many lianas grow together in small areas (Table 2.1).

Many tree species occur at low densities; for example, at the Jengka Forest Reserve in central Pahang (now cleared), Poore (1968) recorded 375 species in a plot of about 23 ha; of that number, 143 (38 per cent) were represented by only a single mature tree and 307 (81 per cent) by only 1–10 individuals (see also Wong and Whitmore 1970; Whitmore 1971). Clearly, common species are generally rare and rare species are common (Ng and Low 1982).

Much debate has focused on the problem of how to account for tropical high diversity[9] (e.g. Connell 1978; Rosen 1981; Mabberley 1983; Deshmukh 1986). Flenley (1979, after Ricklefs 1973) summarized the basic hypotheses, and Deshmukh (1986, chapter 7) provided an assessment of the empirical evidence in support of the various mechanisms that have been advanced to account for such diversity. Tables 2.2 and 2.3

Table 2.2 Hypotheses concerning high species diversity in the tropics

Non-equilibrium hypothesis
Time—the tropics are older and more stable, hence tropical communities have had more time to develop.

Equilibrium hypotheses
I. Speciation rates are higher in the tropics.
   1 Tropical populations are more sedentary, facilitating geographical isolation.
   2 Evolution proceeds faster due to:
      2a a larger number of generations per year;
      2b greater productivity, leading to greater turnover of populations, hence increased selection;
      2c greater importance of biological factors in the tropics, thereby enhancing selection.

II. Extinction rates are lower in the tropics.
   1 Competition is less stringent in the tropics due to:
      1a presence of more resources;
      1b increased spatial heterogeneity;
      1c predators exercise increased control over competing populations.
   2 The tropics provide more stable environments, allowing smaller populations to persist, because:
      2a the physical environment is more constant;
      2b biological communities are more completely integrated, thereby enhancing the stability of the ecosystem.

Source: Flenley (1979, p. 10) based on Ricklefs (1973).

stand in lieu of a detailed description, which is much beyond the scope of this work. We emphasize that the factors listed in the tables are not mutually exclusive. Another point to keep in mind is that 'definitive studies of tree populations in tropical rain forests are rare and extremely difficult because of potential longevity and the enormous life time reproductive output of individuals' (Deshmukh 1986, pp. 216–17).

In 'the ultimate analysis', as Whitmore (1984a, p. 238) observed, 'a region contains only those plant species which migration and evolution, plus survival, have enabled to be present'. Historical (geological and evolutionary) mechanisms are therefore important. We have noted that plate tectonics and plant migrations resulted in the intermingling of taxa of Gondwanic and Laurasian origin, thereby contributing to the richness of the Malesian flora, and that sea-level fluctua-

tions during the Quaternary alternately facilitated migrations or speciation resulting from isolation.

It would appear that Malesia has enjoyed a humid tropical climate since Tertiary times (Flenley 1979, p. 23), although there is evidence (as already noted above) that quite extensive areas may have been much drier during the Pleistocene (Morley and Flenley 1987). Today, the largest number of species are in those areas that have remained the most humid, including Malaysia (Whitmore 1984a, p. 240). Such a benign climate of long duration may have dampened rates of extinction (see Tables 2.2 and 2.3) although, as Deshmukh (1986, pp. 204–5) noted, the biotic environment is far from being benign as 'higher diversity means that many species are at low density, making their extinction more likely than high-density populations on a purely probabilistic basis'. Over a very long time, however, species

Table 2.3 Assessment of evidence in support of various mechanisms proposed to account for high tropical diversity

Mechanisms	Evidence for high tropical diversity	Taxonomic groups
Time scales		
Seasonal to a few years	Slight	All groups
Primary succession	Poor	All groups
Secondary succession (including disturbance)	Good	All groups
Geological and evolutionary		
Specialization	Fair	Mainly animals
Coevolution	Fair	All groups
Refugia	Fair	All groups
Higher speciation rate	Slight	All groups
Lower extinction rates	Poor	?
Spatial heterogeneity		
More biomes	Good	All groups
More habitats	Good	Mainly animals
Competition and species packing	Fair	Mainly vertebrates
Predation	Fair	Plants and invertebrates
Production		
High production	Poor	—
Year-round production	Good	Vertebrates

Source: after Deshmukh (1986, p. 217).

could be expected to evolve and accumulate and it has been argued that speciation rates are higher in tropical than they are in temperate areas (see Tables 2.2 and 2.3). It has been suggested that there are more species of trees and other plants in the humid tropics than there are in temperate areas because the former were not covered by ice sheets and glaciers during the Pleistocene; in the latter areas, plants migrated equatorwards, there were many extinctions, and in the relatively short time since the end of the Pleistocene only a limited number of species have been able to re-establish themselves.

Evolutionary diversification has been related to assumptions concerning the prevalent role played by one or other particular plant breeding system, to the alleged seasonal and geological stability of the climate (which is thought to have allowed time for niche differentiation and specialization), and to the great age of tropical rain forest (Federov 1966; Ashton 1969, 1977; P. W. Richards 1973). On the other hand, there is now evidence of climatic change in the tropics and of the retreat of the rain forest into *refugia*, although scholars differ in their interpretation of the role that such areas played in diversification (see Chapter 1 this volume). 'Shorter-term evolutionary interactions', according to Deshmukh (1986,

p. 204), 'are undoubtedly a major feature of
tropical diversity. The prevalence of plant
defenses has led to specialization in some herbi-
vores, and this may have increased speciation
through coevolution, although direct evidence is
lacking' (see also Whitmore 1984a, p. 239).

Species diversity has been related to several
environmental and biotic factors. The former
include spatial heterogeneity and disturbance.
Contrasts are usually drawn between tropical and
temperate areas and, with that in mind, it has been
argued, for example, that diversity is related to the
existence of more biomes, more habitats, and
more resources in the tropics; to the more heterogeneous abiotic environment of tropical forest
gaps; to local variations in soils, nutrients, topography, drainage, and other site conditions (as
already mentioned in connection with certain tree
species in Malaysia); to the greater range of tree
architecture in tropical forests, which affords
many habitats for climbers, epiphytes, and other
life forms (Tomlinson 1978); and to the set of
niches associated with the forest growth cycle (e.g.
Oldeman 1983; see below). In short, spatial
heterogeneity refers to the variety of habitats
present within an area, and such variety is considered to be greater in the tropics. Disturbance
adds a dynamic dimension to spatial heterogeneity. (As noted in Chapter 1 of this volume and
described more fully below, disturbance results
from fire, floods, landslides, cyclones, and other
processes.) Connell (1978) suggested that diversity is enhanced by 'intermediate' rates of disturbance and gap formation. There is 'good' evidence
that diversity is related to disturbance, more
biomes, and more habitats, according to Deshmukh (1986; see Table 2.3 this volume), who also
suggested that the continuity of primary production throughout the year in the tropics also may be
an important factor (cf. Rosen 1981, p. 123).

Arguments based on biotic considerations
suggest that there are more niches in the tropics
(because there are more habitats or resources,
which means reverting to environmental factors);
that tropical species are more specialized (that is,
the available resources are more finely divided);

that competition for limited resources leads to
finer degrees of sharing (see the discussion of
fauna below); and that there are more predatory
and defensive specializations in the tropics.

Relatively little is known about niche specialization in tropical trees. Although it is known that
species specialize in large or small gaps, it seems
most unlikely, as Whitmore (1982) noted, that
each of the great diversity of tree species has a
well-defined and separate niche. Perhaps, as
Deshmukh (1986, p. 210) suggested, 'many species have more-or-less overlapping niches and
there is an element of chance about which species
actually establishes in any particular gap'. It has
been argued that seed predation results in a
widely spaced distribution of mature trees that
enhances diversity. However, Hubbell (1979) has
countered this argument by noting that many tree
species exhibit clumped rather than widely
spaced patterns of dispersion.

There is no simple explanation for the richness
and diversity of the Malesian flora (Whitmore
1984a, pp. 237–40). Rather, it would appear that
several interacting extrinsic and intrinsic factors
are involved (most of which are probably not
unique to the tropics) and that their relative significance can only be elucidated by detailed
studies of particular forests (see Whitmore 1990,
pp. 78–80, 149–52).

2.3 THE FOREST FAUNA

Malaysia harbours a great wealth of animal life.
This section outlines the origins and current
distribution of the forest fauna and discusses the
ability of many species to coexist in relatively
small areas. Attention focuses on the larger
mammals and the birds because much less is
known about many of the other species.

2.3.1 The regional context

The fauna of both Peninsular and East Malaysia
belong to the Oriental (or Asian) Region of
Darlington (1957). Further eastwards lies the

Australian Region, with its 'extraordinarily different' fauna. This distinction was brought to the attention of the scientific community by Wallace (1859, 1869), who was impressed by the abrupt faunal changes that he found between Borneo and Sulawesi and between Bali and Lombok. It became generally accepted that the straits separating these islands marked the boundary between the Asian and Australian faunas, and the term 'Wallace's Line' eventually acquired almost mystical significance (Keast 1983, p. 370). But not all scholars have accepted the famous line and several other boundaries have been suggested (e.g. I. H. Burkill 1943; Mayr 1944; Schuster 1972; Simpson 1977; Cranbrook 1981).

The migration of various Asian species into the Malay Peninsula, Sumatra, Java, and Borneo was facilitated by lower sea-levels during the glacial phases of the Pleistocene. Fossil evidence from Sundaland (see above) indicates the presence during the early Pleistocene of large mammals similar to contemporary stock in South Asia (Medway 1972; Hooijer 1975), including a number of stegodonts (e.g. *Stegodon trigonocephalus*), several ancestors of the present-day Asian elephant (e.g. *Archidiskodon planifrons*), as well as deer (*Cervus problematicus*), pig (*Sus stremmi*), and large carnivores (*Panthera* spp.). These species were all forest or forest-fringe dwellers, indicating the presence of extensive unbroken tracts of tropical rain forest. During the middle Pleistocene a different but still essentially forest-dwelling fauna invaded western Malesia. Among these were arboreal primates (e.g. the gibbon and the orang-utan) and browsing animals, suggesting a somewhat more diversified vegetation that included both rain forest and more open woodland.

Climatic change was probably the main cause of a number of extinctions during the upper Pleistocene (Table 2.4). Species that disappeared completely included the stegodonts, the large pangolin (*Manis palaeojavanica*), the large tiger (*Panthera palaeojavanica*), and the hippopotamus (*H. sivalensis*). Some species became locally extinct: the Indian elephant disappeared from

Table 2.4 Current and former distribution of some large mammals in Sundaland

Mammals†	Java	Sumatra	Borneo
Orang-utan	o	x	x
Siamang	o	x	—
Tiger	x	x	o
Panther	x	o	—
Bear	o	x	x
Elephant	o	x	x
Tapir	o	x	o
Javan rhinoceros	x	x	—
Sumatran rhinoceros	—	x	x
Banteng	x	o	x

† Extant (x); Extinct (o)
Source: Hooijer (1975, p. 50).

Java and Borneo (existing species are almost certainly human introductions), the tiger from Borneo, and the orang-utan from Java and the Malay Peninsula. Some local extinctions are of relatively recent origin; for example, the tapir (*Tapirus indicus*) survived in Borneo until at least 7000 BP (Medway 1972). Despite the many extinctions that occurred, the extant fauna of Malaysia is remarkably rich and diverse, although how long some of it will survive is already a moot point.

2.3.2 Extant fauna

Malaysia's larger mammals and its birds have attracted the attention of travellers and scientists for more than a century (e.g. Hose 1893; Maxwell 1907/1960; Begbie 1834/1967; Foenander 1952; Locke 1954) and are generally well known. But other animals, particularly the invertebrates, have received relatively little attention. Many species undoubtedly await discovery.

Numbers of species

Two hundred and three species of mammals have been recorded in Peninsular Malaysia and 221 in East Malaysia (Payne *et al.* 1985; Cranbrook

1988).[10] In both areas, bats comprise about half of the total–83 species in Peninsular Malaysia and 92 in East Malaysia. In Peninsular Malaysia, two large mammals, the Javan rhinoceros (*Rhinoceros sondaicus*) and the banteng (*Bos javanicus*), have almost certainly become extinct in this century (Stevens 1968, p. 98).

Malaysia also has a rich avifauna. There are some 460 species of resident birds in Peninsular Malaysia and 388 in East Malaysia (Wells 1976; Smythies 1981; Davies and Payne n.d.). Other animals for which there is information on numbers of species include snakes, with 136 species in Peninsular Malaysia and 166 in East Malaysia (Haile 1958; Grandison 1978); frogs and toads, with 80 species in Peninsular Malaysia and 100 in East Malaysia (Inger 1966); and butterflies, with 1014 species in Peninsular Malaysia (Barlow 1988).

A number of mammal species found in Peninsular Malaysia do not occur in East Malaysia and vice versa; for example, Borneo's best-known creature, the orang-utan (*Pongo pygmaeus*), is not found in the Peninsula, while the tiger and tapir, which are (or were) widespread throughout the region, do not occur in either Sabah or Sarawak. The Asian elephant is found in both Peninsular and East Malaysia, but it is thought to have been introduced into the latter area by humans during the eighteenth century (de Silva 1968*a*).

A number of species are geographically unevenly distributed. In Peninsular Malaysia, for example, the three species of gibbons have different ranges (Chivers 1978): the lar gibbon (*Hylobates lar*) is widely distributed, the agile gibbon (*H. agilis*) is found only in the northern part of the Peninsula, and the siamang (*H. syndactylus*) is found across the central part of the region. Butterflies also are unevenly distributed in the Peninsula, where Barlow (1988) recognized three major faunistic regions: Kedah–Langkawi, Kedah to Johor, and eastern Johor/south-eastern Pahang–Pulau Tioman.

Preferred habitats   The Malaysian land faunas are predominantly forest-dwelling. This is hardly surprising, considering the long history of rain forest in the region. Most such faunas are concentrated in the dryland forests of the lowlands, where certain preferred habitats can be identified. According to Stevens (1968), 53 per cent of the non-bat mammal species of Peninsular Malaysia prefer primary forest and another 25 per cent prefer either primary or tall secondary forest. Some 12 per cent readily live in primary or secondary forest or in cultivated areas, leaving only 10 per cent that are confined to agricultural areas. More than 50 per cent are generally found below 305 m and 81 per cent below 610 m; nine per cent can exist at any altitude.

Not all of the large mammals prefer deep-forest habitats. The seladang (*Bos gaurus*, see Plate 4), for example, is a forest-fringe dweller, sheltering in the forest during the heat of the day and grazing at night near the jungle edge or in forest openings of either natural or human origin (Kitchener 1961). The ability of different species to adapt to habitat disturbance varies considerably. Some, like the seladang and the elephant, thrive in cleared areas (such as those abandoned by shifting cultivators) while others cannot readily adapt to change.

Like the mammals, the resident birds are predominantly forest-dwelling. For example, some 60 per cent of the 460 species in Peninsular Malaysia prefer forest habitats (Wells 1976) and in that region 26 migrant species are exclusively forest-dwellers (Wells 1988). Few species are found in all forest formations and there is a major avifaunal boundary in the Peninsula at an elevation of approximately 910 m—that is, near the junction between the hill dipterocarp forest and the lower montane forest. According to Wells (1988), 282 species occur in the lowland and hill (dryland) forests, compared with 96 species in the montane areas.

Not all of the lowland forests have rich faunas. A case in point is the locally extensive wetland forests, which generally do not contain as many species or numbers of birds and mammals as do neighbouring dryland forests. Their simpler botanic composition and structure presumably do not provide as many food sources or niches as do the more complex dryland forests. Apparently, few species of mammals or birds are restricted to the wetland forests; for example, none is unique to the peat-swamp forests of Sarawak, although that type of forest occupies some 12.5 per cent of the state. Metcalfe (1961) suggested that the swamp forests of Peninsular Malaysia were the preferred habitat of the Sumatran rhinoceros (*Dicerorhinus sumatrensis*), which is known to enjoy sojourning in wallows, and that its present scattered distribution is the result of human disturbance.

The wetland forest with the most distinctive flora and fauna is the mangrove swamps (Berry 1972), although very few non-aquatic species are restricted to this formation. Some 50 species of birds in Peninsular Malaysia frequently use mangrove forests, but, of that number, only 25 are partially restricted to the mangroves and only six are mainly confined to such areas (Nisbet 1968; Wells 1988). Mangroves are important bird habitats because they provide nesting sites for species of heron and storks, many of which do not nest elsewhere; because they provide roosts for migrating species; and because they are visited seasonally, for reasons that are not clear, by many species of pigeons and parrots.

Although there are no terrestrial mammals that are confined to the mangroves, some leaf and proboscis monkeys live there semi-permanently and they are important fishing grounds for otters and mongooses (Berry 1972). The seaward zones of the mangrove forests are rich in aquatic fauna, and trees are often encrusted with snails and bivalves. Tee (1982) recorded 15 401 animals (9199 of which were bivalves) on a single tree with a girth of 22.4 cm. Crabs and mudskippers abound, fish and prawns feed on the decaying vegetation, and juvenile prawns probably depend on the mangroves for shelter (J. E. Ong 1982). An estimate reveals that some 56 per cent of the revenue from fisheries in Sarawak in 1980 was derived from mangrove-dependent fish and prawns (Mahrus Ibrahim 1986). Clearly, removal or alteration of the mangrove forests could have significant economic implications.

Coexistence

A characteristic feature of the Malaysian fauna is that many species often coexist in relatively small areas, particularly in the dryland forests of the lowlands. Species lists based on observations over varying lengths of time have been compiled for a number of sites. For example, 97 mammal species have been recorded in the Ulu Gombak area of Selangor and 89 such species have been sighted in the now isolated Pasoh Forest Reserve in Negeri Sembilan (Kemper 1988). Other mammal counts include 62 species for the Ulu Endau area of the proposed Endau–Rompin bi-state park (Davison and Kiew 1987), 84 species at Bukit Lanjan in Selangor (B. L. Lim *et al*. 1977), and 64 species at Kuala Lompat in Pahang (Medway and Wells 1971). Relatively large numbers of frogs, lizards, snakes, and birds have been shown to coexist in relatively small areas (Wells 1971, 1988; Inger 1979; Davison 1987). Bird counts include 202 species in a 2 km^2 area of the Kerau Game Reserve in Pahang, 196 species in a similar area at Pasoh, and 195 species in Ulu Endau. In contrast, only 25 and 41 species have been recorded on the two offshore islands of Pulau Tioman and Pulau Langkawi, which suggests that the survival of a large number of species in an area depends on setting aside extensive, preferably interconnected, tracts of forest (Wells 1971).

How are large numbers of species able to coexist in relatively small areas of forest? The explanation has been related to a number of mechanisms, including time partitioning, space partitioning, and resource partitioning (e.g. Bourlière 1983; also Whitmore 1990, pp. 61–5). The most obvious type of time partitioning is the fundamental division between those species that are diurnal (e.g. most birds) and those that are nocturnal (e.g. bats and a number of other mammals). With respect to space partitioning, many animals show a preference for a particular layer of the forest. Resource partitioning mainly relates to species that show a preference for a particular type of food, although the differential use of trees for nesting sites, roosts, and refuge can be significant. In the case of birds, the variety of tree architecture in tropical rain forests (Hallé *et al.* 1978; Tomlinson 1978) offers a greater range of microsites for nesting than does the architecture of extratropical forest trees. Some examples illustrate how certain species share space and resources.

Sutton (1979, 1983) has reported a vertical zonation of flying insects in Brunei. He found that Homoptera and Lepidoptera were concentrated in the upper canopy, Ephemeroptera in the middle levels, and Diptera closer to the ground. Similarly, birds are vertically stratified (Wells 1971): for example, hornbills, barbets, pigeons, and sunbirds are found mainly in the canopy layer; trogons, woodpeckers, bulbuls, and flycatchers occupy the middle stratum; and pheasants, pittas, thrushes, and babblers favour the ground layers.

The complexity of coexistence is illustrated by two studies undertaken at Kuala Lompat in the Kerau Game Reserve in Pahang, one of diurnal squirrels and the other of primates (K. S. MacKinnon 1978; J. R. MacKinnon and K. S. MacKinnon 1980). K. S. MacKinnon (1978), who studied seven of the 11 species of diurnal squirrels that have been reported at Kuala Lompat (Payne 1979), concluded that the coexistence of so many species in a relatively small area is possible because they differentially exploit the available resources. Distinct vertical differentiation was observed: the three largest squirrels—the black giant squirrel (*Ratufa bicolor*), the cream giant squirrel (*Ratufa affinis*), and Prevost's squirrel (*Callosciurus prevostii*)—live highest in the canopy; the kampung squirrel (*Callosciurus notatus*) feeds and travels in the lower storeys; and the three-striped ground squirrel (*Lariscus insignis*), as its name implies, is essentially terrestrial. The body form of some species is well adapted to their preferred habitats: for example, the terrestrial *L. insignis* has a short tail that is suited to running; the agile and slim *C. notatus* is well suited to the low bushes and creepers that it frequents; the broad, flattened

form of the slender little squirrel (*Sundasciurus tenuis*) is well adapted for frequenting vertical tree trunks; and the three larger canopy dwellers all have the kind of long tail that is required for balance in such a habitat.

Dietary preferences also vary. The three larger squirrels are predominantly fruit-eaters, but whereas *C. prevostii* eats mainly soft fruit (especially *Ficus* spp.) and some leaves, *R. bicolor* eats various kinds of flowers and a wide range of soft and hard fruits but no leaves, and *R. affinis* eats both soft and hard fruits and some leaves. *S. hippurus* feeds predominantly on large hard fruit, whereas *S. tenuis* appears to feed heavily on bark. The most common of the smaller squirrels, *C. notatus*, has a varied diet. At first glance, it would appear that two of the larger squirrels, *R. bicolor* and *C. prevostii*, occupy the same niche in the upper canopy, where both feed mainly on fruit. However, their different sizes—*R. bicolor* is three times larger than *C. prevostii*—ensure that they do not occupy exactly the same niche. Because of its size *R. bicolor* has a relatively small geographical feeding range, whereas the smaller *C. prevostii* is able to forage much further afield.

A similar situation exists among the six primates at the Kerau Game Reserve (J. R. MacKinnon and K. S. MacKinnon 1980). Observations have revealed differences in canopy use and preferences for particular types of forest. Both species of gibbon—the white-handed gibbon (*Hylobates lar*) and the siamang (*H. syndactylus*)—frequent the middle and upper levels of the forest. In the case of the two species of leaf monkeys, *Presbytis obscura* and *Presbytis melalophos*, the former spends more time in the upper canopy than does the latter. And the other two species, both macaques, also have different preferences: the terrestrial *Macaca nemestrina* favours hill forest, while the arboreal *M. fascicularis* prefers riverine forest.

The dietary preferences of the six species allow them to be divided into three phylogenetic couples: the leaf monkeys, the gibbons, and the macaques. Between them the six species eat some 376 different food items from four food groups (fruits, flowers, leaves, and insects), but only 17 of these are common in the diet of at least five of the six species. Competition for food resources between the three groups is clearly minimal, and it can be assumed that ability to consume different food items is related to differences in the anatomy and in the chemistry of their gastro-intestinal tracts. In addition to leaves, the leaf monkeys also eat chemically defended fruits that other species tend to avoid. The two species are able to coexist because, as noted, they tend to frequent different forest strata.

The gibbons exploit the smallest range of food items and show a distinct preference for figs. As in the case of two of the squirrel species—*R. bicolor* and *C. prevostii*—the two gibbons are able to coexist because of different foraging strategies (Raemaekers 1978). The much larger *H. syndactylus* does not travel as far as does *H. lar*, which exploits more dispersed food sources and sources on smaller branches beyond the reach of its heavier 'cousin'. The macaques' dietary preferences lie between those of the leaf monkeys and those of the gibbons. They are able to coexist because, as already noted, they prefer different habitats.

2.4 THE FOREST SYSTEM

The composition of tropical nature in Malaysia is certainly important but so also is how it 'works and changes'. This section outlines some aspects of forest function (in relation to soils, nutrients, hydrology, and denudation) and dynamics (as expressed in the forest growth cycle). The discussion of these matters sheds light on three major themes that are taken up later in this book, namely resource utilization (Chapter 3), human impact on forest systems (Chapter 4), and conservation (Chapter 5).

There is a rapidly growing literature on how tropical (and other) forests function and change, but we make no attempt to review it (e.g. see the many references in Proctor 1983, 1985, 1986, 1987; Sutton *et al.* 1983; Chadwick and Sutton

1984; Jordan 1987*b*). More detailed accounts of these matters, as they pertain to the tropical rain forest of Malaysia, can be found in a series of papers that report on research undertaken at the Pasoh Forest Reserve in Negeri Sembilan as part of the International Biological Programme (Kira 1978; Leigh 1978*a,b*; M. T. Lim 1978; Yoda 1978; Yoneda *et al.* 1978; Manokaran 1979) and at the Gunung Mulu National Park in Sarawak by members of the Royal Geographical Society's expedition to that area in 1977–8 (Anderson *et al.* 1983; Gautam-Basak and Proctor 1983; Proctor *et al.* 1983*a,b*; Newberry and Proctor 1984). The condensed and selective coverage that follows is in keeping with our limited objectives, which focus more on human/environment relations than they do on function and dynamics for their own sake.

2.4.1 Soils and nutrients

Care is needed not to exaggerate the differences between the soils of tropical and temperate regions. Parent material, erosional and depositional processes, and time are similar or not markedly different between the two regions, hence the term 'tropical soils', which 'can be quantitatively defined only as those soils which lack a significant summer to winter temperature variation' (Sanchez and Buol 1975, p. 598), is of rather limited utility. One difference, however, is that the areal extent of younger soils is greater in (glaciated) temperate regions than it is in tropical regions, where many soils have been formed from parent materials that have been intensively weathered for millions of years.

There is a broad distinction between soils of high base status, such as those that have developed from alluvium or volcanic ash, which are generally fertile, and soils of low base status, such as the leached, acid, and generally infertile soils of regions like the Amazon Basin and Central Africa. Although the latter group of soils—most of them are Oxisols and Ultisols—may have excellent physical properties, their chemical deficiencies pose many problems for agriculture—or at least for continuous production of annual crops (Irion

1978; cf. Sanchez *et al.* 1982 and Nicholaides *et al.* 1985 with Fearnside 1987*a*).

Soils vary with lithology, drainage, altitude, and other conditions and often are more heterogeneous than commonly realized. Forest communities have in some cases been shown to reflect subtle local or micro-patch variations in soils (as already noted in relation to the distribution of certain tree species in parts of Malaysia). A persistent misconception is that most tropical soils are transformed into laterite (plinthite) when they are cleared of vegetation, whereas it has been estimated that laterite at or close to the surface accounts for only about seven per cent of the entire area of the tropics (Sanchez and Buol 1975, p. 599). Soils have been variously classified and named, with generally rather confusing results (e.g. see Sanchez 1976, pp. 64–78; Burnham 1984, pp. 137–9).

In Malaysia, Ultisols (red–yellow podzolic soils) are the most widely occurring soils on well-drained lowland sites, where they tend to be relatively deep, to have thin litter and humus layers, and to be relatively rich in clay because virtually all minerals succumb to chemical weathering (Burnham 1984, pp. 138–9). Depth to unweathered rock varies: in some areas it is many tens of metres, in others little more than a metre—as is the case, for example, at the Pasoh Forest Reserve in Negeri Sembilan and at several localities in central Sarawak (Leigh 1982*a*; Baillie *et al.* 1987, p. 204). Ultisols are characterized by a clay-rich, blocky-textured subsurface horizon and by more sandy upper horizons. They are poor in nutrients and extremely susceptible to erosion once the forest cover is removed (Burnham 1984, p. 142). Over basic igneous rocks, Ultisols tend to be replaced by Oxisols (ferralsols, latosols) that have a high clay content and a characteristic dark-red colour.

Oxisols are highly permeable and do not disperse readily in water, hence they are more resistant to erosion than are Ultisols. In the profiles of both Ultisols and Oxisols, bands of compact, occasionally hardened clays are sometimes found and these, traditionally, have been called 'laterites' and 'lateritic soils'. Laterites in Peninsular Malaysia tend to occur in seasonally drier areas like Kedah and Melaka (Eyles 1967). Where rainfall totals are high over areas of predominantly siliceous rocks such as sandstones and quartzites, Ultisols give way to Spodosols (podzols). Such soils are relatively common in coastal locations over old beach ridges in both East and Peninsular Malaysia and over sandstones and acid volcanics in Sarawak, where they are associated with heath forests (*kerangas*).

Distinctive soils also occur over limestone, ultrabasic rocks, marine and freshwater sediments, and in mountainous areas (Burnham 1984). With increasing altitude, weathering becomes less intense and the clay content of the soil decreases; leaching increases, with the result that there is removal of iron and podzolization at higher elevations; and waterlogging creates anaerobic conditions that often result in the accumulation of peat. Whitmore and Burnham (1969) gave a detailed description of soil changes with altitude on granitic rocks in Selangor.

The nature of nutrient cycling in tropical forests is poorly understood. There is growing evidence, however, that many commonly held assumptions are either incorrect or not universally valid.[11] For example, the long-standing belief that most nutrients are held in the above-ground parts of the rain-forest trees is an unwarranted generalization (Proctor 1983; Whitmore 1984*a*, chapter 10, 1990, chapter 8). Indeed, a substantial proportion of the nutrients may be held in the roots and in the soil. At Gunung Mulu, for example, the nutrient status of the heath- and limestone-forest soils is relatively high, which clearly refutes the suggestion that the low biomass of the heath forest is the result of infertile soils (Proctor 1983).

Two other assumptions that do not always hold are that nutrient cycles are virtually closed and that rates of litter decomposition are more rapid in tropical rain forests than they are in temperate forests. Studies in Malaysia have shown that inputs from precipitation and dry fall-out can be significant (Leigh 1982*a*; Manokaran 1980); that there are losses in stream water (Kenworthy 1971); and that water flowing over the slope and through the soil at the Pasoh Forest Reserve carries nutrients both in solution and in suspension (Peh 1976; Manokaran 1980; Leigh 1982a). Nutrient losses from undisturbed tropical rain forests generally are not as high as one might expect, given the high potential that exists for nutrient leaching. Losses are minimized by a number of nutrient-conserving devices, including large root biomass, concentration of roots near the surface, long-lived and resistant leaves, and thick bark. Some or all of these mechanisms are lost when the forest is disturbed (Jordan 1985, chapter 2).

Further research is needed to identify the nature of nutrient cycling within different rain-forest formations and there is also a need to evaluate the impact that logging operations have on nutrient cycling (Whitmore 1984a, p. 134). It is becoming increasingly clear that there are considerable variations in nutrient cycling within the tropical rain forests and that generalizations must be viewed with caution.

2.4.2 Forest hydrology and denudation

This section describes the hydrological and denudational processes that operate in the rain-forest environment of Malaysia. An understanding of these processes is of critical importance in land-use planning and management. Quantitative data derived from studies of undisturbed forest ecosystems are of great value because they provide a baseline and a yardstick against which the impact of anthropogenic changes can be measured.

Forest hydrology
Not all the rain that falls over the forest reaches the ground, as some is intercepted by the foliage and subsequently evaporated. Rain that does reach the ground percolates vertically downwards, flows laterally through the soil horizons or flows over the ground surface. Surface and subsurface flows feed the stream networks. The passage of water through the forest ecosystem is related to a number of factors, including storm totals and intensities, form of the vegetation cover, litter cover, slope form, soil and regolith properties, and antecedent condi-

tions (for a comprehensive account of tropical forest hydrology, see the excellent review by Bruijnzeel 1990).

Rainfall totals over Malaysia are generally high, most areas receiving in excess of 2000 mm per annum and some areas in excess of 3500 mm (e.g. Dale 1959, fig. 2; Sabah n.d., p. 8). Although rainfall is fairly evenly distributed throughout the year, there are seasonal variations whose magnitude and occurrence vary from place to place (Dale 1959; Y. L. Lee 1965, fig. 5; Jackson 1968a, pp. 25–9; Chia 1977). Both Dale (1959) and Chia (1977) identified five rainfall regions in Peninsular Malaysia, albeit with somewhat different boundaries.

Frequent rainstorms of high intensity and often of long duration are a characteristic feature of the Malaysian climate. The highest rainfall totals and intensities in Peninsular Malaysia occur along the east coast, which lies across the path of the northeast monsoon (Lewis *et al.* 1975). Here exceptionally heavy and prolonged storms sometimes result in widespread flooding. Totals in excess of 610 mm in 24 hours have been recorded (Water Resources Committee 1971).

The generally dense evergreen rain-forest vegetation intercepts a surprising quantity of rain. Estimates and measurements of interception loss for lowland evergreen rain-forest sites in Peninsular Malaysia and Sarawak give average annual losses ranging from 21.8 per cent to 36.0 per cent of above-canopy rainfall (Brünig 1971; Kenworthy 1971; Low 1972; Manokaran 1977). Rates of interception, which vary with storm duration and intensity and also with antecedent conditions, range from 10 per cent to 100 per cent at Sungai Lui in Selangor and from 0 per cent to 99 per cent at Pasoh in Negeri Sembilan.

Leaf litter also intercepts some rainfall but measurements are lacking for Malaysia. In the United States, Helvey and Patric (1965) found that hardwood litter intercepted between two and five per cent of gross rainfall; and in India, Dabral *et al.* (1963) found that litter in tropical hardwood plantations intercepted between 8.9 and 9.1 per cent of gross rainfall.

Most of the water reaching the forest floor soaks into the soil, the relatively open texture of the decomposing litter layer and of the humus layer facilitating the process. Water does, however, flow over the surface when rainfall intensities exceed infiltration capacity or when the water table reaches the surface (Leigh 1978a,b, 1982a). Volumes of surface wash collected on slopes in the Pasoh Forest Reserve and at Kepong in Selangor were found to be closely related to the thickness and completeness of the litter cover (Peh 1976; Leigh 1982a).

Water that is not intercepted or lost through evapotranspiration is lost to the system as streamflow. In the case of mainly forested catchments in Peninsular Malaysia, available evidence suggests that the rainfall–runoff ratio is between 40 and 50 per cent (see Aiken et al. 1982, table 7.13), which is generally lower than for partially forested catchments (see Chapter 4 this volume). During storms of long duration and high intensity, such as those that occurred during December 1926, January 1967, and January 1971, runoff rates were probably similar across a range of land-cover types, hence there was flooding in both forested and non-forested catchments (Toebes and Goh 1975).

Denudation

Eroding hillsides and sediment-laden rivers are not uncommon sights in Malaysia. Evidence of erosion is often apparent on slopes planted with agricultural crops and on cleared slopes awaiting new land uses. Occasionally, massive erosion is seen on cleared slopes. On the other hand, even though regoliths are often very deep and considerable amounts of rainfall reach the forest floor, there is little evidence of bare eroding hillsides and erosion gullies on forested slopes. Clearly, the vegetation and litter cover protect such slopes from the full impact of most storms, and the litter layer inhibits overland flow.

Nevertheless some sediment is removed by the relatively large volumes of water that reach the forest floor, especially during intense and prolonged storms when water can be observed flowing over the ground surface (particularly where the litter layer is patchy) and concentrating in ephemeral rills and stream depressions. Baillie (1976) found erosion tunnels just below the surface in a Sarawak forest.

Rates of sediment transport by surface wash have been measured at the Pasoh Forest Reserve and at Kepong (Peh 1976; Leigh 1982a). The suspended sediment, which was collected in wash traps, showed a positive correlation with the thickness and completeness of the litter layer. Volumetric rates of sediment transport at the two sites were relatively low, ranging from $0.058 \text{ cm}^3 \text{ cm}^{-1} \text{ yr}^{-1}$ to $2.757 \text{ cm}^3 \text{ cm}^{-1} \text{ yr}^{-1}$. At Pasoh, the solute load in surface wash and in throughflow was measured at two pit sites. At one pit, suspended sediments in surface wash accounted for 70.3 per cent of sediment transported, at the other for 49.1 per cent, although these figures probably would have been higher if, as was not the case, inputs of solutes in rain-water had been determined (Leigh 1982a).

Sediment concentrations in streams and rivers draining rain-forested catchments in Malaysia appear to be relatively low. For example, at Pasoh the suspended sediment concentration in the Sungai Maran Kanan, a small stream draining a catchment of some 6.5 km², was found to be only 18.6 mg/litre, with a range from 2.8 mg/litre to 78.0 mg/litre (Leigh 1978a). In contrast, streams draining partially developed suburban catchments in Kuala Lumpur often carry sediment concentrations of several thousand milligrams per litre (Douglas 1972; Leigh 1982b; see Chapter 4 this volume).

Large landslide scars in various stages of revegetation occur in some hill and mountainous areas of Malaysia. Landslides in some parts of the humid tropics are triggered by earthquakes (e.g. Simonett 1967; Garwood et al. 1979; R. J. Johns 1986) but this is not the case in seismically relatively stable Malaysia, where there is evidence that such features are triggered by exceptionally heavy storm events (Windstedt 1927; Fitch 1952; Burgess 1975, pp. 9–13; Day 1980). This was the case following the great storms of December

1926 and January 1971, when there were landslides in the Main and East Coast ranges of Peninsular Malaysia. It is also known that landslides have occurred immediately following heavy storms in the Gunung Mulu National Park in Sarawak.

Slopes underlain by certain types of rock are probably more susceptible to slipping than are others. Landslides in Peninsular Malaysia are more common on granitic rocks, whose unstructured regoliths are particularly prone to movement (Fitch 1952), and it is known that the slaty shales of the Mulu Formation in Gunung Mulu National Park are particularly prone to slipping. It has been suggested, as already noted, that the absence of large trees on steep slopes in the hill forests of Peninsular Malaysia can be attributed to the recurrence of landslides (Burgess and Tang 1972; Burgess 1975).

2.4.3 The forest growth cycle

Forests are dynamic. Flux is of their essence. Change is extremely complex because the processes that bring it about operate at many different temporal and spatial scales. There are, for example, instantaneous, seasonal, and evolutionary changes, and change occurs in areas that range in size from a few square metres up to millions of hectares—from, let us say, a tiny puncture in the canopy to the extent of a regional landscape. This section considers some aspects of change as expressed in the forest growth cycle (Hallé *et al.* 1978, pp. 366–85; Unesco 1978, chapter 8; Whitmore 1978, 1982, 1983, 1984*a*, chapter 7, 1990, pp. 23–5; Brokaw 1985). It is therefore concerned with natural cyclical change (primarily at the local or regional scale) and with such matters as the life cycles of the trees and with competition. Knowledge of how forests change is of fundamental importance in silviculture, in understanding the consequences of anthropogenic change, and in various facets of land-use planning and management (Unesco 1978, chapters 8 and 9).

Trees grow, fall, and others replace them in a growth or regeneration cycle that is usually divided

into three arbitrary phases: a gap phase, a building phase, and a mature phase. The phases of the cycle occur as a mosaic of patches that vary in age, size, shape, and species composition. Tree-fall in the gap phase is followed by colonization and tree growth in the building phase, which eventually attains a mature phase. Tree-fall renews the cycle. Gap size determines the coarseness of the mosaic of forest patches. Disturbance in the form of tree-falls plays a vital role in community function because the 'majority of canopy tree species may depend on growth in a gap to reach maturity' (Brokaw 1985, p. 54). Gaps range in size from small openings in the canopy to huge swaths of destruction. Brokaw (1985, p. 56) collated data on mean gap size, percentage of the forest area in gaps, and gap-turnover rates for several lowland tropical-forest locations.

Two causes of generally small gaps are death of an individual tree and lightning strikes. Old age, decay, and disease are obvious causes of death. Another that has been suggested is epiphytic load (Deshmukh 1986, p. 201). A possibly unique cause of tree mortality was reported from Sarawak, where a large number of trees in the *Shorea albida* peat-swamp forests were killed by a defoliating caterpillar in the 1940s and 1950s. The area affected was in the order of 12 140 ha (see Anderson 1961). Individual trees or groups of trees are killed by lightning strikes (Anderson 1964*b*, 1966; Brünig 1964; R. J. Johns 1986). Lightning is thought to be the main cause of tree mortality in the *Shorea albida* peat-swamp forests of Sarawak, where sometimes a single lightning strike may cause quite large gaps and kill many trees; for example, Anderson (1964*b*) reported 70 tree deaths in a gap of some 0.61 ha.

Wind-throw, cyclones (Dittus 1985), landslides, volcanic activity, conflagrations—all these are causes of generally large gaps. Anderson (1964*b*) reported a wind-throw gap of some 101 ha in a peat-swamp forest in Sarawak. Disturbance resulting from wind-throw may not be completely random because convectional storms and line squalls occur more frequently in some parts of Peninsular Malaysia (and also possibly in East Malaysia) than they do in others (Watts 1954; Dale 1959). Forests in some parts of Malesia are severely damaged from time to time by cyclones, but such events are probably very rare in Malaysia.[6]

In many parts of Malesia gaps in the forest can be related to volcanic activity and to earthquake-triggered landslides (R. J. Johns 1986). Such gaps are not found in Malaysia, which is volcanically and seismically stable, but landslides triggered by heavy-rainfall events are relatively common in the hill and mountainous areas (Burgess 1975; Day 1980). As already noted, certain rock types are more susceptible to landslides than are others: for example, the regoliths developed over granitic rocks in the mountainous parts of Peninsular Malaysia are particularly prone to slipping.

Gaps and intact understorey environments differ. Generally speaking, gaps are characterized by light of greater intensity and duration, higher soil and air temperatures, lower humidity, higher evaporation from soil surfaces, nutrient pulses as dead plant materials decay, and a temporary decrease in root competition. Gaps themselves are complex, heterogeneous environments, and there is evidence that tree-species richness and gap heterogeneity are related (Brandani *et al.* 1988; also Orians 1982). Big and small gaps differ in their longevity and in the amount of light they receive. The large gaps can

only be filled by individuals from a lower storey and the possibility exists for an individual to grow from the seedling stage to the canopy with its crown always receiving uninterrupted direct sunlight. These conditions are therefore stimulated by sylviculturists, who recognize in them the optimal conditions of growth for the seedlings and saplings (Unesco 1978, chapter 8, p. 189).

Saplings already present in the gap will grow if they can survive the new microclimate, and seeds lying dormant or carried into the gap by animals or by wind may germinate. As for the species that become established in gaps, a general distinction can be made between those that are 'shade-tolerant' and those that are 'light-demanding' (see

the important review by Whitmore 1983, also 1990, pp. 102–12). The former tend to grow relatively slowly and to form dense, dark-coloured timber while the latter, which are often referred to as pioneer species, grow rapidly and form less dense, light-coloured timber (Whitmore 1984a, pp. 84–6). Species other than trees that depend on gaps or thrive therein include climbing palms, lianas, herbs, herbaceous vines, decomposers in gap debris, and various species of arthropods, birds, frogs, and toads (Mabberley 1983, pp. 49–50; Brokaw 1985, pp. 63–4).

Two important factors that determine which particular species will become established in newly formed gaps are availability of seeds and the efficiency of their dispersal. Observations in Malesia indicate that the seeds of many rain-forest trees are carried considerable distances by animals and by wind and that the seeds of many pioneer species can remain dormant for months or even for years, whereas most dipterocarp seeds germinate within a few days or not at all.[12] Unlike the seeds of many other rain-forest species, the generally heavy dipterocarp seeds are not readily dispersed either by wind or by animals, and, for the most part, animals appear to find dipterocarp fruits unpalatable (Ashton 1969, p. 183).[13] The slow rates of dispersal of dipterocarps—estimated at about 2 m/yr—and the short time during which their seeds are viable have very important implications for silvicultural management, because it is the dipterocarps that are the most sought-after commercial species and because there is a trend towards shorter rotation periods. Evidence suggests that the dipterocarps (together with many other rain-forest trees) are 'effectively permanently eliminated when large areas are completely cleared' (Unesco 1978, p. 191).

Maximum rates of growth and productivity[14] occur during the building phase of the forest growth cycle, not during the mature phase when the forest appears most luxuriant (Whitmore 1984a, chapter 9). The length of time that it takes various species to reach maturity, or an acceptable size for the sawmill, is clearly of considerable interest to foresters. Such information, which is well known for temperate species from direct observations and from counts of annual growth rings, is extremely limited for rain-forest species. Long-term observations of tree growth in the tropics are few and far between, and annual growth rings are absent or unreliable (Unesco 1978, pp. 197–205).

Some information on tree growth is available for Malaysia, although data based on extrapolations from short-term measurements must be viewed with caution. One Malaysian forest for which an exact age can be given is the (now cleared) Kelantan Storm Forest. There, on average, it took seedlings some 70–80 years to attain a girth of 1.2 m. Extrapolations from measurements taken over 14 years in the Federated Malay States indicated, for example, that to attain a breast-height girth of 1.2 m would require 31 years for a kapur tree (*Dryobalanops aromatica*), 63 years for merbau (*Intsia palembanica*), and 118 years for resak (*Shorea maxwelliana*) (Edwards 1930).[15] There is a much greater potential for rapid growth under plantation conditions (Unesco 1978, pp. 198–9).

Humans attach certain values to the phases of the forest growth cycle. Consider a few simplified examples. Gaps are favourite hunting grounds because the dense patches of succulent vegetation that frequently occur in such places are attractive to browsing and foraging animals like deer and wild pig, whose meat is an important source of protein for many forest- and near-forest dwellers. Activities such as shifting cultivation, plantation agriculture, and ranching attempt to isolate, and thereby to take advantage of, the superior productivity of the building phase of the cycle. Although productivity slows down during the mature phase of the growth cycle, so-called primary or virgin forest harbours the full assemblage of species and materials. It is the kind of forest that is generally favoured by hunter–gatherers, certain shifting cultivators, loggers, and conservationists. There are many visions of the same scene. Conflicting values frequently give rise to disputes.

2.5 CONCLUDING REMARKS

The characteristic rain-forest vegetation of Malaysia embraces several more or less distinct formations. These include mangrove and peat-swamp forests in wetland areas and the 'classic' tropical lowland evergreen rain forest on dryland sites. Other formations are related, for example, to elevation, or to special edaphic or geological conditions. There are many variations within each formation.

Malaysia's forests harbour a remarkably abundant biota (much of which has never been studied), and no doubt there are a good many 'inconspicuous' species that still await discovery. Richness and diversity reach a maximum in the great dipterocarp-dominated forests of the lowlands and the hills. Most species, including humans, are concentrated below about 300 m. Agriculture, logging, and other activities have taken a heavy toll of the Peninsula's lowland forests, and current trends threaten to yield similar results in East Malaysia.

The country's forests are of great real and potential value: they perform many important environmental and biological services; they constitute a potentially renewable source of many economically valuable products (of which timber is only one); they carry scientific, educational, genetic, aesthetic, recreational, and intrinsic values; and they are the ancestral home of several forest-dwelling peoples. Some of the forests are now greatly valued not only by a growing number of Malaysians but also by many concerned and caring people around the world.

All forests are dynamic, and an understanding of how they function and change is a basic prerequisite of land-use planning and management. Although there is still much to learn about the ecology of tropical nature in Malaysia, there already exists a sufficient fund of knowledge on which to base a sound resource-management strategy. The fate of the forests rests on socio-economic and political considerations, not on new scientific discoveries. The next chapter considers how the forests have been utilized.

3 RESOURCE UTILIZATION AND FOREST CONVERSION: PROCESSES AND POLICIES

3.1 INTRODUCTION: SETTING THE SCENE

Physically, South-east Asia consists of two major regions: continental or mainland South-east Asia, and the structurally complex festoon of islands that comprise the Malay Archipelago or insular South-east Asia.[1] The modern state of Malaysia straddles this basic division: Peninsular Malaysia (hereafter the Peninsula) is part of the mainland, whereas Sabah and Sarawak (East Malaysia) belong to the Archipelago. More revealing, however, is what the two parts of Malaysia share: both lie on the Sunda Shelf; both are in the floristic province of West Malesia; and both are part of a larger island world, for the Peninsula is a near-island.

A certain degree of unity can be discerned in the diverse and far-flung archipelagic world of which Malaysia is a part: the vast majority of the region's inhabitants belong to the Mongoloid world of East Asia; almost all of the indigenous population is Austronesian-speaking; trade—especially trade in forest products—has linked different island groups and forged long-distance connections with other cultural realms for millennia; though they are not unique to the region, two major agricultural systems, swidden and wet rice, have long been the preferred methods of utilizing the land; and, in addition, there are certain similarities in architectural styles, in having been colonized, and in the problems posed by economic development.

Within this great island world, Malaysia belongs primarily to Sundaland (see Chapter 2). With the important exception of 'inner Indonesia',[2] Sundaland is a region of limited seasonality, of ever-wet tropical rain forest, of soils that are prone to rapid leaching, and of widespread swidden agriculture. Little is known about the human presence in the archipelago prior to about 40 000 years ago, when sea-levels were substantially lower. Hunting and gathering probably prevailed everywhere until about 5000 BP, when the slow expansion of Austronesian-speaking agricultural groups began to introduce new ways of life.[3]

It has sometimes been argued that the tropical rain forest is a rather parsimonious environment for humans and that this is especially the case from the standpoint of foraging populations because, among other things, most productivity occurs high up in the canopy where it is generally inaccessible, most potential foods are thinly and unevenly distributed, many plants are poisonous, the ratio of animal to plant biomass is often low, and starchy foods (although available) are often not particularly abundant. On the other hand, there have been many adaptations to such conditions, including the blow-pipe, ability to climb, processing of poisonous plants, nomadism, and, especially, possession of remarkably detailed knowledge of plant and animal life.

Constraints are certainly present (where are they not?), but so too are opportunities, not the least of which is the occurrence of many species of plants whose productivity can be readily manipulated. A great many rain-forest plants have been domesticated in South-east Asia, where they are tended not only in swiddens and permanent fields but also in house gardens.[4] It would be wrong, however, to assume that only farmers manipulate their surroundings, because so too do hunter–gatherers, who 'interfere with their environments in a variety of ways, many of which result in some increase in the productivity of particular plant and animal species' (Hutterer 1983, p. 173).

Malaysia reveals a variety of examples of

different modes of rain-forest exploitation or environmental appropriation. For example, those of pre-modern origin include hunting and gathering, swiddening, collection of forest products for trade, and sedentary farming, while those of modern provenance include plantation agriculture, mining, and timber extraction. The transition to modern resource-use regimes (see Chapter 1 this volume) resulted in mounting human impact on natural systems, including more rapid deforestation. Much of the initial impetus for change came from expanding colonialism, and in this regard Malaysia's experience was by no means unique. This being the case, let us look briefly at some selected aspects of the historical picture, focusing mainly on environmental change.

The perspective of history reveals many centuries of very slow anthropogenic forest change (e.g. Maloney 1985; Delcourt 1987), a speeding up of the process under the impact of colonialism and the emerging world-economy and, with many national and regional variations, a great acceleration of the process in recent decades (see Table 1.2).

The emergence of the European world-economy in the sixteenth century, penetration of the outside world by European powers, incorporation of most agricultural production into a single interdependent world system, and tropical deforestation—all these were related phenomena. As far as agriculture was concerned, incorporation required control of trade, movement of crops, and command of labour. For example: in the case of spices, which had long been traded by Chinese, Arabs, and Indians (Curtin 1984), Europeans came to dominate trade and production; sugar production was moved (as were slaves) to new locations in the Caribbean and parts of Latin America; and new industrial uses were found for plants like rubber and fibres such as sisal, which were then transferred to suitable climatic locations where labour could be acquired (Crow and Thomas 1983, pp. 28–30).

Many plants were transferred via metropolitan botanic gardens, where they were 'improved' and developed by scientists in the service of colonial expansion (Brockway 1979).[5] Excluding the staple foods (e.g. maize, tapioca, peanuts), most such transfers in the tropics occurred in conjunction with the plantation system, which is often regarded as a hallmark of colonialism. Industrialization and rising living standards in Europe and North America during the nineteenth century greatly enhanced the demand for a wide and expanding range of industrial raw materials (e.g. indigo, jute, sisal, cotton), foodstuffs (e.g. sugar, rice), and quasi-narcotics (e.g. coffee, tea, tobacco). As a result, the area devoted to plantation crops expanded sharply, while a reduction of the tyranny of distance and a closer integration of the global economy was achieved by improvements in transportation and communication systems.

There was a marked tendency for plantation crops to replace one another, depending on such factors as shifting metropolitan demand, competition between colonies, and crop failures owing to pests and diseases. But for each crop, 'whether indigenous or newly-introduced, whether grown by peasant or sharecropper or directed by plantation owner, the primary means of increased production to meet the new market demand was to increase the area under cultivation' (Tucker and Richards 1983b, p. xii). One result was widespread depletion of forests and woodlands, a process that was increasingly abetted by a steadily mounting world population.

While there is abundant evidence that the imposition of alien rule or control led to widespread deforestation and considerable ecological disruption, it should be noted that the forces unleashed by colonialism, which itself was not a homogeneous system, had effects that varied over space and through time; and that one of the many contradictions of the system was that it stimulated interest not only in how to exploit resources but also in how to conserve and protect them (Blaikie and Brookfield 1987, chapter 6).

This chapter outlines the processes and policies that have resulted in anthropogenic forest change in the three regions that now comprise the modern state of Malaysia: the Peninsula, Sabah, and Sarawak. Our coverage begins in the late eighteenth century, when Britain acquired its first permanent foothold in the region. This temporal bias is not meant to imply that there was no forest change in earlier times, or that the long and complex history of Malaysia prior to the beginning of colonial rule is somehow less interesting or important. On both counts, we would reject such a view. Put simply, we begin the story at the time when rain-forest change began to pick up speed.

3.2 THE PENINSULA IN THE COLONIAL PERIOD

3.2.1 Introduction

Humans have lived in or alongside the forests of the Peninsula since the early Holocene, and perhaps for a good deal longer (see Bellwood 1985, pp. 159–61). For almost all of that time, however, their numbers were very low, amounting to probably no more than about a quarter of a million in the late eighteenth century. Making up the majority of the total were the village-dwelling Malays, most of whom lived in river-bank clearings where they tended rice and other crops; some lived in busy little entrepôts, for maritime trade had long been a Malay way of life. Roaming the coastal waters of the region were the boat people, the Orang Laut, whose livelihood was oriented to the sea.

Living in or on the fringes of the forest were groups of aborigines whose descendants are now called the Orang Asli. Most ethnologists have divided these peoples into three main units based on physical, linguistic, and cultural criteria and on mode of ecological adaptation: Semang (or Negritos) opportunistic foragers, Senoi swidden farmers, and Proto-Malay horticulturalists (Rambo 1982, also 1988a,b). Unlike the Malays, who peopled the forest with evil spirits, most Orang Asli groups felt at home in the security of their green world. Among their ranks were specialized collectors of aromatic woods, resins, medicinal plants, kingfisher feathers, rhinoceros horn, and a great assortment of other forest products—products that for many centuries had made the Malay Peninsula an important link in a complex network of regional and international trade whose tentacles ultimately extended from Europe to China.

Great forests clothed virtually all of the Peninsula until well into the nineteenth century. It would be a mistake, however, to assume that they were pristine or virgin forests, because thousands of years of human agency on the part of the Orang Asli had selectively altered their genetic resources (see Chapter 4 this volume). It was into the margins of this forested realm that Britain, or more precisely the East India Company, began to penetrate in the late eighteenth century.

The Company acquired Pulau Pinang (Penang Island), Singapore, and Melaka in 1786, 1819, and 1824, respectively (Fig. 3.1). Subsequently known as the Straits Settlements, these were Britain's first colonial bases in the Peninsula, and their acquisition was eventually to have far-reaching consequences for the region's forests. As far as the Company was concerned, however, the major purpose of the three settlements was to protect and foster the China trade. In short, since trade, not territory, was of paramount concern, very little interest was shown in the heavily forested mainland, large parts of which remained *terra incognita* to Europeans until the late nineteenth century. When the Company lost its monopoly of the China trade in 1833, it also lost most of its interest in the Straits Settlements, which eventually became a crown colony in 1867 (Turnbull 1972).

Although Britain continued to pursue a policy of non-intervention in the affairs of the Malay states until the 1870s, her presence in the Straits acted as an economic magnet to a swelling tide of mainly Chinese immigrants, many of whom soon gravitated to parts of the western lowland region in the early decades of the 1800s. The activities of the Chinese included tin mining, which devastated parts of Perak and Selangor, and a resource-depleting form of shifting plantation agriculture

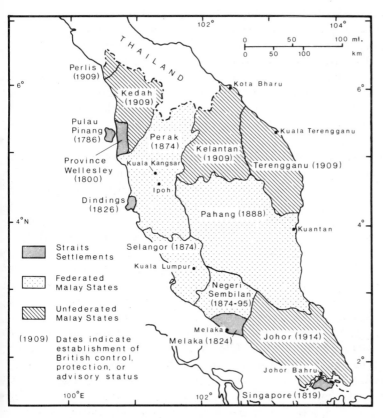

Fig. 3.1 British rule in the Malay Peninsula. (From Turnbull 1981, p. 185.)

that left behind a discontinuous hollow frontier of degraded soils and forests. From the early decades of the nineteenth century, economic activity was increasingly concentrated in the central-western lowland region facing the Strait of Melaka, which rapidly emerged as the most dynamic and developed part of the Peninsula.

Britain adopted a new forward policy in the 1870s, and between 1874 and 1914 colonial control was extended over the entire Peninsula, whose landscape and life were thereby eventually transformed. Behind this new departure were probably several motives, these including strategic considerations, new ideas and values regarding Britain's place in the world, and a growing demand at home for industrial raw materials. British rule, or the 'Residential system' as it was called, was introduced into Perak, Selangor, Negeri Sembilan, and Pahang between 1874 and

1895, and in 1896 all four states were formally united to form the Federated Malay States (hereafter FMS), with the mining town of Kuala Lumpur as the capital (Fig. 3.1). The remaining five states, which were known by the negative title of Unfederated Malay States, eventually accepted British advisers, thereby completing the system of so-called indirect rule (Fig. 3.1; see Emerson 1937; Sadka 1968; Turnbull 1981, chapters 12 and 14).

The spread of British rule over the Peninsula set the stage for the first major assault on the region's forests. After 1874 the colonial government began to establish the framework for a profitable export economy based on tin and plantation agriculture: an efficient legal and administrative apparatus was established, a modern system of communications was laid down, the Torrens system of land tenure was adopted (this providing undisputed title to land, thereby stimulating land development and resource exploitation), liberal taxation policies were implemented, a trigonometrical land survey was started, cheap labour was made available to planters and miners, and law and order were upheld. These various measures were, in due course, rewarded 'by a vast infusion of European capital, enterprise and management skills without which the rapid economic development of Malaya in the twentieth century would not have occurred' (Andaya and Andaya 1982, p. 209; see H. C. Chai 1964).

Chinese-dominated tin mining was already well established in the western lowlands (see above) and it was here, especially in the tin-rich states of Perak and Selangor, that economic development proceeded most rapidly. Until about 1914 most of the government's revenue was derived from the export duty on tin, and the bulk of this was ploughed back into building railways and roads both within and between the main mining areas, thus stimulating existing mining enterprises and also serving to attract an increasing inflow of European capital, management, and technology. Conditions were also becoming more favourable for the development of commercial agriculture: peace and security, better communications, support for botanic gardens, generous loans, a light tax burden, improvements in public health, favourable land laws—all these combined to place agriculture on a firmer basis. In short, in the two or three decades following British intervention in the Malay states, the colonial government erected an edifice in which private enterprise could flourish. One result was a rapid increase in the pace of deforestation.

This section focuses on the main economic activities and policies that resulted in forest loss in the Peninsula during the colonial period (up to 1957); it also has something to say about forest administration and changing forest policy.

3.2.2 Malay peasant agriculture

The Malays lived in coastal, estuarine, and deltaic flats, along river valleys, and in parts of the interior foothills. The basic unit of settlement was the permanent village (*kampung*), usually linear in form, featuring wooden houses on stilts, each set in its own compound and surrounded by a vegetable garden and an orchard (*dusun*). The river was 'the high-road, the water supply, the bath and the drain' (Gullick 1958, p. 28), and each state was centred on a river valley or a system of rivers.

There are no accurate population figures for the Malay states prior to British intervention, when the number of Malays (excluding those in the Straits Settlements) was perhaps in the order of 200 000–300 000 (see Sadka 1968, pp. 3–5; Ooi 1976, p. 111). The population began to grow rapidly after *c*.1880, when immigrants from Sumatra,[6] Sulawesi (the Celebes), and elsewhere in the Archipelago arrived to open up new areas for padi[7] cultivation and later for rubber; in addition, natural increase within the indigenous community also contributed to swelling numbers, though to what extent remains difficult to estimate (Bahrin 1967; Ooi 1976, pp. 122–6). Population growth was accompanied by forest removal and expansion of settlement: following roads and railways, inland along rivers, in a broad front from the coast towards the interior as swamp forests were cleared and drained for padi, and at an advancing distance from lines of communications as land was brought under padi or rubber cultivation (Dunn 1975, pp. 107–8; Ooi 1976, pp. 154–8).

Rice was the basis of the subsistence economy. Three main types of cultivation can be distinguished: as one of many crops grown under shifting (*ladang*) cultivation in the hills; as dry padi in sloping or lowland areas where water could not be retained; and as wet rice grown in bunded or embanked fields featuring either traditional rainfed rice or rice grown with the aid of supplementary water, including water from irrigation systems. There were many local and regional variations (see Anon. 1901; Gullick 1958, pp. 28–31; Hill 1977).

Ladang cultivation was widespread, especially in the foothills of Kelantan, Terengganu, Pahang, Kedah, and parts of Negeri Sembilan (Wyatt-Smith 1958*a*). Foresters in particular were critical of the practice; for example, Arnot and Smith (1937) estimated that it had rendered some 20 per cent of Terengganu useless for forestry by the turn of the century. Rotation periods were shorter than those practised by Orang Asli shifting cultivators because, unlike the latter, the Malays lived in permanent settlements. The result, it appears, was extensive areas of secondary forest or scrub (*belukar*).

There were large areas under wet-padi cultivation in Kelantan, along the Pahang river valley, in Province Wellesley, and on the southern Kedah Plain by the mid-nineteenth century (Logan 1851; Zaharah 1966). Cultivation of irrigated rice was probably best developed in the central valleys of Negeri Sembilan, where the necessary expertise was provided by immigrant Minangkabaus from Sumatra (Grist 1922). As for lowland dry padi, this was probably most extensive in Kelantan and Terengganu.

Important dietary supplements to rice included fish, forest game, coconuts, kitchen garden and orchard products, and at least some peasants had a few cattle, ducks, and chickens (Winstedt 1961, chapter 7). Building poles, bamboo, rattan, *atap* (or thatch, usually made from woven leaves of *nipah* palm), and a variety of other wild products came from the rain forest, which around many settlements closed in like the walls of a prison. It was not, as Sadka (1968, p. 6) noted (referring to the period 1874–95), an entirely self-contained economy because

there was usually a local exchange of surplus food-crops and livestock, and there was an export of jungle produce (rattans, *gutta*, firewood, *atap*, precious woods), and for centuries there had been an important export of tin and gold. This sustained a trade cycle, in which the tin export was balanced by the import of trade goods, mainly textiles, ironware, tobacco, salt and opium.

While it is true that the Malays, especially the local-born Malays, were generally less affected by the colonial economy than were other ethnic groups, they were nevertheless increasingly drawn into a larger monetized world of market relations and changing life-styles. Some Malay peasants

combined subsistence agriculture with the commercial production of padi or the sale of vegetables, fruits, and meat in nearby urban centres, while others, especially after *c*.1910, grew rubber together with subsistence food crops; yet others eventually became fully specialized commercial farmers, growing mainly rubber on smallholdings (T. G. Lim 1977).

After the turn of the century, colonial officials began to give some rather desultory attention to irrigation, and, in some areas, irrigated rice cultivation spread at the expense of the rain forest. At Krian in Perak, swamp forest covering some 20 250 ha was cleared and drained between 1899 and 1906 (Short and Jackson 1971), and an area of similar size was likewise developed for rice production in southern Perlis and on the Kedah Plain during the 1920s. A Rice Cultivation Committee was established in the FMS in 1930 to consider ways and means of encouraging rice production in the Peninsula. An important recommendation of the Committee was acted upon in 1932, when the Drainage and Irrigation Department was established.

By 1937, according to Winstedt (1961, p. 126), 'the total area in British Malaya with special irrigation schemes for rice-planting was just over 142 000 acres' (57 368 ha). Several swamp-forest sites were cleared and drained for padi production between the 1930s and 1960, these including the Sungai Manik scheme in lower Perak (8100 ha), the Panchang Pedena area of Selangor (22 275 ha), and parts of northern Kedah/south-central Perlis (10 125 ha).

Other extensive tracts of wetland forest in the western lowlands were cleared for agriculture (or destroyed by mining activities) during the first half of this century. In addition to rice, which did well in areas of reclaimed freshwater-swamp forest, pineapple cultivation in Johor expanded at the expense of large areas of peat-swamp forest.

3.2.2 Tin

Tin had been mined in the Peninsula since ancient times, and it was the chief support of Melaka's trade during the fifteenth century. The Portuguese, and later the Dutch, attempted to secure a monopoly of the trade, but neither attempted to control the mines, which remained in the hands of Malay royal families and district chiefs for whom they were an important source of revenue. Mining was conducted on a modest scale, featuring many small shallow mines in narrow river valleys and along hillsides where the ore was obtained by panning (*dulang*) and by separating the ore in a water race (*lampan*). After centuries of working the 'tin lands', the Malay miners had made hardly a dent on the omnipresent rain forest.

The scene began to change in the 1820s and 1830s, when growing numbers of full-time Chinese miners began arriving in the west coast states, and soon most management and labour had passed out of Malay hands. Financial backing came from Chinese and European merchants in the Straits Settlements, and more efficient mining techniques were introduced.[8] Remarkably rich tin deposits were discovered in Perak and Selangor during the 1840s and 1850s and by the 1870s, when Britain began to implement the Residential system in the protected states, the main tin mining areas were at Larut and Kinta in Perak, around Kuala Lumpur in Selangor, and on the upper Linggi in Sungai Ujong (later merged into the Negeri Sembilan confederation). Change was now rapid: Malay rulers invited in large numbers of Chinese miners, an open-door immigration policy fostered mass migration from southern China, transportation facilities were expanded and much improved, and tin production, which was greatly stimulated by the rapid growth of the European tin-plate industry from the 1860s, began to soar. By 1904, the Malay states produced half of the world's tin (see Fermor 1939; Allen and Donnithorne 1957; L. K. Wong 1965; Yip 1969).

The Chinese continued to dominate the tin-mining industry, and in 1913 some 74 per cent of production came from Chinese mines. After the formation of the FMS, however, mining became increasingly capital-intensive and European enterprises gradually took over the larger share of

production: 'By 1930 European mines were producing 63 per cent and the Chinese only 27 per cent' (Ooi 1976, p. 332).

Tin mining scarred the landscape, leaving behind denuded and eroded hills and a chaos of pits, ponds, and mounds of sterile tin tailings; sediment clogged many streams and rivers and flooding in some areas became more frequent; and the forests surrounding the mines were degraded or completely destroyed. Large quantities of timber were required for buildings and machinery, for firewood, and especially for charcoal, which was used to smelt the ore. In the Larut district of Perak, where mining had been more or less continuous since the 1840s, 'all the best timber within fifteen to twenty miles of the mines had been destroyed by 1879', according to L. K. Wong (1965, p. 160).

A new form of mining machinery, the floating dredge, first appeared in 1912, and by 1937 dredging accounted for 50 per cent of tin output (Allen and Donnithorne 1957, p. 152). Dredging meant that swampy alluvial flats could be mined, consequently extensive tracts of wetland forests began to disappear. The tin-mining economy stimulated the growth of new urban centres, enhanced the demand for foodstuffs, and was accompanied by major improvements in transportation systems, all of which meant that its impact on the rain forest was not confined to the immediate vicinity of the mines.

3.2.4 Early plantation agriculture

European and Chinese planters in the Straits Settlements experimented with a variety of cash crops, notably pepper, nutmegs, cloves, sugar, coffee, cotton, indigo, cinnamon, and coconuts but, with the exception of sugar in Province Wellesley and pepper and gambier in Singapore, commercial agriculture was a dismal failure and a great disappointment to the East India Company (Anon. 1893; Ridley 1905; Turnbull 1972, pp. 140–60). Most such agriculture was speculative from the outset, and there was a tendency for crops to replace one another depending on factors such as price fluctuations, soil and nutrient exhaustion, ignorance of the local environment, ravages of pests and diseases, competition from other colonies, and changing government policy. Most forms of export agriculture were dominated by the Chinese during the nineteenth century, when there were two kinds of plantations: permanent estates, which were run by both Chinese and Europeans, and 'shifting plantations', which were operated and financed almost exclusively by the Chinese.

Export-oriented agriculture spread from the Straits Settlements to the mainland Malay states with the Chinese: pepper and gambier cultivation diffused from Singapore to Johor in the 1840s and eventually to Negeri Sembilan and Selangor in the 1870s; also in the 1870s, tapioca spread from Melaka to Negeri Sembilan and sugar from Province Wellesley to Krian in northern Perak. Meanwhile, successful European planting was largely confined to sugar cane in Province Wellesley.

The development of commercial agriculture in the protected states was a high priority of the colonial government (see above). Assistance and encouragement came in several forms: generous loans were made to pioneer planters; experimental agricultural stations were established; land was alienated on very liberal terms; many new roads were extended into outlying areas; and large numbers of Indian and Chinese labourers were imported. European planters in the Malay states gained valuable experience in the last quarter of the nineteenth century, but that was about all they had to show for their efforts until rubber appeared on the scene.

Commercial agriculture spread at the expense of the rain forest. Although most early plantations were relatively small, it should be kept in mind that there were many of them in some regions and that they were accompanied by new roads, tracks, settlements, swelling human numbers, and expanded food production. The progress of the main crops conveys some idea of the impact of the nineteenth-century plantations on the natural scene (see Jackson 1968*b*).

Pulau Pinang was heavily forested when Francis Light acquired it on behalf of the East India Company in 1786. Wright and Reid (1912, p. 84) commented that 'Light found Penang a jungle: he left it a garden'. But for the most part it was not a very productive garden. Pepper (*Piper nigrum*) was introduced from Aceh in 1790, and in the early nineteenth century cloves and nutmegs were brought in from Maluku (the Moluccas) by the East India Company botanist (Ridley 1905, p. 295). Most of the island's early spice plantations were developed on land newly cleared of forest, often on steep, erosion-prone slopes. Pepper was already in the doldrums by the 1830s, and although Pulau Pinang at mid-century had 'hundreds of small estates . . . and it was predicted that within five or six years the whole island would be cleared and planted with nutmeg' (Turnbull 1972, p. 145), blight and drought had ravaged most of the plantations by the 1860s.

Cultivation of pepper and gambier—the extract from the boiled leaves of gambier (*Uncaria gambir*) was used in the dyeing and tanning industries (see Ridley 1892)—expanded rapidly in Singapore in the 1830s and 1840s, eventually spilling over into Johor where there were some 200 plantations in 1848 (Turnbull 1972, p. 152; see L. Oliphant 1860, I, pp. 23–33; Humphrey 1982, pp. 335–9). Pepper–gambier (the two crops were often growth together) and tapioca (*Manihot esculenta*), which became commercially important in Melaka and Negeri Sembilan in the 1860s and 1870s respectively, were grown by Chinese on shifting plantations. Tapioca exhausted the soil in about five years, pepper and gambier after about 15 years (Thomson 1850; Ridley 1892), and large quantities of wood were needed for boiling the gambier leaves. The practice of clean weeding must have produced high rates of erosion, and abandoned plantations were soon overrun with tough grasses such as *lalang* (*Imperata cylindrica*), which inhibited forest regeneration (Balestier 1848; Anon. 1895; Lim 1908),[9] or were colonized by dense scrub (*belukar*). Competition from rubber, declining prices, and official restrictions on the wasteful practices of the planters led to a decline of shifting plantation agriculture at the turn of the century.

Sugar cane and coffee were two other crops of some importance in the nineteenth century. The former was the main commercial crop in Province Wellesley, where it had been introduced by Chinese planters before 1820. Europeans opened up large estates in the 1840s, and by c.1860, when they controlled four-fifths of production, some 4050 ha were under cultivation (Khoo 1972, p. 93; Turnbull 1972, pp. 144–5). Chinese planters began moving into Krian in the 1870s, and in 1882 the large and important Gula estate was opened up in that district by Europeans (it was the only European estate in Perak until 1893).[10] Labour shortages and competition from other colonies resulted in a sharp decline in the area under cultivation in the early 1900s, and in 1913 the last sugar refinery was closed.

Coffee was grown mainly by Europeans, who introduced the crop into Perak, Selangor, and Sungai Ujong between 1879 and 1881. Many of the planters were from Ceylon, where a blight had destroyed the coffee plantations during the 1870s. The area devoted to coffee in the 1880s was insignificant, but by the mid-1890s, thanks largely to rising prices and an adequate supply of labour provided by Indian immigration, cultivation of the crop reached boom proportions—especially in the Kelang Valley of Selangor, which emerged as the hub of the industry. But the boom years were short-lived, and around the turn of the century the industry collapsed because of competition from Brazil.

3.2.5 Rubber (and other crops)

The Malayan rubber industry has been described as 'one of the greatest achievements of Western colonial enterprise' (Allen and Donnithorne 1957, p. 106). Together with tin, it formed the backbone of the Peninsula's economy, and its impact on the landscape was profound.

Seedlings of Brazilian rubber trees were sent to the Singapore Botanic Garden from Kew in 1876, but they soon died. A second consignment was successfully established in 1877, thus marking the origin of the Malayan rubber industry.[11] The father of the new enterprise was Henry Ridley, who was appointed director of the Singapore garden in 1888. Known to his critics as 'mad Ridley' or 'rubber Ridley', he was a tireless experimenter and the first person to recognize rubber's great commercial potential (Eaton 1935).

The new tree crop got off to a rather slow start, and the area devoted to it in 1897 was only 140 ha. By that time, however, planters had recognized that rubber—particularly Para rubber[12] (*Hevea brasiliensis*)—was well suited to a variety of site conditions, and they began to plant it more extensively following the collapse of the coffee industry in the late 1890s. The timing of the transition from coffee to rubber was fortuitous because it coincided with the initial growth of the American automobile industry, which greatly stimulated demand for pneumatic tyres and other rubber products. As prices soared, the area devoted to rubber increased dramatically in the first two decades of this century: from about 4500 ha in 1903 to some 405 000 ha in 1916, and then to more than 810 000 ha in 1921 (Drabble 1973, p. 215), when the crop represented about half of the world's rubber production and some three-fifths of all cultivated land in the Peninsula.

Cultivation spread along lines of communication that had been developed to service the tin mines, and then into other areas as they were opened up. Although initially concentrated in Perak, Selangor, and Negeri Sembilan, planting soon spread to Melaka, Johor, and southern Kedah, eventually spilling over, albeit at a much reduced pace, into the more isolated states east of the Main Range.

Government policies that encouraged greater production included giving planters access to cheap land and generous loans, facilitating large-scale immigration of Indian labour, building roads, tracks, and bridges, and launching attacks on malaria and other diseases of the expanding settled area. Into this politically favourable late nineteenth-century environment was inserted that 'extraordinarily flexible, permanent, imper-

Plate 1. Tropical lowland evergreen rain forest in the Pasoh Forest Reserve, Negeri Sembilan. Great lofty forests of this type formerly clothed most of the country up to an elevation of *c.* 750 m. (C. H. Leigh).

Plate 2. Heath forest is extensive in Borneo, where it is known as *kerengas*. Much of it, as here in Bako National Park, Sarawak, has been converted to grassland with scattered shrubs and remnant woodlands. (K. Rubeli).

Plate 3. Tropical upper montane rain forest near the summit of Gunung Brinchang in the Cameron Highlands; the elevation is *c.* 2000 m. Note the low, even-height canopy and the absence of emergents. (C. H. Leigh).

Plate 4. Seladang (*Bos gaurus*) or gaur are wild cattle that live in herds of up to 25 individuals. They shelter in forest by day, emerging at nightfall to graze in clearings and along grassy river banks. (Tourist Development Corporation, Kuala Lumpur).

(a)

(b)

Plate 5. FELDA-sponsored rural development, Jengka Triangle, Pahang. (a) Cleared area awaiting planting. Note the extent of bare ground. (b) FELDA settlement in 'desert-like' setting. (C. H. Leigh).

Plate 6. Shifting cultivation along the Melinau River (Sungai Rajang basin, east of Kapit), Sarawak. (K. Rubeli).

Plate 7. Abandoned tin mine, Kinta Valley, Perak. The revegetation of such areas is hampered by the sterile nature of the tailings. (C. H. Leigh).

Plate 8. Cutting in deep regolith and weathered bedrock: Main Range between Seremban and Kuala Pilah, Negeri Sembilan. Such cuttings are prone to erosion and slumping (especially if not properly drained), resulting in high sediment loads in nearby streams. (C. H. Leigh).

Plate 9. Part of the Genting Highlands hill station. The scar along the hill in the middle distance is the road up to the resort. Potential threats to montane forests include further development of hill stations and an ill-conceived ridge-top road. (C. H. Leigh).

Plate 10. Log landing area, Ulu Gombak, Selangor. Logging activities often leave behind extensive areas of bare ground and cause considerable damage to residual stands. (K. Rubeli).

Plate 11. Fallen two-horned Sumatran rhinoceros, shot in 1914 on the Main Range above Kuala Lumpur. The species is now highly endangered. (Courtesy of the Bombay Natural History Society. Photograph by T. R. Hubback).

Plate 12. The Great Hornbill (*Buceros bicornis bicornis*). Hornbills are wide-ranging forest-dwellers. In many areas their survival is threatened by forest clearance and by loss of mature trees with nesting holes. (K. Rubeli).

Plate 13. Many of the Orang Asli now live in 'pattern settlements' like this one in the Cameron Highlands, Pahang. (C. H. Leigh).

Plate 14. Sumatran rhinoceros: symbol of the 'Save Endau-Rompin National Park' campaign.

sonal yet supremely competent institution, the corporation or joint-stock company' (Fryer 1979, p. 70), which applied industrial forms of organization to its various operations.[13] Increasingly, ownership and control of agricultural enterprises passed out of the hands of individuals and into those of corporate businesses. In due course the modern, industrial-style plantation became the model for planned settlement and land development, its most notable exponent being the Federal Land Development Authority (see below).

Most planting was on land newly cleared of forest, which was 'burnt as a sort of preliminary offering to the goddess Rubber' (Wright and Reid 1912, p. 292). A drab greenery soon took hold in the charred wake of the retreating rain forest, leaving behind an orderly but depauperate landscape of unrelieved monotony. Although the large estates were controlled by Europeans, Malay and Chinese farmers began to plant rubber in their gardens and orchards, on old tapioca and gambier lands, and in newly cleared areas, so that by 1921 smallholdings accounted for 33.4 per cent of the total area under rubber in the FMS (T. G. Lim 1977, p. 143).

Falling prices after World War I and the impact of the Great Depression resulted in the introduction of restriction schemes that curtailed the alienation of new land for rubber production, but in spite of these impediments the area under cultivation expanded from 0.93 million ha in 1922 to 1.4 million ha in 1940, when rubber covered about 11 per cent of the entire area of the Peninsula. There were two reasons for this advance: development of mainly estate lands alienated prior to the introduction of the restriction schemes, and an increase in smallholder cultivation on lands previously devoted to or intended for other crops (Ooi 1961). The industry suffered a reversal during the Japanese occupation and in 1957, when independence was proclaimed, the area devoted to the crop differed little from that of 1940.

The colonial government gave some attention to agricultural diversification during the generally lean inter-war years. Considerable success was achieved with oil palm (*Elaeis guineensis*),[14] and the planted area increased from some 5050 ha in 1926 to 31 900 ha in 1940. The cultivated area continued to expand after World War II, and by 1960 oil palm covered some 54 675 ha (Tan 1965). Much of the planting occurred on former rubber and coconut land but, in contrast with the case of rubber, colonial officials permitted alienation of forest lands for cultivation.

Also of some commercial importance were the coconut and the pineapple. The former was a traditional Malay subsistence crop, although there were some plantations of 40 ha or more in Selangor prior to the turn of the century (Sadka 1968, p. 345). Drainage of several western coastal areas during the 1920s and 1930s facilitated the spread of the crop, and a number of estates were established in Selangor and southern Perak. The area devoted to coconuts in 1961 was 210 000 ha, with smallholdings making up 85 per cent of the total. The area given over to the pineapple, which initially was interplanted with immature rubber, increased from 1200 ha in 1929 to a peak of 15 000 ha in 1956 (Ooi 1976, pp. 284-5). Depletion of the west coast peat-swamp forests continued after World War II, when large areas in the Pontian district of western Johor were cleared for pineapple cultivation.

3.2.6 Forest administration and policy

During the colonial period, forest management systems were established and forest-related legislation was introduced. In addition to the needs of day-to-day forest management, foresters also undertook research and promoted the reservation of forest areas for watershed and hill-land protection and for sustained production. The evolution of forest administration in the Peninsula and attempts to formulate forest policy are briefly outlined in this section. These are important themes because they shed light on the course of forest clearance and alteration during the first half of the present century and because the legacy of past decisions still reverberates today.

A forest department under the director of the

Singapore garden was established in the Straits Settlements in 1883, and one of the early directors, Henry Ridley (see above), had the foresight to reserve most of the remaining forests in the state of Melaka (Troup 1940, pp. 375–6; Watson 1950, p. 64). A broader interest in forest management was revealed in 1901 when a forest department was created for both the Federated Malay States and the Straits Settlements (hereafter FMS/SS), although one of its first tasks was to help to organize the orderly removal of the forest for rubber cultivation (Kumar 1986, p. 65)! The department's duties and responsibilities were given legislative support between 1907 and 1918,[15] and after World War I it became progressively more involved in forest botany, silvicultural practice, policy formulation, and forest reservation (see e.g. Watson 1934; Strugnell 1938; Troup 1940, pp. 374–83; Wyatt-Smith and Vincent 1962; Menon 1969). During the 1920s and 1930s the FMS/SS model of forest administration and legislation was adopted by all the unfederated states (with the exception of Perlis).[16]

It was already apparent in the late nineteenth century that extensive and permanent forest reserves would be required to maintain a self-sufficient supply of timber and other forest products and to protect steepland areas from erosion. Although some reserves were established prior to the turn of the century (see Wyatt-Smith and Vincent 1962, p. 210), it was not until the FMS/SS forest department was formed, and not until departments were etablished in the unfederated states, that systematic forest reservation programmes were introduced. For example, immediately after his appointment in 1901 the first Chief Forest Officer of the FMS/SS department ordered the reservation of mangrove forests in Perak and Selangor in order to prevent their overexploitation,[17] and his successor, who took office in 1915, reserved tracts of forest throughout the FMS that were stocked with the general utility keruing (Dipterocarpus spp.) and meranti (Shorea spp.) timbers, whereas previously the policy had been to reserve forests containing the more valuable

cengal (Neobalanocarpus heimii) and merbau (Intsia palembanica) (Watson 1950, pp. 65 and 67).

Forest reservation proceeded apace in the inter-war years, and by 1936 the forest reservation programme had been completed in the Straits Settlements and was nearing completion in Kedah, Perak, Selangor, Negeri Sembilan, and Johor. At the end of 1937, productive and protective forest reserves covered some 27 110 km^2 (or about 20 per cent of the land area of the Peninsula), the latter covering most of the mountainous areas (J. N. Oliphant 1932, p. 240; Troup 1940, p. 382). In addition, 7500 km^2 had been reserved for wildlife and other purposes (see also Chapter 5 this volume).

The inter-war years also witnessed greater interest in devising a forest policy that would lead to a permanent forest estate for the FMS and, eventually, for the whole Peninsula (Watson 1950, p. 67). Unfortunately this idea was dealt a severe blow in the early 1930s when a policy of administrative decentralization was adopted in the FMS, with the result that, among other things, mining, agriculture, and forestry were placed under individual state control.[18] Foresters were opposed to this development, one subsequently commenting that '[a]nything less suited than Decentralization to the needs of forestry or more subversive to adequate control of exploitation, utilization or protection it would be difficult to imagine' (Watson 1950, p. 70). But given the administrative complexity of British Malaya, it is hardly surprising that a Malaya-wide forest policy failed to materialize. On the other hand, a certain degree of administrative uniformity was achieved by the secondment of FMS forest officers as advisers to the unfederated states, where forest reservation programmes were introduced in the 1930s (Troup 1940, p. 378; Watson 1950, p. 70).

In the years following the Japanese occupation, foresters again pressed for a country-wide forest policy, and in 1952 the federal government released an Interim Forest Policy Statement containing the recommendation that up to 25 per cent of the land area of the Peninsula should be set aside for sustained timber production (Wyatt-Smith and

Vincent 1962, p. 206).[19] This was accepted in principle by most states but formally adopted by none. Meanwhile, control over forests continued to rest with the states, as indeed it still does.

'Envoi'

Stimulating economic development was the main goal of the colonial government. A basic requirement was land uses that were 'more productive' than the apparently interminable forest, which was viewed as largely empty, a virtual *tabula rasa* awaiting replacement and development. Compared with plantation agriculture and tin mining, unexploited forests had little worth, a perception that tended to ignore their scientific, genetic, and other values, to underrate their ecological and environmental services, and to overlook their vital importance to the forest-dwelling Orang Asli. The myth of vast emptiness has long worked against both the forests and their human inhabitants.

Rubber in particular made substantial inroads into the forest, but even by 1935, according to Troup (1940, p. 380), about 80 per cent of the Peninsula 'was still under forest of some kind or another', and in 1958—one year after independence and 172 years after Francis Light's occupation of Pulau Pinang—74 per cent of the land area of the Federation of Malaya (now Peninsular Malaysia) remained forested (Wyatt-Smith 1958b, p. 40). Probably only about 20 per cent of the forest cover was eliminated during the entire colonial period (it is a lot more difficult to estimate the area that was variously degraded). In 1966, when a survey of the Peninsula's land resources was conducted, close to 70 per cent of the area was still under forest, most of it so-called primary forest (Ooi 1976, p. 89). Much has changed since then.

3.3 THE BORNEO TERRITORIES IN THE COLONIAL PERIOD

3.3.1 Introduction

We turn now to a discussion of forest change in the ex-British Borneo territories of Sarawak and North Borneo (later renamed Sabah). This section is less comprehensive than the previous one because both territories were still heavily forested at the end of the colonial period (1963). We begin with a brief sketch of the scene today, which in turn leads into a short account of the origins and spread of British rule in the region. We hope this procedure will be of use to the general reader.

Sabah and Sarawak (East Malaysia) occupy the north-western part of the great island of Borneo. Covering 73 940 km^2 and 124 449 km^2 respectively, they together make up about 60 per cent of the total land area of Malaysia but embrace only about 18 per cent of the nation's population. The East Malaysian region comprises three main topographical units: a mangrove-fringed, low-lying, and often swampy coastal plain that narrows towards the north, a broad belt of hill country, and a zone of interior highlands. Sabah is the more mountainous of the two states but, in general, possesses better soils and more accessible plains. The extensive peat-swamp forests of Sarawak contain valuable timber but the acid soils of the swamps preclude almost all crops (other than the sago palm), and most of the hilly interior of the two states is unsuitable for permanent-field agriculture. East Malaysia is rich in oil, gas, and timber and these resources are being rapidly exploited to boost the nation's export economy. Most of Malaysia's remaining forests are found here (see Table 3.3 below). Fig. 2.2 shows the distribution of the major rain-forest formations.

As noted in the Preface, East Malaysia is characterized by great ethnic and cultural diversity. The Malays are in a minority here, and most of the indigenous population is non-Muslim. Both states have substantial Chinese minorities. The great majority of the region's inhabitants are concentrated in coastal and riverine areas, and many of the non-Muslim indigenes are still engaged in shifting cultivation (see below). Some 80 per cent of the total population of the region (about 2.8 million in 1985) live in rural areas, overall population densities are low, communications are still poorly developed in most areas, primary-producing activities dominate the economy, and poverty is widespread.

3.3.2 The creation of new political units

North Borneo and Sarawak were carved out of the territories of the once-great Brunei sultanate. The first half of the sixteenth century was Brunei's golden age, and at that time it claimed suzerainty over the whole west and north Borneo coast as well as the Sulu Archipelago. Thereafter, however, its power ebbed away and by the early 1840s, when the first serious British penetration of the region began, the domain over which it could lay claim was largely confined to the district around Brunei town, the Sarawak river area (in the future First Division), and a strip of the north-western coast (Turnbull 1981, pp. 55 and 155).

Most of northern Borneo at this time was very thinly populated and large areas were essentially uninhabited. There were a few pockets of thriving economic activity—in the lower reaches of the Mukah and Oya rivers (in the later Third Division), where Melanaus began producing sago for the Singapore market as early as the 1840s, in and around Brunei town, and in the Sarawak river basin, where Hakka Chinese were engaged in gold mining and the Bidayuh (Land Dayak) in working antimony ore—but otherwise forests prevailed virtually everywhere, clothing the plains and the hills and extending up into the interior mountains. Since the mid-sixteenth century, however, growing numbers of pioneer Iban swidden farmers had been migrating into the region from the Kapuas Valley of Kalimantan, and by the mid-nineteenth century they were the largest ethnic group in the territory that forms modern Sarawak (see Sandin 1967; Pringle 1970; Turnbull 1981, chapter 13). It was against this background that the middle-class English adventurer James Brooke (1803–68) made his unexpected advent in 1839.

Brooke rule—the rule of the three 'white rajahs', James, Charles, and Vyner—was established in 1841 and lasted until World War II. James Brooke and his nephew Charles (he ruled

during 1868–1917) pursued a policy of territorial acquisition that extended the Sarawak Raj all the way to Brunei Bay (see Fig. 3.2). After 1890, when the Limbang Valley was included in Sarawak, Brunei was split into two parts. Forts played an important role in this territorial expansion because they were used to deny Iban 'pirates' access to the sea, to control river traffic, and to provide safe havens for European administrators in dangerous outlying districts (Fig. 3.2; Pringle 1970, pp. 84–94).

The main priority of the paternalistic Brooke administration was to protect the native peoples from undesirable outside influences; consequently, very little encouragement was given to Western-style commercial activity. This is not to say, however, that there was no economic development, but rather to emphasize that the Brookes, who 'saw outside investors as a potential threat to [their] political dominance' (Reece 1988, p. 32), opted for a policy of slow and gradual change. Charles Brooke left Sarawak relatively prosperous and in the period after c.1910 it progressively moved towards an export-oriented cash economy (Pringle 1970; Criswell 1978; Reece 1988).

A different form of colonial expansion was set in motion in the extreme northern part of the island, over whose coastal areas both Brunei and Sulu exercised at least nominal overlordship. Probably fewer than 100 000 people lived in this area in the nineteenth century. An American trading company working out of Hong Kong attempted to establish a settlement on the west coast in the 1860s, but after this failed the company was sold to an Austrian aristocrat, Baron von Oberbeck, who went into partnership with a British businessman, Alfred Dent. In the late 1870s, Dent and Oberbeck obtained title to a vast tract of territory from the sultans of Brunei and Sulu, and in 1881 Dent, who bought out Oberbeck, was granted a royal charter to form the British North Borneo (Chartered) Company whose territories became known as 'North Borneo'. Additional territories were acquired in the period down to about 1901, when the Company's realm was more or less conterminous with

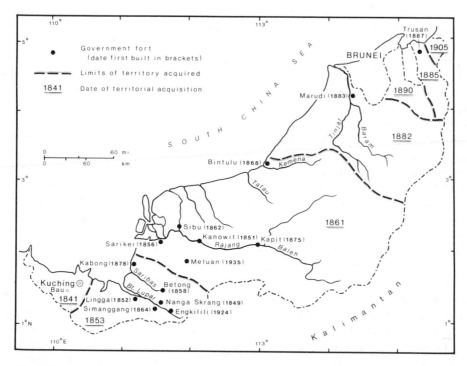

Fig. 3.2 Growth of the Sarawak Raj, 1841–1905. (From Jackson 1968a, p. 59.)

that of the modern state of Sabah (see Black 1983, map 1, p. 2).

3.3.3 North Borneo: economic activities and forest conversion

Although it was never more than an administrative body, and hence never directly involved in trade, the Chartered Company sought, through encouraging the efforts of others, to make North Borneo into a profitable commercial venture. So, with this goal in mind, labour was brought in from China and Indonesia (mainly Java), government stations were established, some attention was given to improving transportation facilities, and law and order were upheld. In the early years of its existence, however, the Company struggled merely to survive, and it was not until after about 1909 that it began to pay any dividends to its shareholders.

Except in the handful of towns, social improvements were minimal.

The Company was initially sustained by revenue derived from the exploitation and export of forest and sea products; in particular, edible birds' nests, which were shipped directly to Hong Kong, proved to be a dependable source of revenue. Among the forest products that were collected and sent to Singapore, where they were cleaned, graded, and re-exported to world markets, were rattan, gutta-percha, damar, kapok, illipe nuts, camphor, and elephant tusks. Forest and sea products accounted for one-third of North Borneo's entire export trade in 1885 (Y. L. Lee 1965, p. 28). It was these so-called minor forest products that saved the Company from bankruptcy.

The government sponsored or encouraged the cultivation of a wide range of crops: those that failed completely included sugar, tapioca, opium, silk, soya beans, pineapple, and orchids; others that met with only very modest or temporary success included pepper–gambier, jute, and kapok; yet others, among them the coconut (the area devoted to its cultivation increased from 4733 ha in 1914 to 22 855 ha in 1940) and Manila hemp (the planted area was 2354 ha in 1940), were somewhat more successful. Other activities that contributed to the diversification of the economy included an export trade in dried fish and sea-pearls, and coal mining on a small scale (see Tregonning 1965, chapter 5). None of these crops or activities had much impact on the rain forest and they all remained subordinate to the three great bulwarks or mainstays of the colonial economy: tobacco, rubber, and timber.

Tobacco was the only crop of any real significance in nineteenth-century North Borneo. A boom crop during the 1880s and early 1890s, it was grown mainly by Dutch and German planters in the east-coast region. By 1890 there were 61 estates listed as growing tobacco. These

stretched down the east coast, from the German-managed plantation on Banggi Island, round Maruda Bay, and far up the long remote rivers, the Labuk, Sugut, Kinabatangan and Segama. There were some [estates] near the government outposts at Lahad Datu and Tawau, as well as round Sandakan Bay (Tregonning 1965, p. 86).

Although the boom was short-lived—more than half of the estates had closed down by 1895, in part owing to an American tariff on imported tobacco-leaf—it was tobacco that first brought large numbers of Chinese to North Borneo, it was tobacco (along with jungle and sea products) that sustained the Company in its early days, and it was tobacco that led to the initial penetration of the interior along the main rivers. The value of tobacco exports peaked in 1902, which saw the beginning of a long-drawn-out decline. There were only 162 ha of land devoted to the crop in 1940. Rain forest was cleared to make way for tobacco, but it would appear that abandoned areas reverted to forest or were (later) converted to rubber.

Rubber was the main agricultural crop after the

turn of the century. Unlike tobacco, which was concentrated in the east, rubber was grown primarily in the west-coast region, where transportation facilities were better. Between 1896 and 1900 the Company had constructed a railway between Weston and Beaufort and this was later extended northwards to Jesselton (Kota Kinabalu) and inland to Tenom and Menlap. It was along this rail line that rubber cultivation began to spread in the boom years prior to World War I. Eager to emulate the Peninsula's success with the new crop (actually it had been first introduced in 1882), the Company offered land to estates and smallholders on very generous terms, with the result that the planted area expanded rapidly— from about 40 ha in 1902 to 1305 ha in 1907, increasing to 14 091 ha by 1917 (Tregonning 1965, pp. 88–91).

Following the slumps and restrictions of the post-World War I period, the planted area in 1928 consisted of 27 200 ha of estate rubber and 11 655 ha of smallholdings; by the end of the 1930s, however, the area of the latter exceeded the former. At the close of the colonial period (1963) roughly 70 000 ha of land were under rubber cultivation, some three-quarters of this in the west-coast region. Unlike the situation in Sarawak, where smallholdings made up the great bulk of the planted area, Company policy emphasized the development of estates.

The adoption and spread of rubber cultivation was accompanied by a substantial influx of Chinese and by the emergence of the western coastal zone as the core area of North Borneo. This was already the more settled part of the country, and large areas of forest had been repeatedly cleared by shifting cultivators. It therefore seems reasonable to conclude that most of the area into which rubber advanced was probably clothed in secondary forest or even scrub.

Timber was the third mainstay of the colonial economy. Like tobacco, whose export value it exceeded after 1921, it was concentrated in the sparsely populated and heavily forested coastal and riverine areas of the eastern region. Exploitation of timber for export began in the mid-1880s and continued to grow in importance thereafter. The timber-rich dipterocarp forests were the basis of the industry, and 'during the 1930s they represented about 95% of all timber exported in log or converted form' (John 1974, p. 55). However, it was not until 1959 that timber replaced rubber as North Borneo's most valuable export. Down to 1922, when Tawau began exporting logs to Japan, Sandakan was the only port of export. Although its impact on the settlement pattern was less than that of rubber, the timber industry contributed to population diversity by drawing in substantial numbers of Indonesians and Filipinos.

The early industry was concentrated along the rivers leading into Sandakan Bay, and from there exploitation spread northwards into the Labuk and Sugut districts and southwards into the Kinabatangan River, Darvel Bay, and Cowie Harbour areas. From the outset the rivers were the main lines of communication between the logging camps and the coastal towns and it was along these arteries that virtually everything moved—men, infectious diseases, supplies, and logs. The main market for timber during the early decades of the industry was Hong Kong, which supplied the expanding Chinese railway network with railroad sleepers (ties), but after World War I other markets were sought.

The early industry was dominated by four enterprises, of which the largest was the British Borneo Trading and Planting Company. Beginning in 1920, however, control passed into the hands of the giant British Borneo Timber Company, which obtained a monopoly for twenty-five years

to cut, collect and export timber on all State land. This meant virtual control of all timber cut and exported from North Borneo, and forced the smaller companies to procure licences from, and to be directed by, this new colossus (Tregonning 1965, p. 83).

During the inter-war years Japan replaced Hong Kong as the main market for North Borneo timber, and by that time Sandakan had become one of the world's great timber ports. In addition to timber, the export of cutch—a brown, astrin-gent extract derived from the bark of mangrove trees and used as a tanning agent—remained a profitable activity during the colonial period.

Wet-rice cultivation and shifting agriculture featuring hill or dry padi also made inroads into the forests during the colonial period. The former was concentrated along the west coast and on parts of the interior high plains, the latter in the north-west, on the hill slopes of the Crocker Range, and on the hillier parts of the interior plains. In some of the long-settled areas shifting cultivators had replaced the forests with broad expanses of bare ground or with *lalang* and *belukar* (Y. L. Lee 1961). Only about four per cent of North Borneo was under settled or permanent-field agriculture at the end of the colonial period, whereas shifting cultivation covered 12 per cent. More striking is the fact that 'forests (excluding secondary forests resulting from shifting cultivation) cover[ed] 80% of North Borneo . . .' (Y. L. Lee 1965, p. 68).

3.3.4 Sarawak: economic activities and forest conversion

Sarawak was very thinly populated during the colonial period: in 1905, when it had attained its present limits, the total population of the territory was only about 400 000, and in 1960 it was still only 744 529 (see Jones 1966, p. 18). Down to about 1910, jungle produce—mainly rattan, rhinoceros horn, camphor, birds' nests, damar, and gutta percha—regularly accounted for about one-third of total exports. Prior to the adoption of rubber, which began to be planted in earnest in 1905, almost all attempts to grow plantation crops—these included tobacco, sugar cane, coffee, and tea—were dogged by failure. Western-style private enterprise, whether European or Chinese, was discouraged, and only one European concern, the British-owned Borneo Company, was permitted to operate in Sarawak during the nineteenth century. More Chinese were allowed to enter the country after about 1880, but steps were taken to ensure that they did not dominate the economy.

Clearly, in these various respects Sarawak and North Borneo had much in common. In addition, both were similar in that they failed to keep up with the rapid pace of economic growth in the western Malay states. Both shared more with the rural, poor, largely isolated, and generally underdeveloped eastern part of the Peninsula than they did with the dynamic core area of that region, which remained firmly planted in the lowlands facing the Strait of Melaka.

The bulk of Sarawak's population was (and still is) concentrated in the western part of the country, that is, in the First and Second Divisions and the western section of the Third Division, and it was here, predictably, that the human impact on the forests was most evident. (Sarawak was divided into five divisions during the period under discussion. New divisions have been created in recent years.) In the relatively thickly settled First Division the Bidayuh practised a form of semi-sedentary cultivation of hill rice in the upper reaches of the main rivers and, as we have mentioned, worked antimony ore. Over time their numbers increased substantially, resulting in forest depletion and land degradation in some areas. Various other minerals, including gold, iron, mercury, and coal, were at various times mined in the First Division, consequently there must have been at least some temporary forest loss in the mining areas.

The Iban had occupied most of the river valleys of the future Second Division by the eighteenth century, and from the early 1800s large-scale movements into the Rajang Basin (in the later Third Division) were under way. Subsequent migrations in a mainly northerly direction led to Iban settlement in many other river valleys, including those of the Mukah, Oya, Kemena, Baram, and Limbang. Government-sponsored or spontaneous migrations still occur (see Pringle 1970; Padoch 1982, chapter 2).

The majority of the migrants were shifting cultivators of hill rice, and it has generally been assumed by most scholars that they were (and continue to be) 'addicted' to a prodigal, forest-destroying form of land use that was geared to a continual invasion of primary forests. While it is true that primary forests were cleared in largely uninhabited areas, the assumption of prodigality has been challenged by Padoch (1982), who demonstrated that migration to primary forest has not been the sole Iban response to changes in resource availability (see also below). Furthermore, the fact

that there are numerous river valleys in Sarawak which have been continuously settled by Iban for over three hundred years, immediately suggests that land use among all groups is not predicated on constant migration. The further observation that no extensive area of Sarawak colonized by Iban in the past has been voluntarily completely abandoned by them, and all such areas continue to be exploited by shifting cultivation, points to the fact that the natural resources of these areas have not been 'exhausted' (Padoch 1982, pp. 10–11).

Under Brooke rule the Malays were persuaded to give up their involvement in trade and to become farmers, small planters, or to enter government service. Throughout the colonial period they were concentrated in coastal and riverine areas, where they were engaged in such activities as wet-padi cultivation, fishing, and rubber planting. The Melanaus have already been mentioned. Largely confined to the low-lying coastal plains of the Third and Fourth Divisions, they specialized in the production of sago flour, Sarawak's main export during the nineteenth century. In the deep interior were numerous other cultural groups, including Kenyah, Kayan, Kedayan, and Kelabit, all of whom were farmers, and the nomadic Penan.

Greater encouragement was given to Chinese immigration from the 1880s onwards, partly in order to fill the hiatus caused by the Malay withdrawal from trade, and partly in the hope of stimulating agricultural development. Most new arrivals were concentrated in the First Division (where Chinese were present, as we have seen, before James Brooke arrived on the scene). From the early 1900s Foochow agriculturalists began to settle in the Sibu area and later in other parts of the Rajang Basin, where their activities included pepper and rubber planting.

The cultivation of pepper and gambier by Chinese smallholders increased in importance from the 1870s. Most production was concentrated in the First Division and on the south bank of the lower Rajang. Pepper was the more important of the two crops, and gambier faded into insignificance in the early part of this century. Prior to World War II, pepper was grown on a shifting plantation basis, in part, as Jackson (1968*a*, p. 101) explained, owing 'to the liberal use of burnt earth as a fertilizer, because four acres of jungle land were required to supply the burnt earth necessary to support one acre of pepper for ten to fifteen years'.

Rubber was the first commercial crop to have any substantial impact on Sarawak's forests, which remained virtually omnipresent at the turn of the century. *Hevea brasiliensis* was introduced into Sarawak in 1881, but it was not until 1905 that planting began in earnest. Whereas government policy in both the Peninsula and North Borneo called for the establishment of estates, Brooke policy favoured smallholder production. High prices resulted in a marked increase in the area brought under cultivation during the boom years of the World War I period, and again in the decade after 1923. By the mid-1920s the area under rubber in Sarawak had increased to about 40 500 ha, and 'when the Great Depression ushered in the era of restriction the total area devoted to [the crop] exceeded 220,000 acres' (89 012 ha), according to Jackson (1968*a*, p. 90). In 1965, two years after the end of the colonial period, more than 125 000 ha of land were under rubber cultivation, most of this in the lower and middle Rajang and in parts of the First and Second Divisions. All in all, however, rubber covered only about one per cent of the total land area of the state.

Timber played a minor role in Sarawak's export trade until after World War II (see Mead 1925; Smythies 1963). Logging on a large scale began in the late 1940s and by 1964 timber had supplanted rubber as the main export product. The early timber industry was largely confined to the extensive coastal peat-swamp forests, where there was a rapid increase in the exploitation of ramin (*Gonystylus bancanus*, now largely depleted) from the 1950s. Timber exports quadrupled between 1958 and 1965. Mechanization proceeded apace from the early 1960s and since that time, but especially since *c*.1970, the pace of logging in the dipterocarp forests of the interior has greatly accelerated (see below).

Probably less than five per cent of Sarawak had been brought under more or less permanent cultivation by the end of the colonial period. On the other hand, shifting cultivation is thought to have utilized about 18 per cent of the state's land in the mid-1960s, and this activity had resulted in numerous tracts of secondary forest in such long-settled regions as the First and Second Divisions and parts of the Rajang Basin. Almost all of the remaining 77 per cent or more of the state was clothed in largely undisturbed or primary forests. Clearly, then, the combined total of secondary and primary forests was well in excess of 80 per cent (see Y. L. Lee 1968).

3.3.5 Forest administration and policy

A Forest Department was established in North Borneo in 1914, but its early policy 'was one of exploitation rather than of conservation and sustained yield . . .' (Troup 1940, p. 389). Timber exploitation remained largely uncontrolled until 1930, when a policy was initiated to assure a sustained yield of timber and minor forest products, although the Department's small staff reduced the effectiveness of the policy. A forest reservation programme was launched in 1920, but progress was slow in the inter-war period. Yet by 1941, thanks largely to the efforts of the Conservator of Forests, H. G. Keith, the Forest Department comprised 'a well organized team with clearly defined responsibilities in timber administration, silviculture and research' (Tregonning 1965, p. 85).

A Forest Department headed by J. P. Mead was established in Sarawak in 1919 (see Mead 1925). No comprehensive forest legislation was introduced under the Brooke administration until 1920, when a code of Forest Rules came into effect to regulate 'the taking of timber, firewood, and certain other forest products, as well as the tapping of jelutong trees and the collection of gutta percha' (Troup 1940, p. 392). Forest reservation got under way in 1920 but, as in North Borneo, progress was slow. During the inter-war years Orders were introduced that provided for the constitution of three types of permanent forest: Forest Reserves, Protected Forests, and Communal Forests (see Smythies 1963, pp. 237–8).

As noted below, North Borneo and Sarawak introduced new legislation and policies in 1953–4 (for details, see Kumar 1986, appendix A, pp. 207–13). The important matter of forest reservation is taken up again in Chapter 5 of this volume.

'Envoi'

The two ex-British Borneo territories remained largely undeveloped and heavily forested throughout the colonial period. Shifting cultivation, not commercial agriculture or timber extraction, was the major process of forest change, and by the 1960s both territories exhibited extensive tracts of secondary forest. Forests in some areas were replaced by rubber cultivation and food crops but these and other land uses comprised only a small proportion of the total land area. Great primary forests still clothed more than three-quarters of the entire region when North Borneo and Sarawak joined the Federation of Malaysia.

3.4 FOREST EXPLOITATION IN THE MALAYSIAN PERIOD

We now turn to the period since 1963, when the Federation of Malaysia came into existence. The main watershed event of this period was the 13 May 1969 race riots in Kuala Lumpur. Prior to that event, the interracial *modus vivendi* recognized that political power would rest with the Malays, leaving economic power mostly in non-

Malay, especially Chinese, hands. Although steps were taken to improve the lot of Malays in rural areas through a programme of land development, the government pursued a mainly *laissez-faire* economic policy.

Following the riots a New Economic Policy (NEP) was formulated, and this was incorporated in the four five-year national plans covering the period 1971–1990 (see Lim 1982/83). The NEP, which has now run its course (see Tsuruoka 1990), had two main prongs or objectives: poverty eradication among all Malaysians, irrespective of race; and a restructuring of society to correct economic imbalance, 'so as to reduce and eventually eliminate the identification of race with economic function' (Malaysia 1971, p. 1). The NEP had important implications for forests because, among other things, it sought to increase the access of the poor to land and to modernize rural life. One result was a great increase in the pace of land development in the Peninsula; another was a great proliferation of federal and state agencies and authorities (see Chapter 5 this volume).

Malaysia is still heavily dependent on primary production. This is especially the case in East Malaysia, where manufacturing is much less important than it is in the Peninsula (Lim 1986; Bowie 1988). Although still thinly populated and largely underdeveloped, the eastern periphery is rich in oil, gas, and timber, all of which are being rapidly exploited to fuel Malaysia's export economy. Poverty remains widespread in the region and large numbers of the non-Muslim native peoples are still engaged in swidden cultivation (see King 1988). This section focuses on the three main processes of anthropogenic forest change: land development, timber exploitation, and shifting agriculture.

3.4.1 Land development

Many countries have adopted government-sponsored planned schemes or projects to develop and settle hitherto underutilized areas with agricultural potential. The common aim of such projects is to improve the socio-economic status of the landless poor, although specific aims vary considerably. Most planned schemes have proven inordinately expensive, few appear to have met their objectives, and some have merely resulted in heightened social tensions (Oberai 1988). Not a few such projects have been ecological disasters.

Land development schemes have played a major role in Malaysia's programme of rural development and modernization. Largest and most important of the welter of institutions that have emerged to implement such schemes is the Federal Land Development Authority (FELDA), one of the largest and most profitable plantation companies in the world. More than 300 FELDA-sponsored land development and resettlement schemes have been implemented in the Peninsula. Although FELDA has also been active in Sabah (but not in Sarawak), most land development in East Malaysia has been conducted by state agencies, albeit generally in accordance with the FELDA model.

Land development in thinly settled or frontier areas has been the single most important cause of deforestation in the Peninsula, whereas in East Malaysia it has played a much less significant role. We elaborate on this distinction in the following two subsections.

FELDA in the Peninsula

Land development in the Peninsula has been motivated by a combination of social, political, and economic factors: most of the rural poor are Malays and since the mid-1950s the governing Malay élite has stressed the importance of rural development. FELDA was established in 1956 with the aim of improving the socio-economic status of the landless and the unemployed. Now a huge plantation company with a large bureaucracy and several subsidiary corporations, FELDA is charged with opening up new areas, bringing in settlers, and supervising the transformation of undeveloped land into large-scale schemes devoted to the production of mainly oil palm and rubber for export. The various components of the FELDA model of land development

and resettlement—land alienation, selection of settlers, phased development of land, management and administration, and so on—have been described in considerable detail elsewhere, and we do not propose to go over the ground again (see Bahrin and Perera 1977; Aiken *et al.* 1982; Bahrin 1988).

Originally, most FELDA schemes were relatively small, averaging about 1800 ha and accommodating some 400 settler families. It was hoped that these schemes would eventually result in the growth of small urban centres, thereby helping to retain second- and third-generation settlers in the predominantly rural areas. In part because this did not happen, the trend since the late 1960s has been towards ever-larger integrated regional development projects featuring both land development and urban growth centres (Higgins 1982; Majid and Majid 1983). Jengka Triangle in Pahang was the first of the large-scale projects: work began in 1968, and, by 1977, 24 separate FELDA schemes covering 121 700 ha had been established (see Plate 5). Since the 1970s FELDA has been variously involved in several other, even larger regional development projects. Prominent among these are the following (together with the approximate areas covered by the project master plans): Pahang Tenggara (over 1 million ha), Johor Tenggara (0.30 million ha), Terengganu Tengah (0.44 million ha), and South Kelantan (1.17 million ha).

The pace of FELDA-sponsored land development has been very rapid since *c.*1970, this reflecting the government's emphasis on 'the modernisation of rural life' (Malaysia 1971, p. 1) and the increasing size of development projects. All in all, FELDA has developed some 847 705 ha of land, almost all of which was originally heavily forested (Table 3.1). In addition, a considerable amount of land has been developed by state authorities and the private sector.

Much of the frontier region east and south of the Main Range has been transformed by land development schemes. Here, in little more than two decades, vast areas of forest have been

Table 3.1 Land development by FELDA, 1956–1990

	Period	Area (ha)
First Malaya Plan	1956–60	6619
Second Malaya Plan	1961–5	48 269
First Malaysia Plan	1966–70	72 423
Second Malaysia Plan	1971–5	166 847
Third Malaysia Plan	1976–80	216 447
Fourth Malaysia Plan	1981–5	161 600
Fifth Malaysia Plan	1986–90	175 500†
Total		847 705

† Estimate; includes 58 090 ha in Sabah.
Sources: Bahrin and Perera (1977, p. 51); Malaysia (1971, p. 125; 1976, p. 302; 1981, p. 270; 1986, p. 306).

replaced by uniform, serried rows of plantation crops that now march endlessly across the landscape. This great and sudden transformation owes rather little to the initiative and resourcefulness of the settlers themselves, for contractors do the work of clearing and preparing the land for settlement; rather, it is the product of the managerial and technical skills of a giant authority that is at one and the same time a profitable plantation company and an agent of modernization. But as Mehmet (1986, p. 63) has noted, these two roles are inherently conflicting, and in recent years FELDA's

split-personality complex has become more acute as the Authority has tended to divert increasing surpluses to growth and diversification at the expense of its objectives of poverty amelioration.

Land development in East Malaysia

Most deforestaton in the Peninsula has resulted from land development. This has not been the case in East Malaysia, where logging and shifting cultivation have played the major roles in anthropogenic forest change (see below). Over the past 20 years or so, however, increasing attention has been given to land development in the two Borneo states.

Agriculture in Sarawak has long been dominated by smallholders. The Brookes, as we have seen, did not encourage plantation agriculture (see Reece 1988), and the post-war colonial government (1946–63) continued to emphasize smallholder agriculture, albeit with generally limited success (Morrison 1988). Although the importance of agriculture in the state's economy has declined relative to minerals (oil and natural gas) and timber, it is still the principal way of life of the majority of the population.

Lands and forests are of vital significance to the Iban and other peoples, most of whom are shifting cultivators of hill rice. From the government's perspective, however, swiddening is an inefficient and wasteful form of land use that needs to be modernized—or better still, eliminated. Although the controversial issue of native customary land tenure has not been resolved, this has not prevented land development from proceeding apace, whether or not this is what local people want. The official goal of agricultural modernization is to alleviate rural poverty, but one suspects there is much truth in Colchester's (1989, p. 55) observation that the tacit goal of socio-economic change has more to do with removing

the native peoples from the extensive lands that they are using so 'wastefully' so that the State can alienate them for other purposes, be it for 'forestry' (ie logging), hydro-power or the creation of national parks.

Integrated land development schemes featuring crops like rubber, oil palm, cocoa, and tea, and often including provision of facilities such as housing, piped water, and electricity, have been implemented by three major state agencies since 1972: the Sarawak Land Development Board (1972–), the Sarawak Land Consolidation and Rehabilitation Authority (1976–), and the Land Custody and Development Authority (1981–). In addition, the Department of Agriculture (usually in association with other government departments) has been involved in assisting small farmers to improve or diversify their agricultural activities; several areas have been targeted for coordinated development under the Malaysian government's Integrated Agricultural Development Projects programme (see *Borneo Post*, 20 August 1990); and the *Fifth Malaysia Plan 1986–1990* (Malaysia 1986) called for greater emphasis on both *in situ* development of smallholdings and the resettlement and regrouping of communities so as to foster rural growth centres and small-scale industries (see especially King 1986, 1988; also Hong 1987, chapter 5; Masing 1988). Most of the development schemes have followed the FELDA model, and much encouragement has been given to private-sector investment.

By 1981, state agencies had developed some 19 700 ha of land for mainly rubber, oil palm, and cocoa (Malaysia 1981, p. 35), and the area under those crops in 1985 was 40 000 ha according to Colchester (1989, p. 55). However, most data appear to be unreliable or unavailable, and we are not in a position to assess the impact of land development on the rain forest: some schemes have been implemented in newly cleared areas, some in already settled areas, but in neither case do we have accurate data on their extent, nor do we know the area of the forest, either secondary or primary, that has been cleared. What we can say with certainty is that land development remains a minor cause of forest conversion—at least when compared with logging and shifting cultivation.

Land development schemes in Sarawak have met with very little success: development has been slow and costly, numerous financial, managerial, and other problems have been encountered, some schemes have resulted in mounting social tensions, and the key difficulty of customary land tenure—a difficulty that should have been resolved before the schemes were implemented—is far from being resolved. King (1988, p. 294) has referred to the schemes as a 'rather costly mistake', and it is difficult not to agree.

In 1969, Sabah had 3.5 million ha of land 'suitable for agriculture', of which some 202 000 ha were 'under agricultural uses' (Malaysia 1971, p. 131). These are rather interesting 'facts', considering that some 12 per cent of the region, or about 0.89 million ha, was being utilized for

shifting cultivation in *c*.1965 (see below). Few figures reveal more clearly the government's long-standing bias against shifting cultivation.

Prior to the 1970s, relatively little attention was given to rural development in Sabah (previously North Borneo): it was not a high priority of the Chartered Company or of the post-war British colonial administration (1946–63), and the pace of change was rather slow during the period of the first two national plans (1966–75), although the incidence of rural poverty remained very high. A major change occurred in 1976 when the Berjaya Party assumed office, and thereafter poverty alleviation through rural development was given much greater emphasis (see Voon 1981). In addition to implementing various programmes in existing agricultural areas, government policy also called for a speeding up of land development and settlement schemes in underdeveloped areas, and this was to be achieved in large part through resettlement of isolated communities of mostly shifting cultivators.

The Sabah Land Development Board (formed in 1969) soon emerged as the major agency of land development, and by 1978 it had brought some 32 500 ha of land under mainly oil palm, although attracting settlers to the schemes remained a major problem. FELDA (see above) has also been engaged in land development in Sabah: it was responsible for planting some 37 200 ha of new land with oil palm and cocoa during 1981–5 and is expected to develop another 58 090 ha during the *Fifth Malaysia Plan* period, 1986–90 (Malaysia 1986, pp. 180–1 and 320). As in the case of Sarawak, however, we have no accurate information concerning the amount of rain forest that has been cleared.

'Envoi'

FELDA's land development and resettlement schemes have boosted the production of export crops and contributed to higher living standards in rural areas, although the hoped-for growth of urban centres has lagged well behind the pace of land colonization. The environmental cost of rural modernization has been the elimination of a vast area of species-rich rain forest. Ironically, the amount of idle land in already established agricultural areas probably now exceeds the total amount of land that FELDA has brought into production. Surely it is the case, as Bahrin (1988, p. 116) has suggested, that the transfer of farmers to FELDA and other schemes 'contributes to the abandonment of land elsewhere'.

Land development has done very little to alleviate rural poverty among the shifting cultivators of East Malaysia. Here, especially in Sarawak, social tensions are mounting, in large part because the issue of customary land tenure has not been resolved. One has the impression that most programmes of rural modernization in the region fail to recognize that culture and environment are intrinsically interrelated, each affecting the other in complex ways. Indigenous peoples attach not only economic but also social and spiritual value to their lands and forests: habitat is livelihood but it is also, in a sense, a field of care, a centre of meaning (see Chapter 4 this volume).

3.4.2 Timber and forest policy

Malaysia possesses forests of great commercial value. Of premier importance are the great dipterocarp-dominated, timber-rich forests of the lowlands and the hills. Other forests of commercial significance include the extensive peat-swamp forests of Sarawak and the mangrove forests. Although some of the former have been largely depleted of ramin (*Gonystylus bancanus*), others still contain large stocks of valuable *Shorea albida*. Some of the mangrove forests of the Peninsula have been managed for fuel production since *c*.1918 (Cubitt 1920, p. 14), and extensive tracts of East Malaysia's mangrove forests are being felled for wood-chip production.

Prior to the 1950s, the timber industries of Malaya and Sarawak were mainly geared to meeting local demand. Only in North Borneo, where the industry had long been a mainstay of the economy, was any emphasis given to large-scale exploitation of timber for export. Both supply and demand factors combined to restrict the growth of the industry, and forest products other than timber continued to retain some of their long-standing significance. There have been dramatic changes in all three major regions—the Peninsula, Sabah, and Sarawak—since *c*.1960, and in 1988 Malaysia earned some US$2.2 billion from timber exports, her third-largest foreign-exchange earner after manufacturing and oil (W. C. Lee 1990).

This section does two things: firstly, it describes some of the basic characteristics of the logging industry; and secondly, it picks up again a theme that was introduced in earlier sections, namely forest policy. Tables 3.2 and 3.3 provide some essential background information. Some of the human and environmental consequences of broad-scale logging are discussed in the next chapter.

Table 3.2 Log production (million m³) in Malaysia, 1960–1989

Year	Total	Peninsular Malaysia	Sabah	Sarawak
1960	5.6	2.3	2.1	1.2
1970	17.7	6.5	6.5	4.7
1980	27.9	9.3	9.1	9.6
1987	36.1	10.3	12.2	13.6
1989†	39.7	12.0	9.5	18.2

† Estimate.
Sources: Kumar (1986, p. 91); Chin (1989, p. 3); Anon. (1990, p. 24).

Table 3.3 Distribution and extent of major forest types in Malaysia, 1989 (million ha)

Region	Land area	Dipterocarp	Swamp	Mangrove	Total forested land	Per cent land area forested
Peninsular Malaysia	13.16	4.94	0.46	0.11	5.51	41.9
Sabah	7.37	3.93	0.19	0.32	4.44	60.2
Sarawak	12.33	7.04	1.24	0.17	8.45	68.5
Malaysia	32.86	15.91	1.89	0.60	18.40	56.0

Source: Anon. (1990, table 1, p. 9).

Technological change

During almost all of the colonial period, timber extraction techniques were simple and non-destructive: trees were felled by axe, cut into suitable lengths at the stump by hand-saw, and hauled to the nearest road, river, or tramway (narrow-gauge rail line) by human or animal power (Arnot 1929; Desch 1938; Brown 1955; Martyn 1966; Whitmore and Ho 1968). In Johor, Pahang, and North Borneo logs were placed on wooden sleds and hauled by teams of men along wooden pathways. Most extraction was confined to a distance of a few kilometres from roads and rivers and to areas of gentle terrain, although the construction of tramways allowed deeper penetration of forests in some areas (Martyn 1966). As Wyatt-Smith (1987, p. 12) has noted, 'extraction was on an extensive, selective basis, and damage to the site and the natural vegetation was minimal'.

The post-World War II logging industry was revolutionized by the introduction of mechanized equipment, including ex-military vehicles (such as weapons-carriers), the winch lorry, the tractor, and the one-man chain-saw.[20] Armed with this equipment, logging operators were generally no longer constrained by distance and the slope of the land, while the chain-saw not only speeded up extraction but also cut back on manpower. Winch lorries—they were dubbed *san tai wong* ('king of

the hills')—were soon regarded as essential equipment, and increased use was made of tractors for yarding and road construction, although the introduction of the latter (especially in North Borneo and Sarawak) was much slower owing to the high initial capital outlay and the costs and difficulties of maintenance (F. S. Walker 1950; Webb 1950; Brown 1955; Burgess 1973). The trend has been for equipment to become heavier, more powerful, and a great deal more destructive, an example being the use of the crawler tractor instead of the earlier winch-vehicles with wheels (Wyatt-Smith 1987, pp. 12 and 17).

Silviculture

Three different silvicultural systems have been utilized in the Peninsula since the emergence of organized forestry in the early 1900s: the Regeneration Improvement Felling System (RIFS) that was practised until the Japanese occupation, the Malayan Uniform System (MUS) that was used between 1949 and 1980, and the Selective Management System (SMS) that was introduced in 1977. There is a substantial literature on these systems (e.g. Wyatt-Smith 1961*b*, 1963; Ismail 1966; H. S. Lee 1982; Salleh 1982, 1983, 1988; Whitmore 1984*a*, chapter 8; Tang 1987; Chin 1989; Whitmore 1990, pp. 118–27). There is space for only a brief overview.

The RIFS, a polycyclic logging system introduced into the Peninsula from Burma and India, was well suited to ensuring a steady supply of heavy hardwoods, poles, and firewood (Whitmore 1984a, p. 97). Prior to felling, unwanted species were cut or girdled to promote the regeneration of desired species; subsequently, several 'cuts' of desirable species were made, relatively few trees being felled at any one time. The system proved extremely successful, providing some of the best regenerated forests in the Peninsula (Ismail 1966, p. 229).

But cutting only a few trees at any one time was commercially unattractive to logging operators with substantial capital investments in new technology (see above), and so for economic reasons the RIFS was replaced by a new monocyclic system—the Malayan Uniform System (MUS)—that was geared to harvesting relatively large quantities of timber. The essential elements of the MUS were a pre-felling sampling to establish the presence of an adequate stock of desirable seedlings, a removal of all marketable trees in a single operation, and a post-felling girdling of all unwanted species. Regeneration produced an even-aged forest of desirable species that could be logged after some 70–80 years. The MUS was introduced into North Borneo and Sarawak in the 1950s and is still utilized in a modified form in the former (Nicholson 1965; Chai and Udarbe 1977; H. S. Lee 1982, p. 4; Chin 1989, p. 5).[21] The MUS was developed to manage

the lowland dipterocarp forests, but since the 1960s, in the Peninsula, logging has moved progressively into the hill dipterocarp forests—and it is there that most of the permanent forest estate (see Table 3.4) will be concentrated.

In the late 1970s and the 1980s the MUS was replaced in most parts of the Peninsula by a Selective Management System (SMS) with a felling cycle of 30 years (Chin 1989). This new system was introduced mainly for economic reasons—essentially because the much shorter cutting cycle meant that more timber could be harvested. The official government position, however, is that the SMS was introduced:

(1) for silvicultural reasons, because regeneration from seeds was considered to take too long, because seedling regeneration in the hill forests (see above) was patchy and seedlings on steep slopes were often destroyed by logging (Lall and Wan Hassan 1968; H. S. Lee 1982), and because the cheapest and most effective poison-girdling agent, sodium arsenite, was banned;

(2) for ecological reasons, because 'a mixed forest crop offers the best cover for soil and water conservation' and helps to conserve genetic resources (Anon. 1990, p. 18);

(3) only in part for economic reasons, because many species that formerly were not marketable

Table 3.4 Proposed permanent forest estate in Malaysia (million ha)

Region	Productive forests	Protective forests	Total	Per cent total land
Peninsular Malaysia	2.85	1.90	4.75	36.1
Sarawak	3.24	2.42	5.66	45.9
Sabah	3.00	0.35	3.35	45.3
Total	9.09	4.67	13.76	41.8

Source: Tang (1987, p. 43).

can now be utilized (Salleh 1988, p. 135; Anon. 1990, p. 19).

The SMS involves a pre-felling inventory, followed by one of three (hence 'selective') procedures: if adequate adolescent trees are present, a thirty-year bicyclical felling system is followed; if this is not the case, the MUS is used; and if seedling numbers are inadequate to ensure regeneration, enrichment planting is undertaken or 'compensatory' plantations are established (Whitmore 1984*a*, p. 98). The felling cycle in Sarawak's mixed dipterocarp and peat-swamp forests is 25 and 45 years, respectively (Chin 1989, p. 5; Anon. 1990, p. 19).[22] An eighty-year cutting cycle is still employed in Sabah (see above).

A successful polycyclic felling programme depends on ensuring that no more than 30 per cent of intermediate-sized residual trees are damaged and that seedlings and saplings are not extensively destroyed. In practice, however, most logging operations are highly destructive, hence these conditions are often not met (see Chapter 4). Government officials acknowledge that current management problems include an inadequate number of trained staff to enforce harvesting regulations, insufficient knowledge of how to manage the hill forests for sustainable production, and a paucity of data on the impact of different harvesting systems on residual stands. Yet all this has not deterred some local foresters—foresters who know perfectly well that the SMS only works if damage to seedlings, saplings, and poles is minimized—from voicing ingenious justifications for practices that are unsustainable. It is clear that the main rationale behind the sequence of silvicultural systems—RIFS, MUS, SMS—has been steadily mounting economic pressure on the forests.

Production for export

Exports of tropical hardwoods have increased dramatically since World War II. Between 1946–50 and 1976–80, for example, exports grew by a factor of 24: from an annual average of 2.8 million cubic metres to 66.6 million cubic metres (Laarman 1988, p. 151). Insular South-east Asia has been the major source area of tropical hardwoods and Japan has played the dominant role in the international tropical timber trade (Nectoux and Kuroda 1989).

There are several reasons for the rapid growth of the tropical hardwood timber trade: growing prosperity and demand in Japan, Europe, and North America; mechanization of the logging industry (see above); new pulping techniques that permit the use of a wider range of species; development of grading rules that made tropical timbers more acceptable on foreign markets (Salleh 1988, p. 130); and policies of major development agencies and governments, leading to greater involvement by multinationals (Hurst 1987, p. 173). Owing to depletion of commercial stands, it is predicted that harvests, exports, total revenue, and the number of exporting countries will all decline sharply in the future (Repetto 1988*b*, p. 10).

In recent years Malaysia has been the world's major exporter of tropical-hardwood logs, sawnwood, and veneer (Laarman 1988, table 7.2). The great bulk of Sabah and Sarawak's exports are in the form of logs, whereas virtually all of the Peninsula's exports since *c*.1978 have been in the form of processed sawnwood, plywood, and veneer. Vincent (1988) has noted that there are several reasons why processing industries are less developed in East Malaysia: energy is expensive and wage rates are high; small domestic markets have a limited capacity to absorb the low-grade by-products of export-oriented manufacturing; and relative location and poor communications mean higher transportation costs. In Sabah, however, sawmilling and pulp-and-paper production have been given more emphasis in recent years (Gillis 1988, pp. 135–8). Most of Malaysia's log exports go to Japan, Taiwan, and South Korea, while Japan, Singapore, and Western Europe are among the more important importers of sawnwood and veneer.[23]

The relative importance of log production in the three main regions has changed: the Peninsula was the main timber producer in 1960, a position that it shared with Sabah in 1970, but by 1980 Sarawak had advanced ahead of both (Table 3.2). Sarawak remains the major producer (and it has been here that environmental disputes and social tensions have been most pronounced in recent years). Timber extraction in the Peninsula has often gone hand-in-hand with land development (see above), whereas this has been of minor importance in East Malaysia.

Log production in Malaysia increased from 5.6 million cubic metres in 1960 to an estimated 39.7 million cubic metres in 1989, East Malaysia in the latter year accounting for 70 per cent of the total (Table 3.2). On the basis of the 1987 log production, Chin (1989, p. 3) estimated that 863 000 ha were logged in Malaysia in that year, primary forest accounting for 760 00 ha: 146 000 ha in the Peninsula and 614 000 ha (rounded) in East Malaysia. Only some 5–6 million ha of 'productive' primary forest remain in Malaysia (most of it in Sarawak); if present rates of depletion continue, virtually all of this will be gone by c.2001 (Chin 1988, 1989). The Peninsula, which was already experiencing a shortage of prime logs in the 1970s, is expected to become a net importer of timber by the turn of the century (Thang 1985; *Borneo Post*, 15 June 1990). Sawmills in most parts of the Peninsula are already facing shortages (Rachagan and Bahrin 1983, p. 495; Chin 1988).

Although the timber industry is a major source of foreign exchange and a very important source of revenue for certain state governments (especially those of Sabah and Sarawak), it is a much less significant source of employment: for example, it accounted for 0.1 per cent of the Peninsula's labour force in 1984 and for 7 and 9 per cent of Sabah and Sarawak's labour force in c.1980 and 1984, respectively (Gillis 1988, pp. 117, 124, and 127). While the latter two figures are substantial, it should be kept in mind that East Malaysia's population makes up only some 20 per cent of the national total and that employment in the industry is often of a casual or temporary kind, mainly involving unskilled and low-paid workers (King 1988, p. 273). In Sarawak, where logging is

proceeding at an unprecedented pace, the main beneficiaries of the timber business are a coterie of concessionaires and their favoured contractors; very few benefits accrue to local communities (see Hong 1987; Ngau *et al*. 1987; Jomo 1989; Pearce 1990).

Government policy

Certain provisions of the Malaysian constitution have fundamental significance for forests and forestry. We emphasize in particular that the constituent states of the federation have jurisdiction over land, agriculture, and forestry. Although the federal government may introduce legislation relating to these matters for the purposes of ensuring uniformity of law and policy between the states, such laws cannot be enforced unless separately introduced by the state legislatures (see Mohamed Suffian 1976). The states own and have control over their forests: they gazette (or degazette) reservations, they issue logging concession licences, and they collect royalties from logging operators;[24] on the other hand, federal responsibility in this area is limited to research and to providing services, advice, and training (see Salleh 1983).

No attempt was made to formulate a national forest policy in the immediate post-colonial period, although foresters in particular continued to stress the need for such a strategy (e.g. Setten 1962; Anon. 1967; Anon. 1968). Land development in the Peninsula continued apace during the 1960s (see above) and it was not until the end of the decade that the federal government began to recognize the need for a rational land-use policy and for a better knowledge of the forest resource base. This awareness had two immediate results (although both were confined to the Peninsula): a forest inventory was undertaken during 1970–2 and a land capability classification (LCC) was completed in 1970 (P. C. Lee 1968; Rachagan and Bahrin 1983, p. 499; Salleh 1988, pp. 127–30).

Predictably, the LCC was heavily biased in favour of agricultural development, with the result, in short, that areas of mostly poor soils and steep terrain were assigned to forests, leaving the lowlands for agricultural development. The adoption of the LCC has had two fateful results: firstly, forests on land deemed better suited to agriculture were rapidly degazetted and cleared; and secondly, logging progressively moved into the hills, although no proven or suitable management system has yet been devised for the hill forests (Tang 1987).

The need for a permanent forest estate (PFE) became increasingly apparent during the late colonial period, which witnessed a considerable expansion of timber operations. An Interim Forest Policy that provided for a PFE comprising protective and productive forest reserves was introduced in the Federation of Malaya in 1952 (Kumar 1986, p. 205). In North Borneo and Sarawak, new legislation and policies were introduced in 1953–4, provision being made for the establishment of PFEs—Sarawak's PFE to comprise Forest Reserves, Protected Forests, and Communal Forests, North Borneo's to comprise Protection, Communal, Domestic, Amenity, and Mangrove forests (Anon. 1979a, pp. 311–12; Anon. 1979b, pp. 288–9 and 294; Kumar 1986, pp. 71, 73, and 207–13; Hong 1987, pp. 73–8).

After much negotiation, a National Forest Policy[25] for Peninsular Malaysia was endorsed by the National Land Council in 1978 (Anon. 1979c, p. 329; Muhammed Jabil 1980). This provided for a PFE comprising protective, productive, and amenity forests. The federal government considered it to be desirable to incorporate the new policy in legislation, hence a *National Forestry Act* (Laws of Malaysia, Act 313) was proclaimed in 1984 to provide 'for the administration, management and conservation of forests and forestry development within the States of Malaysia and for connected purposes'; the Act applies throughout Malaysia and provides for the establishment of a PFE divided into many types of forest reserves (Table 3.4).[26] But note, however, that the Act is not binding on the states unless gazetted as a state law, and it appears that several states have not been prepared to do this (*Star* 10 June 1985).

'Envoi'

Logging in Malaysia is proceeding at a pace that is greatly in excess of sustainable yield, and a sharp decline in log and timber exports is inevitable in the near future. The once great timber-rich forests will soon be all but gone. More attention is being given to compensatory plantations, and a programme is in place in the Peninsula that calls for the establishment of some 188 000 ha of fast-growing species by the year 2000; there are plans for similar programmes in Sabah and Sarawak. But plantations are not without their problems, not least of which is lack of knowledge regarding the most suitable species. Meanwhile the amount of investment in management of logged forests is inadequate, with the result that the area of treated forest falls far short of the area that has been logged. Although Malaysia's forests provide a wealth of products, perform many vital ecosystem services, and have many non-economic values, they have been managed almost exclusively for industrial wood production. There is an urgent need to give greater attention to multipurpose management, including much greater attention to the needs of indigenous forest-dwelling peoples. The economic value of so-called minor forest products has been grossly underestimated, whereas calculations of timber's worth are inflated by excluding environmental costs.

3.4.3 Shifting cultivation in East Malaysia

Shifting cultivation is often considered to be one of the major causes of tropical deforestation (Lanly 1982, pp. 80–4; Myers 1984, chapter 8). Nowadays most of those engaged in this activity are recent arrivals in the forest: they are the so-called 'shifted cultivators' (Myers 1989). This part of the chapter attempts to clarify the role that shifting cultivators play in deforestation and land degradation in Malaysia. This is quite a task because the scene is rather murky.

Shifting cultivation: west and east

Shifting (or swidden) cultivation (see Plate 6) is of minor importance in the Peninsula, although it was formerly widely practised (Wyatt-Smith 1958*a*). In 1966, when a land-use survey was conducted, it covered only 0.1 per cent of the region's total land area (Ooi 1976, p. 201). Today, shifting (*ladang*) cultivation in the Peninsula is almost entirely confined to the slopes of the Main Range, where it is still the principal subsistence technique of the Senoi-speaking Orang Asli—notably the Semai and the Temiar (see Cole 1959*a,b*; Dentan 1968, chapter 4; Rambo 1982, pp. 258–9, 266–72). But even in these less accessible areas the practice appears destined to disappear because most of the aborigines are being 'encouraged' to move into new settlements and to adopt permanent-field agriculture (Endicott 1979; see Chapter 4 this volume).

Most of the peoples of interior Borneo have long been shifting cultivators of hill (or dry) padi, and swiddening is still an important way of life among most of East Malaysia's cultural groups— including, for example, the Iban, the Bidayuh, and the Kayan of Sarawak, and the Kadazan (Dusun) and the Murut of Sabah. In addition to padi, many other crops are often grown, and most shifting cultivators also engage in hunting and fishing as well as in collecting forest products for sale (Geddes 1954; Cramb 1985; Hong 1987, chapter 3). Cramb (1988, p. 106) has pointed out that Iban farmers in some of the longer-settled areas 'have been practising an established system of shifting cultivation, supplemented and to some extent replaced by commercial agriculture, for the best part of this century'. Most of Sarawak's swidden farmers have long been permanently settled, hence it is important to note that they do not belong in the category of so-called 'shifted cultivators'.

Shifting agriculture is well suited to the forested but niggardly environment of East Malaysia: here most soils are infertile and many are acid, and most of the region behind the coastal plain is rugged, hilly, and steep. Very little of the interior is suitable for wet-rice cultivation, although some farmers engage in downstream cultivation of *padi paya* in naturally swampy areas (see Padoch 1982, pp. 67–71). In short, over most of East Malaysia, at least as far as farming is concerned, there are few if any sustainable alternatives to shifting cultivation—or to some other, related form of agroforestry.

A truism must be kept in mind, namely that shifting cultivation in the humid tropics is forest-based:

(1) stored nutrients in the vegetation are released by burning the forest; secondary succession replenishes lost fertility;

(2) forests generally assure supplies of good-quality drinking water;

(3) forests (including fallows) provide building materials, fuel, fibres, dietary supplements, medicines, and other items of household consumption;

(4) forests are sources of valuable products that can be collected for sale.

As Chin (1987, p. 8) has noted, the forest is the shifting cultivators' life-support system, hence it 'is absolutely essential that any plans for development, improvement or change must take this into consideration'.

Estimates

In this section we try to throw some light on four basic questions regarding shifting cultivation in East Malaysia. Firstly, how many shifting cultivators are there? Secondly, how much land is currently involved in the shifting cultivation cycle? Thirdly, how much so-called primary, as opposed to secondary, forest is cut each year? And fourthly, who has the greater impact on the forests: shifting cultivators or loggers? Satisfactory answers to these questions are bedevilled by inadequate data. It is possible, however, to provide some rough estimates. Much of the discussion focuses on Sarawak, where a heated and polemical dispute has raged in recent years over the answer to our fourth question.

Numbers involved Estimates of the number of shifting cultivators vary widely. The only figure we have for Sabah comes from a major FAO report on tropical forests (see FAO/UNEP 1981c, p. 86), which claimed that there were some 390 000 swiddeners in the state in 1980 (or about 39 per cent of Sabah's 1980 population of 1 million). There are various estimates for Sarawak:

1. In the mid-1960s, according to Jackson (1968a, p. 81), approximately 40 per cent of Sarawak's total population (then about 840 000) was engaged in shifting cultivation; this would yield a figure of about 336 000.

2. Hatch (1982) estimated (probably on the low side) that about 50 000 households were involved in shifting cultivation in the late 1970s, or about 250 000 to 300 000 persons.

3. FAO (FAO/UNEP 1981c, p. 86) placed the number at 280 000 for 1980.

4. Some official figures go as high as 500 000 (see Colchester 1989, p. 48), and a noted observer of Sarawak appeared willing to accept a figure of about 450 000 (King 1986, p. 96).

Clearly, there is a need for more accurate data! This is by no means a trivial matter because, among other things, a stated goal of government policy is to alleviate poverty among the shifting cultivators. A head-count would be a useful place to begin.

Areas involved It would appear that roughly 12 per cent of Sabah (or about 0.89 million ha) and approximately 18 per cent of Sarawak (or about 2.24 million ha) were being utilized for shifting cultivation in the mid-1960s (see Y. L. Lee 1965, p. 68 and fig. 17; 1968, p. 282 and fig. 1). Estimates for 1980 by the FAO (FAO/UNEP 1981c, p. 86) revealed a considerable expansion: 19 per cent in Sabah (or 1.4 million ha) and 26 per cent in Sarawak (or 3.3 million ha), presumably representing a substantial increase in the total number of shifting cultivators (see above). Hatch (1982) reported a very similar figure for Sarawak (25.7 per cent). We have no other data for Sabah. Estimates for Sarawak of the amount of land 'in use' at any particular time (as opposed to the area under fallow) vary from 73 589 ha to 158 900 ha (according to Colchester 1989, p. 49). It appears safe to conclude that the total area of forest within the shifting cultivators' gamut has increased considerably since the end of the colonial period (1963).

Forest primary or secondary? Lau (1979) calculated that the rate of forest 'destruction' by swiddeners in Sarawak during 1966–76 was 100 000 ha per annum: 60 000 ha in primary forest, the remainder in secondary forest. These (controversial) findings were based on a total of 36 000 households, each clearing or utilizing some 2.8 ha/yr. Lau's calculations have been rejected out of hand by Chin (1985, 1987), who noted that his studies (mostly, but not exclusively, of a Kenyah community) revealed an average rate of clearing and planting of 2 ha/yr, and that only about five per cent (3600 ha) of new swiddens were in primary forest. Lau has also been taken to task by Hong (1987, p. 137), who concluded that probably 'less than 18 000 hectares of new primary forest are cleared by swidden farmers each year'.

Shifting cultivation versus logging Chin (1989) estimated that the total area logged in Malaysia in 1987 was in the order of 863 000 ha, of which 760 000 ha (88 per cent) was in *primary* forest: 146 000 ha in Peninsular Malaysia, 292 000 ha in Sabah, and 325 000 ha in Sarawak; and Hong (1987, pp. 128–9) calculated that some 270 000 ha were logged in Sarawak in 1985, or about 2.8 per cent of the region's forest area. We now have the following:

(1) primary forest logged/yr: 270 000 ha (1985) and 325 000 ha (1987), according to Hong (1987) and Chin (1989), respectively;

(2) primary forest utilized by shifting cultivators/
yr: 3500 ha (Chin 1987), less than 18 000 ha
(Hong 1987), 60 000 ha (Lau 1979).

Clearly, then, logging is in the first place.

Most selective logging in Malaysia leaves
behind extensive areas of damaged residual forest
and often much bare ground (see Chapter 4 this
volume). In other words this is a process of
degradation, leading to both qualitative and
quantitative forest loss. It must be emphasized,
however, that this is not outright elimination,
which, strictly speaking, is deforestation. Similarly, it is often alleged that shifting cultivation
causes much deforestation in East Malaysia (e.g.
FAO/UNEP 1981c; Gillis 1988). While there is
no doubt that swiddening has produced considerable land degradation in parts of the region (see
below), it is equally true that extensive tracts of
forest in many areas have been utilized for
generations. It would appear that three points
should be underscored: firstly, logged forest that
has been left to recover is still forest; secondly,
shifting cultivation initially destroys the forest,
but the process of secondary succession usually
produces forest on 'abandoned' swiddens (see
Chapter 1 this volume) after about 20 years; and
thirdly, continual cultivation on a cycle of less
than about 20 years is unlikely to yield vegetation
that can be regarded as 'forest' (see Lanly 1982,
pp. 74–7; also L. S. Hamilton 1987a).

Images
Malaysia's shifting cultivators are held in very low
regard by many government officials and politicians, a perception that is generally shared by the
local press. According to the official line shifting
cultivation results in severe soil erosion, depletes
valuable timber (although, as we have suggested,
most swiddening occurs in secondary forest), and
causes rural poverty, whereas, so the argument
continues, logging creates jobs, brings other
commercial benefits, and is less wasteful of forest
resources (King 1988, pp. 270–3). Although, as
we noted, logging is generally destructive, Sarawak's Minister of Environment and Tourism is

reported to have claimed that logging operators
attempt to minimize logging damage, whereas the
shifting cultivators permanently destroy the forest
(see *Borneo Post*, 5 September 1987). Lurking
behind these rather formalized and commonplace
official images of shifting cultivation is a good deal
of prejudice and vested interest.

Shifting cultivation is widely misunderstood.
Officials (and some scholars) still tend to view it
as a 'primitive', uniform, static, and prodigal form
of land use, steeped in tradition and resistant to
change, whereas studies by anthropologists and
geographers have shown that it is usually diverse,
flexible, generally resource-conserving, and
responsive to changing circumstances. Thus, for
example, Padoch (1982) demonstrated that
migration to new areas of primary forest is not the
sole response of the Iban to changes in resource
availability; rather, the Iban adopt a range of
'alternative behaviours', including

shifts in agricultural practices, particularly the use of
more conservative cropping–fallow regimes and more
labor intensive farming patterns, as well as increases in
the rates of borrowing of land, of temporary wage labor
migrations, and lower rates of human fertility and
population growth (pp. 11–12).

Land degradation
Some of the long-settled parts of East Malaysia
(and Brunei) have long shown signs of abuse by
shifting cultivators: for example, Y. L. Lee (1961,
pp. 106–7, 1965, pp. 10–12) commented on the
extensive tracts of *belukar* in Brunei and drew
attention to the eroded and *lalang*-infested hillsides of both north-western Sabah and parts of
the First and Second Divisions of Sarawak (also
Spurway 1937; Jackson 1968a, pp. 81–5).
Apparently, the main cause of these ill effects was
population pressure and shorter fallows, although
the introduction of commercial crops and the
practice of clean-weeding (of pepper gardens for
example) must have played a role.

Shifting cultivators continue to cause land
degradation in some areas—in part because their
numbers appear to have increased, in part
because the total amount of land available to them

has diminished (owing, for example, to competition from logging, creation of forest reserves and conservation areas, and implementation of land settlement schemes), and in part because they tend to move into areas already opened up by new logging roads (Colchester 1989, pp. 51–2). It is this last process, namely the invasion, clearing, and planting of previously logged-over areas, that results in some of the most severe forms of land degradation. While not denying that shifting cultivation can (and sometimes does) result in environmental deterioration, evidence suggests that logging is a much more damaging process. As an experienced observer of the Sarawak scene has observed, it is

... obvious to anyone who has travelled through regions subject to large-scale timber exploitation, that it is this activity more than cultivation which has resulted in serious soil damage and erosion, long-term forest destruction, siltation and water pollution, dramatic fluctuations in river levels and increased flooding in down-river settlements (King 1988, p. 272).

A greater emphasis on careful road construction would eliminate many of these problems.

'Envoi'

The future of the shifting cultivators is clouded in doubt. Most land development schemes in the region have met with limited success (see above) and the 'problem' (for development agencies) of customary land tenure has not been resolved. Timber generates considerable revenues for both the federal government and the state governments and it can be expected to receive priority over the needs of the mainly non-Muslim shifting cultivators, who are generally perceived to be 'traditional', resistant to change, and in need of modernization. Like the region in which they live, most shifting cultivators are poor. Most have cash crops, but by and large these are low-yielding, producing relatively little in the way of income. On the other hand, their generally flexible and adaptable ways of life create other opportunities.

What appears to be needed includes giving assistance to the poorest groups, improving or intensifying agricultural production, recognizing the significance to many groups of so-called minor forest products, reforesting the degraded areas, and devising mechanisms that would allow local groups to participate in decisions that affect their lives. The responsiveness of many shifting cultivators to changing circumstances should stand them in good stead, for although their style of agriculture is well suited to much of the region, it is doubtful that it will be tolerated for much longer. There is still timber for the cutting.

3.5 CONCLUSION

The colonial economy of the Peninsula was dominated by tin and rubber. Mining created barren lunar-like landscapes in some areas, but it was the rapid adoption of rubber that provided the momentum for the first major assault on the forests. The generally reckless pace of deforestation highlighted the need for a Peninsula-wide forest policy, but no such strategy was adopted; instead, forest policy and administration remained incohesive, reflecting the fragmented nature of British rule in the region. Economic development was heavily concentrated in the western lowland region, most of which was transformed into an extensive humanized landscape. The pace of change was much slower in the region east and south of the Main Range, and here great forests continued to prevail. At the end of the colonial period, some three-quarters of the Peninsula remained forested.

Much of the Peninsula has been deforested in recent decades, and our estimates suggest that probably little more than 40 per cent of the region remained under forest cover in *c*.1990 (Aiken and Leigh 1988, fig. 2, p. 294; see Table 3.3 this volume). Land development featuring mainly rubber and oil palm has been the major cause of forest loss. But the human impact on the region's forests has been much greater than even their current extent would suggest because:

(1) the genetic resources of the remaining forests have been selectively altered by thousands of years of human agency;

(2) most of the surviving lowland and hill forests have been logged;

(3) most forests now occur in upland areas (because the great species-rich lowland forests are all but gone).

The pace of economic development was much slower in the two Borneo territories, and probably 80 per cent or more of their respective areas remained forested at the end of the colonial period. Rapid selective logging and the (further) spread of shifting cultivation have been the major processes of change in recent decades. For the period 1974–89 there is some evidence that forest cover as a percentage of total land area declined from 76 per cent to 68 per cent in Sarawak and from 86 per cent to 60 per cent in Sabah.[27] But these figures must be accepted with caution because, among other things, it is not always clear what is meant by 'forest' and in one of the sources we have drawn on (Myers 1989) logging is viewed as 'deforestation'. In Sabah and Sarawak, logging is almost entirely in primary forests, virtually all of which will probably be gone by c.2001. As we have already noted, however, this is not deforestation (see Sayer and Whitmore 1991).

4 THE HUMAN IMPACT

Environmental change spans all of human time. Even so-called primitive peoples have left their mark on the earth. Fire, the oldest of anthropogenic landscape modifiers, has long been implicated in the origins and maintenance of tropical savannahs and mid-latitude grasslands, and big-game hunters of the Stone Age may have caused or at least contributed to late Pleistocene mammalian extinctions. Peoples with domesticated plants and animals have shaped natural systems even more profoundly. Among their creations are extensive humanized landscapes, some of them of great productivity and beauty; widespread soil erosion and soil salinization are two other results.

There has been a great increase in both the scale and the intensity of environmental change since about 1700. Processes such as industrialization, urbanization, colonialism, expansion of cultivated land, population growth, new social relations of production—all of these have variously combined to produce profound changes in the global environment. Clearly, the growth and integration of the world-economy and the widening circle of environmental degradation are related phenomena. Recent decades in particular have witnessed a great proliferation of generally more complex and refractory environmental ills, some of which are now truly global in extent: pollution of the oceans, the 'greenhouse effect', and extinction of species are three examples. No longer are humans simply modifying the face of the earth; they are transforming it.

Malaysia is being transformed. Here, as elsewhere in the humid tropics, the clearest expression of ecological change is the demise of large tracts of tropical rain forest. Behind the now familiar images of the throbbing chain-saw, the toppling forest giants, and the relentless march of monocultures are national planning goals and policies, promises of future rewards that fuel rising expectations, more complex networks of communication and the demands of distant markets, the ever-present jockeying for control over natural resources, and a set of Western-style attitudes and values that reveal a desire to shape and control the natural world solely for the benefit of humans.

Now an upper-middle-income country, Malaysia's economic performance has been very impressive and its general standard of living is the envy of most of its neighbours. But there are accumulating costs to pay for this success—costs in the form, for example, of an alarming impact on a growing number of species of both plants and animals. In Malaysia, as in virtually all other countries, the environmental and biological costs of economic policies are invariably absent from the debit side of national account books. A more rational accounting requires an assessment not only of economic growth rates but also of the value of what is being wasted, degraded, or destroyed.

This chapter deals with the human impact on the natural systems of Malaysia. It aims to outline and document certain of the environmental, biological, and human consequences of the processes and policies that were discussed in the previous chapter. Our coverage is necessarily selective, focusing primarily on the contemporary scene, although we do make an effort to range as widely as possible in search of examples of change in past time periods. The subject matter of this chapter warrants book-length discussion, and we are aware that what follows is merely a kind of prologue.

4.1 SOME CONCEPTS

Our discussion falls under the very general rubric of human/environment relations or, more grandly, of culture/nature relations, on which subject there is a voluminous literature, especially in the disciplines of anthropology and geography. Although there is no space here to discuss the theoretical and methodological issues surrounding such enquiry, it is perhaps incumbent upon us to clarify our own particular stance, if only in general terms.

On the environmental side of the equation we note that:

1. Nature is entirely indifferent to our survival, and natural processes do not operate because we will them to do so.

2. While all natural systems possess a certain degree of resilience in the face of human intrusion, there are limits to what they can sustain, and some systems are more fragile than others. It may be true, as Hare (1980) has suggested, that the *physical* environment is more resilient than we normally assume, but many biologists believe that the biosphere is fragile and delicate.

3. *Homo sapiens* is only one of a great many species on earth, and our indefinite future is not assured. Extinction is a natural process (see below).

On the human side of the equation we note that:

1. Since no organism can exist apart from its environment it follows that humans are part of nature. In that they possess culture, however, humans differ from other species. While it can be argued that other life-forms possess 'rights' or 'interests', only humans reflect on their actions and only humans bring about large-scale, purposeful, or intended (as opposed to natural) changes in the biophysical environment. Such actions and abilities generate moral responsibilities.

2. Humans everywhere and at all times have brought about selective changes in the natural world. Such changes—they have been both positive and negative, adaptive and maladaptive—are part of human history. Viewed in the long term, human/environment relations reveal that nature has been increasingly incorporated into culture. Many find this

undesirable or unfortunate, but the facts of the matter cannot be ignored.

3. Although there may be much to learn about sustainable ecological relations from so-called tribal or indigenous peoples, it should be kept in mind that such cultures are not static, that they do not necessarily possess a 'conservation ethic', and that they are

only pauses in the overall historical tendency toward exponential increases in environmental use and impact. This fact must always be remembered lest we commit the 'ecological fallacy' of locating desirable remedies for our environmental ills in the specialized systems of tribal societies (Bennett 1976, p. 12).

4. The prevailing Cartesian dualism between nature and culture has had the most unfortunate consequence of reducing all non-human species to the status of mere objects or things (see Chapter 5 this volume).

A good deal of this chapter is concerned with the more specific matter of the human interaction with other species, including the growing danger of plant and animal extinctions. Some framework or structure is required for the subsequent discussion of this matter, and that is what we attempt to provide in the remainder of this section.

Extinction is a natural process. Frankel and Soulé (1981, pp. 10–30) provided a classification and a discussion of the factors that contribute to extinction. These, they noted, include biotic factors (competition, predation, parasitism, and disease), isolation (insular species are especially prone to extinction), and habitat alteration (resulting from slow geological change, climatic change, catastrophic events, and human activities). One of their main conclusions is stressed here, namely that 'in historical time, not a single species of plant or animal is known to have become extinct except by the direct or indirect hand of man' (p. 29).

The earth's present complement of species—some 5–30 million—is only a small fraction of the estimated 500 million species of organisms that have ever existed. Calculations suggest that the so-called 'background rate' of extinction over the past 200 million years has been in the order of one species every one and one-ninth years. Many (though not all) observers believe that the recent, human-induced rate of extinction is hundreds or even thousands of times higher than the background rate. There is general agreement that most of the loss is occurring in tropical forests (Myers 1979; WCED 1987, chapter 6; Raven 1988; Wilson 1989). Extinction is not the only threat. A related matter of concern is the loss of genetic variability within species, because this greatly reduces a species' ability to adapt to changing circumstances.

As biophysical conditions change—and they are changing rapidly in many parts of the humid tropics—certain kinds of adaptations can make a species vulnerable to extinction. In the animal world, according to Humphrey (1985, pp. 11–12), the adaptations include narrow habitat requirements related to dietary preferences, conservative population performance adjusted to apparently stable environments (i.e. so-called K-selected species), large body size, and high trophic level. Humphrey (pp. 12–18) also noted that three biological circumstances lead to extinction:

1. Initial rarity. In tropical rain forests, as we saw in Chapter 1, few species are common and many are rare; the latter, when compared with initially abundant species, are more likely to become extinct with the passage of time.

2. Smallness of habitat. Fragmentation of tropical forests is proceeding apace, resulting in a variety of edge- and area-related effects. The smaller remnants, in particular, cannot contain the necessary array of habitats that associations of species require, hence they cannot support a region's full complement of species.

3. Inbreeding depression. This, as already intimated above, refers to loss of genetic varia-

bility. Small populations that are genetically isolated are likely to suffer deleterious consequences from inbreeding. It is therefore important to 'have some knowledge of the genetic diversity of a species and the number of individuals necessary for the long-term preservation of a population in order to make species' conservation a realistic business' (p. 18).

Finally, we note that tropical rain-forest species are increasingly threatened or endangered by two broad categories of human activities: firstly, those that have a direct impact, such as depredations resulting from hunting and collecting; and secondly, those whose impact is indirect or unintentional, the most significant of which involve habitat destruction or degradation. Among the many processes that are involved in this second category are the following: urbanization, rural land development, logging, soil erosion and siltation, disposal of industrial wastes, spraying with pesticides and herbicides, and vacationing in sensitive habitats. The two categories of human activities are not mutually exclusive; rather, their spatial distributions may overlap and their effects are often cumulative (see Ehrlich and Ehrlich 1983, pp. 155–212).

4.2 ENVIRONMENTAL IMPACT

As described in Chapter 2, rain-forest vegetation intercepts a considerable amount of rainfall: humus and leaf litter protect slopes from the erosive impact of rain (or throughfall), and the open-textured soil humus layers encourage infiltration rather than surface runoff. Processes that replace or significantly alter the forest cover result in changes to hydrological and denudational regimes. Some changes are observable and dramatic, as when hillsides are left denuded for any length of time; others are more subtle and, in the general absence of long-term measurements, difficult to forecast.

Contrary to popular belief, removal of the rain

forest does not necessarily lead to massive erosion and more frequent flooding. Generally more important is the quality of subsequent land management. This is the case because the potential for significant erosional and hydrological changes is always present owing to the occurrence of high storm totals and high rainfall intensities and because deeply weathered and loosely structured regoliths are widespread. It should also be borne in mind that landslides and floods occur in forested as well as in non-forested catchments and that land use may have little or no impact on flood magnitudes during extreme meteorological events.

4.2.1 Hydrological changes

The limited available experimental evidence indicates that rain-forest vegetation intercepts more rainfall than do stands of rubber trees, although the results are not conclusive (see Aiken *et al.* 1982, table 7.3). Average annual interception rates in lowland evergreen rain forest in Malaysia range from 21.8 per cent to 36.0 per cent of above-canopy rainfall, whereas average rates in tracts of rubber trees range from 13.6 per cent to 29.8 per cent (Teoh 1971, 1973, 1977). The highest rates for rubber trees, however, were recorded from sites where planting densities were from two to four times greater than those recommended by the Incorporated Society of Planters (Edgar 1958). Rates of between 13.6 per cent and 18.4 per cent were recorded from sites with planting densities within or close to the recommended range, and these values are probably more representative. Rates of interception by oil-palm trees are not known, although their form might be expected to yield lower rates than those from rubber trees.

On crop-covered slopes, rates of overland flow are probably generally higher than under forest cover, this reflecting lower interception rates, greater soil compaction owing to rain-splash and human activities, and loss of absorbant soil humus layers. Experiments in Peninsular Malaysia indicated that volumes of overland flow were some two to four times higher on slopes planted with rubber

than they were on forested slopes (Aiken *et al.* 1982, p. 171).

Studies of forested catchments in Peninsular Malaysia indicated that the rainfall/runoff ratio ranged from 30 per cent to 50 per cent (see Aiken *et al.* 1982, table 7.13). As might be expected, ratios for partially forested catchments tended to be higher; for example, a ratio of 63.8 per cent was reported for a catchment that was almost equally covered by rain forest and rubber in Selangor, and another study reported a value of 56.2 per cent for a catchment in the Cameron Highlands that was approximately two-thirds forested, with the remaining area under tea and vegetable cultivation (Shallow 1956; Goh 1971). Similarly, paired catchment experiments at Sungai Tekam in Pahang (as cited by Bruijnzeel 1990, table 4, p. 86) indicated that total stream flow (water yield) increased after conversion of forest to agricultural crops, as follows: during the first four years after conversion to cocoa and oil palm, annual increases ranging from 94 to 158 per cent were recorded for the catchment planted with cocoa and increases ranging from 85 to 470 per cent for the catchment planted with oil palm. But these findings should be viewed with caution, because considerable variation in stream flow may occur depending on annual rainfall totals and on age of replacement crops (see Bruijnzeel 1990, p. 85 and table 4, pp. 86–8).

Evidence suggests that peak discharges may also be higher in catchments in which rain forest has been replaced by crops (Hunting Technical Services 1971, p. 5; Bruijnzeel 1990, table 5, p. 101). It does not necessarily follow, however, that floods will become more frequent in such catchments, although devastating floods in humid tropical areas are popularly attributed to deforestation (L. S. Hamilton, 1983, pp. 1–2). The particularly severe and widespread floods that occurred in the Peninsula in December 1926, January 1967, and January 1971 can be attributed to unique meteorological events; they showed little variation with land use (Toebes and Goh 1975; Leigh and Low 1978).

4.2.2 Erosional changes

The increased rates of soil loss, river sedimentation, and higher turbidity levels that usually follow forest replacement or alteration have been of concern for more than a century. The excessive soil losses that occurred in the tin-mining areas of the Peninsula during the last century and from rubber plantations in the early years of this century were readily apparent to contemporary observers, just as massive soil losses from newly developed urban areas are apparent today. Erosion of agricultural land and of logged-over forests is continuing apace.

Erosion of agricultural land

Some colonial planters recognized the erosive power of tropical rain and the need to maintain a ground cover. Writing in 1836, James Low noted that the hills of Pulau Pinang were well suited to agriculture, but warned of the potential danger of soil erosion 'unless pains be taken by encouraging the growth of binding grasses' (Low 1836/1972, p. 3). Chinese pepper–gambier and tapioca planters, however, generally practised clean weeding, consequently rates of soil loss must have been high (see Chapter 3). When soils became exhausted, the planters simply moved on. Abandoned plantations were soon overrun with *lalang* (*Imperata cylindrica*), but as a contemporary writer noted, the water 'flowed between the lalang tufts sweeping away decaying litter and soil' (Anon. 1895, p. 76).

The implications of observations such as this went unheeded by both Chinese and European coffee and rubber planters,[1] who considered clean-weeding (the work was done by gangs of coolies) to be indispensable to the well-being of trees.[2] The inevitable soil losses were often extreme, especially on steep slopes (Ridley 1910; Hunter 1928; Durant 1936). Trees on many estates and smallholdings died owing to loss of nutrient-rich topsoil (Malaya 1928), and eroded sediment caused siltation and flooding as well as abandonment of padi land downstream (Drainage and Irrigation Department 1937). According

to Fermor (1939), between 1905 and 1939 an average of 33.5 million tonnes of sediment were added annually to the rivers of the Peninsula from lands devoted to rubber, and some 7.6 cm of topsoil were lost.

Among the factors that led to the gradual abandonment of clean-weeding were the following: introduction of legislation (see Chapter 5); extension work by government agencies and research undertaken by the Rubber Research Institute of Malaya (it was founded in 1925); changing attitudes among the planting community; and the slump in prices following World War I, which made labour-intensive practices unattractive (Anon. 1921; Berenger 1922; Anon. 1938). Measures that were introduced to prevent erosion included bunds and silt pits on established estates and cover crops on newly cleared land or land that had been replanted (Dakeyne 1929; Haines 1929; Anon. 1938). Terracing, which is now a common practice, was not widely adopted until after the Japanese occupation, when mechanized earth-moving equipment became available.

On private rubber and oil-palm estates and government land development schemes, cover crops are now universally used, and steep slopes are invariably terraced. Nevertheless, rates of soil loss in these areas are almost certainly higher than from rain-forested slopes. Measurements made on a well-maintained hill planted with rubber on the University of Malaya campus in Kuala Lumpur indicated that rates of erosion were some 16 times higher than those on rain-forested slopes in the Peninsula (Aiken *et al.* 1982, table 7.5).

Rates of soil loss from areas of tea and vegetable cultivation are also much higher than from forested slopes. Shallow (1956), who conducted research in the Cameron Highlands, estimated that the rate of soil loss from forested slopes was in the order of 336 kg ha^{-1} yr^{-1}, compared with 6720 kg ha^{-1} yr^{-1} from slopes planted with tea and 10 080 kg ha^{-1} yr^{-1} from slopes planted with vegetables. We have no information regarding soil loss from areas planted with oil palm. It is likely, however, that the rate is lower than for rubber because oil-palm trees are generally not planted on steep slopes, for two reasons: they are less tolerant of poor soil conditions, and in such areas it would be difficult to harvest the heavy fruit bunches.

Rates of soil loss vary with the quality of land management and with soil type (see Soong *et al.* 1980, pp. 14–19), and R. P. C. Morgan (1974) showed that some parts of the Peninsula are potentially more susceptible to erosion because rainfall is seasonally concentrated, thereby giving rise to high erosivity values, and because the drainage texture is relatively dense. Experiments by the Rubber Research Institute of Malaysia (Malaysia 1974) indicated the importance of maintaining ground cover even under mature rubber trees. Soil losses from a low-angle slope (4°–5°) under 15-year-old rubber trees over a 15-month period were found to be in the order of 103 tonnes/ha from bare ground, 44 tonnes/ha from an area with grass cover, and almost negligible from a site with a ground cover of ferns.

4.2.2.2 Erosion of mining areas

The tin-mining activities of the latter half of the nineteenth century were environmentally disastrous. Large quantities of soil were eroded from hillsides and massive siltation occurred along many rivers. Hillsides were denuded, and the regolith containing the tin ore was hoed into water races.[3] The heavy ore remained in the race and the lighter material was washed into tailing dumps, large quantities eventually entering the drainage network. The process was mechanized in 1892, when high-powered water jets were introduced. A contemporary description gives a clear picture of the denudation caused by the new technology, which washed away

... everything from the level of the valley drainage system to the topmost points of the hills ... The whole mass of the hill, rich and poor, hard and soft, is served alike; all is removed and passed through the sluice-boxes. The surface remaining after the removal of the hill almost resembles a ploughed field (Warnford-Lock 1907, p. 133).

In addition, extensive areas of forest around the mines were cleared to provide wood for charcoal, which in turn was used to smelt the tin ore.

Storms washed huge amounts of sediment into watercourses from exposed hillsides and tailings dumps and miners discharged sediment into rivers and streams. Frequent flooding resulted from massive sedimentation along many rivers, and adjoining land was blanketed with sterile quartz sand and silt. Rich agricultural land along the Sungai Riai in Perak was '... reduced to stretches of dead and dying rubber trees, dismal lagoons of muddy water and patches of *belukar* and swampy grass' (Drainage and Irrigation Department 1936, p. 78). The town of Kuala Kubu on the Sungai Selangor was so frequently flooded that it had to be abandoned in the 1930s, when a new town, Kuala Kubu Bahru, was built on higher ground nearby. A typical abandoned tin mine in the Kinta Valley of Perak is shown in Plate 7.

The introduction of mining regulations at the turn of the century curbed most of the worst environmental excesses. Even today, however, rivers draining mining areas still carry considerable loads of sediment, partly because little attention has been given to rehabilitating worked-out mining lands.[4]

Erosion of logged-over areas

We know of only two quantitative studies of soil losses from areas of selectively logged forest. One was conducted by Liew (1974), who established two plots on steep slopes (32° and 35°) in the Tawau Hills Forest Reserve in Sabah immediately after logging operations had been completed; both slopes were almost completely denuded of vegetation. The reported net soil losses from the two plots over a 20-month period were 175 m³/ha and 266 m³/ha, respectively. Such heavy losses may be relatively widespread, considering the amount of ground disturbance (see below) that occurs during logging operations.

Another study, this one conducted at Bukit Berembun in Negeri Sembilan, investigated suspended sediment yield from three small catchments before and after logging (Baharuddin Kasran 1988). Catchments BC1 and BC3 were logged using standard procedures, but with these important differences: in the latter, logging was more closely supervised, greater attention was given to proper road construction and alignment, and use was made of cross drains on steep logging roads. Catchment BC2 served as a control. It was shown that after logging, sediment yield from BC1 and BC3 increased by 97 and 70 per cent, respectively, and that the highest weighted suspended sediment concentration increased from 386.0 to 844.5 mg/litre from BC1 and from 158.3 to 318.2 mg/litre from BC3. Clearly, carefully supervised logging operations yield lower amounts of sediment.

4.2.2.4 Road construction and soil losses

Several major roads have been constructed through mountainous areas of Peninsular Malaysia in recent years (see Plates 8 and 9). They include the road to the Genting Highlands resort, the Karak Highway linking Kuala Lumpur to Karak on the eastern side of the Main Range, and the East–West Highway linking Grik in Perak to Jeli in Kelantan. Soil losses must have been heavy during the construction of these roads, and landslides on the steep slopes above and below the roads are an ongoing problem. It is anticipated that work will commence during 1991 on a highway linking Simpang Pulai in Perak and Kuala Berang in Terengganu (*New Straits Times*, 16 June 1990). A federal government proposal to construct a 240-km long ridge-top road linking the Genting Highlands with Fraser's Hill and the Cameron Highlands is of concern to conservationists, who believe that its construction would inevitably result in large quantities of sediment entering nearby watercourses (B. H. Kiew 1988; see below).

4.2.2.5 Cumulative effects

Evidence indicates that rates of soil loss from agricultural land and from areas that have been logged are higher than from forested areas, and it is reasonable to assume that river sediment loads

such areas will also be higher. Estimates for the large Sungai Pahang catchment support this premise. Prior to 1900, when most of the catchment was forested, the estimated sediment load was 2.7 tonnes $\times 10^6$/year, whereas in 1975, when land development was at its height, the estimated load was 8.5 tonnes $\times 10^6$/year (AUSTEC 1974, p. 108).

Such high river-sediment loads have potentially significant long-term and short-term implications. Deposition could reduce the ability of watercourses to pass floodwaters, thereby necessitating costly 'improvement' works; and it could shorten the life-span of HEP installations, more of which will be built if current federal government policy is maintained (Malaysia 1986, pp. 463–4). Of immediate concern is the fact that high turbidity levels are polluting drinking-water supplies, depleting fish populations, and causing hardship in many rural communities (Aiken and Moss 1975, pp. 219–20; Colchester 1989, p. 37). In Sarawak in particular, water quality has deteriorated markedly in recent years in a number of rivers draining areas that are being logged.

4.3 BIOLOGICAL IMPACT

The history of anthropogenic impact on Malaysia's biota spans many millennia. Countless generations of hunter–gatherers, shifting cultivators, and specialized collectors selectively altered the genetic resources of the forests in pre-modern times, and contemporary forest-dwellers continue to deplete certain plant and animal species. Forest removal and alteration during the colonial period exerted ever-greater pressures on a growing number of species, some of which were almost certainly eliminated.

The human impact in recent decades has been profound, and Malaysia is now in the throes of a biodiversity crisis: more species of plants and animals are threatened or endangered than ever before; many species have undoubtedly been expunged for ever; and a spasm of future extinctions appears inevitable.

Especially vulnerable to anthropogenic extinction are species whose characteristics include, for example, small populations, thinly scattered or clumped distributions, and narrow habitat requirements. Malaysia has many such species. The vulnerability of certain plant species is further enhanced by the occurrence of dioecy: that is, there are 'male and female plants' and both are required for reproduction.

An awareness of the increasing threat to certain of the faunas (notably the larger mammals) can be traced to the early 1900s. Interest in the plight of threatened plant species is growing. This is timely, because many such species, including numerous endemics, are now vulnerable or endangered.[5]

4.3.1 Anthropogenic impact on the flora

Many species of plants have apparently been eliminated by human activities, and many others are endangered. In addition to forest removal, which is obviously the major threat to most species, forest alteration and commercial collecting are also important endangering processes. Even where forests remain, alterations to the structure and composition of their vegetation may have adverse consequences for genetic resources.

Forest removal

As discussed in some detail in Chapter 3, it has been the dipterocarp-dominated lowland and hill forests that have borne most of the brunt of human activities. Land development and logging have been pursued with such vigour that almost all of the great species-rich lowland forest of the Peninsula has already gone, and the same processes (together with shifting cultivation) are making major inroads into the lowland and hill forests of East Malaysia. Forests that at one time appeared virtually illimitable are in danger of being reduced to a few protected fragments. Many species are destined for extinction, and even certain forest types may disappear.

The wetland forests have also been extensively cleared. For example, large tracts of freshwater swamp forest on the Kedah Plain had been cleared

for rice cultivation by the mid-nineteenth century. Logan (1851, p. 55) compared the area to

... one of the wide plains of Bengal, for there is nothing like it in the rest of the Peninsula. The whole area is an immense paddy field, broken at great intervals by clumps and belts of trees ...

In this century, tracts of peat- and freshwater swamp forest have been cleared for rice cultivation along parts of the west coast of the Peninsula and on the Kedah Plain; and similar kinds of forest have been cleared for pineapple cultivation in the Pontian district of south-western Johor (Wee 1970; see Chapter 3 this volume).

Mangrove forests have come under heavy attack in recent years. In East Malaysia, extensive tracts of this distinctive formation have been licensed for wood-chip production to feed Japan's rayon industry. Mangroves in Peninsular Malaysia have been replaced by several other land uses: agriculture, industry, airport and port development, saltpans, and aquaculture ponds. The development of the Prai industrial estate in Province Wellesley eliminated an entire mangrove community that included the vulnerable mangrove date palm *Phoenix paludosa* (Ong 1982; R. Kiew 1988*a*).

Instead of protecting the remaining mangroves, several states have excised them from their forest reserves; in 1988 alone, some eight per cent of the mangroves within the Peninsula's forest reserves were degazetted. More than 20 per cent of the Peninsula's mangrove forests have been cleared since independence, and if the current rate of reduction continues, very little of the formation will remain by the century's end (Ong 1982; *New Straits Times*, 12 October 1988).

Several tracts of wetland forest have been destroyed to make way for ill-conceived projects. Two examples will have to suffice. Firstly, 1600 ha of mangrove forest were cleared for salt production in the South Banjar Forest Reserve in Selangor, only to find that local evaporation rates were too low to permit profitable production. Secondly, wetland forest was cleared for rice production along the Sungai Merbok in Kedah, but the project had to be abandoned when the acid sulphate soils proved completely unsuitable for such an enterprise. Surely both failures could have been predicted (Bennett and Caldecott 1986).

Most of the upper hill and montane forests are still intact. The steeply sloping terrain has prohibited most kinds of development, although there are a few extensive clearings where upland resorts have been established. There are three hill stations of colonial origin along the mountainous spinal column of the Peninsula (and another on Pulau Pinang): Bukit Larut (formerly Maxwell Hill), Cameron Highlands, and Fraser's Hill. There is also a post-colonial hill station at Genting Highlands (Reed 1979; Aiken 1987). At Cameron Highlands, the largest of the hill stations, extensive cleared areas are given over to tea plantations and vegetable cultivation (*New Straits Times*, 9 November 1988).

Road construction (see above) in the Genting Highlands (see Plate 9) has eliminated several extremely rare and endangered species, and conservationists are justifiably alarmed by a federal government proposal to construct a 240-km long ridge-top road linking Genting Highlands to Fraser's Hill and Cameron Highlands (B. H. Kiew 1988). The federal Works Minister is reported to have said that the road will enable the three resorts to become 'a mini Switzerland', that there will be 'a transmigration of farmers from the lowlands' to develop some 4500 ha for vegetables, fruits, and flowers, that an International Islamic University with 9000 students will be built, and that residential development will include the provision of summer houses for Japanese visitors (*Star*, 6 July 1988; *New Straits Times*, 1 and 3 November 1988). The proposal was shelved when the estimated cost escalated from M $600 million to more than M $2 billion, but in mid-1990 the government announced that survey work would recommence (*New Straits Times*, 16 June 1990). Some of the likely consequences of building such a road include widespread forest loss, massive soil erosion, and a new round of extinctions.

Forest alteration

Humans have been exerting selective pressures on the forests since the dawn of prehistory. Age-old practices that have altered the forests include hunting and gathering, collecting-for-trade, and swidden agriculture; of much more recent provenance is the now widespread and generally very destructive practice of selective logging. There has been a dramatic increase in the pace and scale of forest alteration in recent decades, with the result that very little so-called primary forest has survived.

Even peoples who are generally regarded as having always lived in near-perfect harmony with nature have left their mark on the forest biota. Take, for example, the Orang Asli, who have been living in or on the margins of the forests for thousands of years (see Chapter 3 this volume). Although endemic disease, low population numbers, and weak technology combined to restrict the ability of the Orang Asli to bring about large-scale changes in the forest, it would be a mistake to assume that their activities have had no appreciable impact on the forest biota. Quite the contrary is the case according to the anthropologist Terry Rambo (1979a; see also 1979b, 1982), who showed that countless generations of selective hunting and plant gathering, habitat modification, dispersal of seed, and domestication have profoundly altered the genetic resources of the forests.

Contemporary Orang Asli groups continue to exploit a wide range of plant (and animal) species, and the same is true, for example, of Sarawak's forest-dwelling Penan. Species are often ruthlessly exploited: for instance, the nomadic Semang foragers are not averse to 'strip mining the resources of the forest' (Rambo 1982, p. 277) and, if need be, they are prepared to chop down a whole tree to get at its fruit.

The seeds of petai, rambutan, and other fruits are deliberately dispersed by some groups, while others reset the tops of wild yam tubers to ensure regeneration. Certain species have been severely depleted by the ancient practice of collecting-for-trade (see below). On the other hand, traditional swiddening, which creates favourable conditions for species that could not otherwise survive in undisturbed forest, has contributed to increasing the total diversity of forest ecosystems (see Rambo 1982, pp. 279–81). There can be little doubt that the forest-dwelling aborigines have had a profound impact on the evolution of numerous species.

Since the 1950s, many Orang Asli have been brought together in government-built settlements (see below). One result of concentrating formerly dispersed bands has been to stimulate overexploitation of rattans and other local resources (Rambo 1982, p. 279; also Wyatt-Smith 1958a; Endicott 1979). Governments are unwilling to recognize customary land tenure rights, and pressure on the remaining forests is depriving indigenous peoples of habitat. Given these circumstances, it can be expected that overexploitation of certain plant and animal resources will continue.

Vast areas of the timber-rich dipterocarp forests have been degraded by selective logging. As we have noted elsewhere (see Chapter 3), silvicultural practices were designed to promote the regeneration of desired (commercial) species and to suppress others. An inevitable consequence of selective logging operations is that the proportions of species present in regenerated forests will often differ from those in virgin stands (Whitmore 1984a, p. 272). Because loggers generally remove only the best trees, it has been suggested that the quality of the residual stands will decline over time, leaving behind the smaller, often malformed and genetically inferior trees to provide the seed for the next crop (Kartawinata 1981, p. 195).[6] Although evidence is rather limited, it would appear that overexploitation has resulted in many species becoming endangered; for example, two formerly prized heavy hardwoods—cengal (*Neobalanocarpus heimii*), which is restricted to Peninsular Malaysia, and belian or Borneo ironwood (*Eusideroxylon zwageri*)—are now very scarce (B. H. Kiew *et al.* 1985, p. 4).

Many species of trees and other plants are also threatened by destructive logging practices.

Damage is often extensive. Burgess (1973, p. 134) reported that in the hill forests of the Peninsula some 10.4 kilometres of roads were constructed for every square kilometre of logged forest; this represented 6.25 per cent of the total area, and up to an additional six per cent of the area was occupied by log landings and dumps (see Plate 10). Wyatt-Smith and Foenander (1962, p. 41) observed that nine per cent of a logged area in Pahang was occupied by roads. Fox (1969, p. 45) noted that 41.7 per cent of a logged area in the Segaliud–Lokan Forest Reserve in Sabah had been variously disturbed by tractor operations.

In addition to soil erosion and compaction and other damage to the ground surface, felled trees cause severe damage to seedlings, saplings, neighbouring trees, and other plants. Here are two cases in point: firstly, only 10 per cent of the trees in a lower hill forest coupe in Terengganu were felled, but 55 per cent of the residual stand was seriously damaged; secondly, logging operations in the Lungmanis and Segaliud–Lokan forest reserves in Sabah damaged 45 and 64.4 per cent of the residual stands, respectively (Nicholson 1958; Fox 1968; Burgess 1971).[7]

Logging and other activities can render the forests more combustible, thereby indirectly contributing to changes in forest structure and composition and to loss or depletion of species. The great Borneo fires of 1982–3 clearly demonstrated that the probability of fires, whether caused by lightning or by human carelessness, is greater in logged forest than it is in primary forest. Thus, of the approximately 1 million ha of forest that burnt in Sabah, 85 per cent had been logged, and in Kalimantan, where about 3.5 million ha of forest were burnt, about one-third had been logged (Beaman *et al.* 1985, p. 23; Whitmore 1990, p. 117). There is evidence that only 'about a third (1.35×10^6 ha) of the destroyed Kalimantan forest was primary and part of that (1×10^5 ha) was killed by drought not by fire' (Whitmore 1990, p. 117). Apparently, the recently logged forests of both Sabah and Kalimantan contained large quantities of readily combustible dry dead

wood. Fires (and droughts) are natural phenomena, but human activities exacerbate their consequences.

Woods (1989) investigated three forest sites in
Sabah that had been burnt in 1983: one in primary hill dipterocarp forest, a second in hill
dipterocarp forest that had been logged 14
months before the fire, and a third in lowland
dipterocarp forest that had been logged six years
prior to the fire. It was concluded that the
prospects for recovery of pre-fire forest structure
were good in burnt primary forest (although
species composition would probably continue to
be different from that of the original forest),
whereas the opposite was true in the case of the
two logged forest sites, both of which suffered
severe canopy loss and were invaded by vigorous
grasses and creepers. Woods (p. 297) noted that
the recovery of the latter forests 'will depend in
the first instance on the ability of the secondary
forest trees to cast sufficient shade to suppress the
grasses and woody creepers which blanket the
soil surface'. Repeated burning impedes the
process of secondary succession, with the result
that the end-product is likely to be unproductive
grassland. Clearly, observations such as these
have important implications for forest management and conservation.

Finally, it should be kept in mind that shifting
cultivation in East Malaysia has also resulted in
widespread forest alteration (see Chapter 3).

Commercial collecting

Collecting forest products for trade is an ancient
practice. Prior to the nineteenth century, the
collectors and primary traders were exclusively
or primarily forest peoples, namely the Orang
Asli in the Peninsula and various indigenous
peoples in Borneo (see Dunn 1975; Hoffman
1986). Malays were long important as middlemen
and resident exporters, although the latter function passed progressively into local Chinese
hands (Dunn 1975, pp. 117–19). During the
nineteenth century in the Peninsula, some Malays
and Chinese became involved in what appears to
have been a rather ruthless phase of collecting in

conjunction with timber cutting and other activities (Dunn 1975, pp. 108–9).

The export of forest (and sea) products to
China extends back to at least the early centuries
AD (Wang 1958; Wheatley 1959), the prized
items including, for example, beeswax, gaharu
wood, camphor, *damar* (resin), lac, ivory, and
rhinoceros horn (Ranee of Sarawak 1913, p. 255;
Hose 1929, pp. 171–9; Wheatley 1959, 1961,
p. 74; Dunn 1975, pp. 111–12). Camphor, which
was used for medicines, incense, and embalming,
was a particularly important export from Sarawak
(Han 1985).

Europe cornered an increasing share of the
trade in forest products during the nineteenth
century. Behind this trend was industrial development, which stimulated demand and created new
uses for a variety of tropical rain-forest resources.
These included jelutong (wild rubber), gutta
percha (a rubber-like forest latex used for, among
other things, cable insulation), illipe nuts (a source
of vegetable oil), cutch (a tanning agent extracted
from the bark of mangrove trees), rattan (mainly
for furniture), and various gums (Rutter 1922,
chapter 8; Hose 1929, chapter 7; Andaya and
Andaya 1982, pp. 133–5).

Although there is no evidence that the collection of these and other products resulted in plant
extinctions, the numbers of individuals of certain
species were severely depleted at times in some
localities. Here are two examples. Firstly, the
collection of the much prized Borneo camphor
crystals was achieved by felling many individuals
of the tree *Dryobalanops aromatica*. Secondly,
the rapid rise in demand for gutta percha resulted
in the overexploitation, death, or felling of many
trees in Selangor, Johor, Singapore, and elsewhere (Gullick 1975, p. 41; Andaya and Andaya
1982, p. 135).

Commercial collecting is still practised. Rattans,
for example, command increasingly high prices on
the world market, either in unprocessed form or as
furniture (R. Kiew 1989*a*; Pearce 1989*a*).[8] Production in Sarawak jumped from 533.7 tonnes in
1979 to 2574 tonnes in 1987. Some 20 species are
commercially important in the Peninsula, three

being especially prized: *Calamus manan*, *C. tumidus*, and *C. caesius*. In addition to the rattans, many other species of palm are widely utilized for such things as food wrappers, rice baskets, fish traps, and mats.[9] In Sarawak, locally woven mats and baskets are becoming popular with tourists, and this could lead to increased production— possibly to overexploitation of certain species.

Many of the rattans are already scarce. For example, the Peninsula's premier species, *C. manan*, has been much depleted in logged-over areas, and virgin stands elsewhere are becoming harder to find, as too are other species (R. Kiew 1989*b*, pp. 4–5). Pressure on the rattans could be relieved by establishing commercial plantations; this is an urgent matter, because an indispensable fund of genetic material is being rapidly depleted.

Plant hunters in search of specimens for their own collections or for sale are active in Malaysia. Both local and foreign residents are involved. Orchids and pitcher plants are especially prized. The white slipper orchid (*Paphiopedilum niveum*), which is confined to Pulau Langkawi, has been plundered by collectors (B. H. Kiew *et al*. 1985, p. 4). Pitcher plants—particularly *Nepenthes rajah*, which can fetch up to US $1000 per plant on overseas markets—have been ruthlessly stripped from the mountains of Sabah and Sarawak, some collectors using helicopters to reach remote peaks (Briggs 1985).[10] Malaysia's palms are of growing interest to foreign plant hunters, some of whom are also collecting seeds from threatened or endangered plants—including one of the country's most endangered species *Johannesteijsmannia magnifica* (R. Kiew 1989*b*, p. 5).

Cumulative impact

What, then, has been the cumulative effect of forest removal, forest alteration, and commercial collecting? The short answer is that many species have almost certainly been lost and a large and ever-increasing number of others has been rendered vulnerable or endangered.

Numerous recorded species of trees from the tin-rich alluvial soils of the Kinta and Larut districts of Perak appear to have vanished. Ng and Low (1982, p. 1) reported that the majority of 66 such species (probably local endemics) have not been sighted since 1940. Two other plant species that have probably disappeared for good are *Didymocarpus perdita* and *Chionanthus spiciferus*: the former was last reported in 1923, when two individuals were found at Selatar in Singapore; the latter is known from only one site, at Rawang in Selangor, but little forest remains there now (R. Kiew 1983*b*, p. 2).

A few palms may have become extinct. Ten species in the Peninsula and 13 in Sarawak have not been sighted for more than 25 years (R. Kiew and Dransfield 1987, table 2; R. Kiew 1989*b*, table 1; Pearce 1989*b*, p. 33). Two species of begonia are also thought to be extinct. *Begonia rajah* collected from a single site in Terengganu won a Royal Horticultural Society medal in 1894; it has not been seen since and it is no longer in cultivation in either the Singapore or the Kew botanic gardens. Likewise, *Begonia eiromischa*, a Pulau Pinang endemic that was illustrated in Ridley's *Flora of the Malay Peninsula*, is no longer in cultivation and has not been seen in the wild for many years (R. Kiew 1988*b*, pp. 74–5).

Many species are vulnerable or endangered. Of the 654 endemic species of trees that they identified in the Peninsula, Ng and Low (1982, pp. 5–6) considered 343 to be endangered, and the plight of the endangered species is further underscored by the fact that 216 of them were found in only one state. Many palms are considered to be threatened. Only nine of the 194 species recorded in the Peninsula are believed to be in no danger. In Sarawak, where 213 species have been recorded (of which at least 56 are endemic), only 16 are considered to be safe (R. Kiew 1989*b*, appendix 1; Pearce 1989*b*, table 1). A similar situation probably prevails in Sabah, although the conservation status of 113 of the 131 species recorded there has not been established (Dransfield and Johnson 1989).

Table 4.1 presents a list of the plant species that are considered 'most endangered'. Several species on the list are known from only a few plants, and one is represented by a sole known survivor.[11] It would appear, however, that several other species share a similarly precarious status.[12] For example, *Maclurodendron magnificum* is known from only eight plants growing by the side of the Genting Highlands road (at the same site as the sole survivor mentioned above), and *Ilex praetermissa* is known from only two sites along the narrow Klang Gates ridge in Selangor (R. Kiew 1983*b*, p. 7; D. T. Jones 1984). Both are dioecious, thereby all the more vulnerable.

The increasing pace and scale of the human impact on natural systems and the fact that plants are generally given little consideration or adequate protection means that most of the endangered species will probably not survive. Scientific surveys similar to the one conducted by the

Table 4.1 Malaysia's most endangered plant species

Plant	Comments
Acrymia ajugifolia	Monotypic species. Known from one site a few hundred metres square in the Kanching Forest Reserve, Selangor.
Melicope suberosa	Known from only one plant alongside the Genting Highlands Road.
Nepenthes rajah	Restricted to Mt. Kinabalu. Under pressure from collectors. Only Malaysian plant in Appendix I of CITES.
Botrychium daucifolium	Celery-leaved palm; known from a few plants found in a small area in the Cameron Highlands.
Johannesteijsmannia magnifica and *J. lanceolata*	Umbrella palms; known from only a few sites in Peninsular Malaysia.
Peperomia maxwelliana	Known from only a few plants on a single limestone hill in Peninsular Malaysia.
Paphiopedilum niveum	White slipper orchid; endemic to Langkawi. Threatened by collectors.
Rafflesia hasseltii	Known from 14 sites in Peninsular Malaysia. Parasitic on vines. Threatened by collecting (for medicine) and by vine cutting prior to logging.
Ficus albipilus	Rediscovered after 40 years in 1980 at Janda Baik, Selangor.
Neobalanocarpus heimii	Cengal; rare by the 1950s owing to overexploitation. Threatened by continuing removal of lowland forest.

Source: B. H. Kiew *et al.* (1985, pp. 3–4).

Malayan Nature Society in the Endau–Rompin wilderness (B. H. Kiew *et al.* 1987; see Chapter 5 this volume) may locate some hitherto unknown representatives of endangered or vulnerable species or even rediscover species that were thought to be extinct. More such surveys are sorely needed because time is running out for many species. The rediscovery after 40 years of *Ficus albipilus* along the Bukit Tinggi road at Janda Baik in Pahang was a piece of welcome news in an otherwise generally bleak scene (R. Kiew 1983*b*, p. 2).

4.3.2 Anthropogenic impact on the fauna

Pre-colonial times

Many cultural groups hunted forest animals for food and because parts of certain creatures could be readily sold or bartered. Birds and small mammals were trapped or killed with poisoned darts from blowpipes; larger animals were caught in nooses and camouflaged pits or dispatched with spears and iron-tipped arrows (e.g. Maxwell 1907/1960; Rambo 1978). Forest-dwelling groups hunted primarily for food; both archaeological evidence and historical accounts indicate that a wide range of species was caught, including deer, pig, serow, monkeys, squirrels, porcupines, and seladang (Rambo 1979*a*, pp. 60–1). Birds and other small, generally plentiful, animals were regularly taken by lone hunters. Occasionally, hunting groups were formed to capture more elusive and dangerous species: for example, Malay villagers drove deer towards rattan lines hung with nooses (*sidins*). Some of these lines could be more than two hundred metres long, with hundreds of nooses (Maxwell 1907/1960, pp. 37–60). Deer were hunted for food, for sport, and because they damaged padi crops. Hunting of the larger mammals was fraught with danger but the rewards for elephant tusks, rhinoceros horn, and the flesh and bones of the tiger were sufficiently high to attract hunters both from within and from without the forest.[13]

The impact of traditional hunting practices on animal populations was probably not great,

although locally it may have depressed the numbers of certain species (Rambo 1988*a*, p. 282). Human populations were small and many of the more frequently caught species bred relatively quickly. It has been claimed that the tapir, the tiger, and the giant pangolin became extinct in Sarawak as a result of human activities (see Rambo 1979*a*, p. 60), but Medway (1972, p. 68), referring to the period prior to the introduction of the gun into the Malesian region, contended that 'there is no evidence that man critically reduced the population of any animal by direct slaughter'.[14]

Traditional attitudes towards conservation appear to have been somewhat ambivalent. Rambo (1988*a*, p. 282) found no evidence of a strong conservation ethic among the Orang Asli of Peninsular Malaysia; on the contrary, he observed that certain groups 'will kill any creature that they consider to be edible with no regard for preservation of future breeding stocks'. The Batek Semang, for example, will readily cut down a tree to reach a nesting hornbill with young or eggs. On the other hand, there is evidence of practices that promoted sustainable exploitation of forest resources. Orang Asli and Dayak groups had clearly defined 'territories', which ensured that several groups did not exploit the same tract of forest, and the Tambunan Kadazans of Sabah established riparian reserves (*tegah*) where fishing was prohibited for certain periods to allow fish stocks to recover (Cole 1959*a*, p. 193; Davies and Payne n.d., p. 13; Lian 1988, pp. 119–21).

The colonial era

The human impact on certain species of animals increased dramatically during the colonial period. This was mainly the result of habitat loss and the introduction and widespread use of firearms. During the nineteenth century most Malay villages in the Peninsula probably acquired at least one firearm, and from the late 1800s it would appear that Orang Asli groups were also able to obtain guns (Rambo 1978). The impact of firearms in the hands of skilled hunters must have been considerable. Both the Orang Asli and the

Malays were now able to hunt the larger animals more successfully. Considering the naturally low densities of these species, and in some cases their slow breeding rates, population structures and dynamics must have been altered. Rambo (1978, p. 214) has suggested that current densities of certain species may be considerably lower than optimum owing to hunting by aborigines.

During the nineteenth century and early decades of the present century, resident and visiting Europeans hunted the larger mammals for sport. L. Oliphant (1860, I, pp. 24–5), who visited Singapore, Johor, and Melaka in the 1850s, commented that

[t]he sportsman who has exhausted every variety of game to be found in the jungles of India, will derive a fresh excitement here in hunting the rhinoceros, or watching for the wary tapir; while on the muddy banks of sluggish rivers he may surprise the seladang or wild ox, a species peculiar to the forests of the Malay peninsula, and which has not yet been described by naturalists.

Big game was also hunted in North Borneo, where, according to Rutter (1922, p. 372), the banteng, rhinoceros (see Plate 11), and elephant provided sport as good as that to be found anywhere in the world.[15]

Forest animals damaged plantation crops and in some cases endangered life. Tigers were particularly feared—wounded and old 'rogue' tigers being particularly dangerous. In Johor in the 1850s, for example, tigers killed several Chinese on newly opened pepper–gambier plantations, some of which had to be temporarily closed (L. Oliphant 1860, I, pp. 28–9). Most of the damage to estates was caused by elephants and sambar—a type of deer. The former were a recurring problem where herds had been virtually isolated by agricultural development and where estates had been opened up in the vicinity of or cut across pathways leading to salt licks. The damage caused by elephants and deer was considerable, and the Wild Life Commission of Malaya described it in some detail.[16]

The magnitude of the problem can be gauged from two examples: at the Palawan Estate in Perak between November 1927 and March 1929, elephants destroyed some 2937 rubber trees, representing 30 per cent of a newly planted area; and on the Jenderak Planting Syndicate's estate in Pahang some 12 000 rubber trees were destroyed by elephants during 1927, representing approximately a quarter of the trees planted (Wild Life Commission of Malaya 1932, I, pp. 338–9; III, pp. 330–1).

Planters waged war on the animals that they considered to be a nuisance. Managers employed armed guards to protect the plantations and some joined together to form patrols to hunt the larger animals in the adjoining forest. One such group in Perak, the Plus Valley Elephant Damage Protection Society, killed 36 elephants in the period between 1911 and 1930 (Wild Life Commission of Malaya 1932, I, p. 17). The planters were encouraged and directly helped by the colonial authorities. In the 1850s, the Chinese pepper–gambier planters in Singapore were rewarded with 100 Spanish dollars for every tiger's head they delivered to the government (Thomson 1864/1984, pp. 229–30), and in later years game wardens throughout British Malaya destroyed large numbers of nuisance animals. It would appear that many game wardens spent as much time destroying wildlife as they did protecting it! Between March 1925 and October 1929, for example, two successive game wardens in Johor shot at least 48 elephants (Wild Life Commission of Malaya 1932, I, pp. 78–9).

Many of the larger mammals that were shot but not killed often became rogues. One of the Johor game wardens told the Wild Life Commission that most of the 30 or more elephants he had killed carried gunshot wounds. Rogue elephants caused extensive damage and were dangerous if confronted, and wounded tigers preyed on village livestock and occasionally killed villagers and plantation workers (see Locke 1954, pp. 113–14; Foenander 1952, p. 109).

Hunting on the part of Orang Asli groups, Malay villagers, Chinese and European planters, and colonial officials combined to have a dramatic cumulative impact on the larger mammals. A progression of witnesses informed the Commission that species that had been plentiful were now scarce or were never seen. The evidence indicated that two of the larger mammals, the banteng (*Bos javanicus*) and the one-horned Javan rhinoceros (*Rhinoceros sondaicus*), were either greatly depleted or on the verge of extinction in the Peninsula, and it is likely that both vanished from the region before the end of the colonial era (Wild Life Commission of Malaya 1932, II, pp. 24–5; Hubback 1926, p. 40; Metcalfe 1961, p. 185). The situation deteriorated during the Japanese occupation and the subsequent Emergency, when many of the larger mammals, particularly seladang, were slaughtered for their meat (Kitchener 1961, pp. 198–9). By the late 1950s, the Sumatran rhinoceros was close to extinction, with an estimated population of only 50 (Metcalfe 1961, p. 185). It is almost certain that most of the other large mammals were either endangered or highly vulnerable.

In North Borneo and Sarawak, both of which remained virtually undeveloped throughout the colonial period, the main pressure on the fauna came from native hunters. With the notable exception of the Sumatran rhinoceros, it would appear that the human impact on most species was not very great until firearms became widely available in the 1950s and 1960s. The extent to which the rhinoceros was hunted in Sarawak can be judged from the following facts: between 1929 and 1931, 79 rhinoceroses were known to have been killed in the Baram area and between 1932 and 1940 some 1820 kg of rhinoceros parts were exported from Sarawak (Caldecott 1987, table 6.2). Firearms became available after the Japanese occupation and imports of these increased from the late 1940s. The scene was now set for an unprecedented onslaught on the forest fauna.

The post-colonial period

The threats to the fauna that were in place during the colonial period have persisted to the present-day, and new ones have emerged. In the Peninsula,[17] loss of habitat has undoubtedly been the main threat to the survival of many species since independence, while in East Malaysia the greatest threat has probably come from hunting with firearms. New threats include the ever-increasing compartmentalization of the remaining forests and widespread selective logging, although the impact of the latter may not be as deleterious as is often believed.

The extensive land development programme in the Peninsula and the further fragmentation of the region's forests by roads such as the East–West Highway and large dams such as Temenggor and Kenyir have put certain mammals under considerable pressure. It was estimated that the absolute numbers of six primate species fell dramatically between 1958 and 1975, the relative losses ranging from 23.4 per cent in the case of the long-tailed macaque (*Macaca fascicularis*) to 56.8 per cent in the case of the siamang (*Hylobates syndactylus*) (Mohd. Khan *et al*. 1983, table 2).

The impact of land development schemes on certain animal herds is illustrated by the fate of seladang herds in Ulu Lepar, Pahang: prior to land development there were 90 animals in seven herds, whereas three years after development only 61 individuals were counted and these were spread between nine herds (Zulkifli Zainal 1983). The size of many elephant herds has also been reduced. Elephants appear to have a particular liking for oil palm and, as in earlier years, many 'nuisance' animals have been shot; for example, between 1966 and 1976 the state game departments shot some 120 elephants (Mohd. Khan 1977, p. 28).[18] A change in policy has resulted in a number of elephants being trapped and relocated in recent years (*New Straits Times*, 13 June 1988).

In East Malaysia, hunting with guns is a major threat to the survival of many species and one that is becoming more serious as remote areas are opened up by timber companies (Davies and Payne n.d., p. 192; Caldecott 1985, 1987, p. viii). There are 61 500 registered guns in Sarawak and two million rounds of ammunition are sold there annually. Virtually every other family in the interior has a gun, the single-barrelled 12-bore shot-gun being the most popular. Some 60 per cent of kills are bearded pig and deer—rusa, kijang, pelandok—but a wide range of species is shot. Caldecott (1987, p. v) noted that just about 'any animal larger than mouthful-sized is liable to be captured and eaten in Sarawak'. Most kills are for food but some species are hunted for trophies: clouded leopards for their skins, hornbills and argus pheasants for their feathers, and rhinos for their horns.[19]

Selective logging has taken a heavy toll of Malaysia's dipterocarp forests (see Chapter 3). The impact of this activity on forest fauna has been studied at Sungai Tekam (Pahang) and elsewhere by A. D. Johns (1981, 1983a,b, 1985, 1986, 1987, 1988a,b; also Wilson and Johns 1982), who showed that some species disappear (at least initially), that most survive, and that certain others flourish and multiply. Surprisingly few animals appear to be killed during logging operations. It may be that the pangolin and the orang-utan are exceptions. A. D. Johns (1983a, p. 207) observed that pangolins, which roll into a ball when frightened, were crushed by machinery; and orang-utans, which are slow-moving and tend to 'freeze' when alarmed, are sometimes killed when trees are felled (C. C. Wilson and W. L. Wilson 1975, p. 272; Davies and Payne n.d., p. 177).

Species with specialized diets appear to be the most adversely affected when the forest is logged. In the case of birds, insectivores such as trogons suffer because the numbers of leaf-eating insects are greatly reduced. It has been suggested that such displaced specialists are unable to settle in adjoining undisturbed forest because the required niches are occupied (McClure and Hussein 1965). Species that are sensitive to changes in microclimate, such as frogs and toads and some understorey birds, are also adversely affected by logging, as are birds that nest in tree hollows, such as hornbills (Plate 12), parrots, and owls.

Elephant, tapir, deer, and seladang are some of the mammals that prefer logged-over forest to primary forest because grazing opportunities are generally much better in the former. Other spe-cies prefer undisturbed forest but can adapt to logged areas. The wide-ranging hornbills return although numbers may be reduced (A. D. Johns 1987) and most of the primates can adapt. A. D. Johns (1986) reported that *Hylobates lar* and *Presbytis melalophos* adapted by changing their foraging and dietery habits; in the general absence of fruit, these two species ate more leaves and, probably because of the lower nutritional intake, spent more time resting and less time travelling. The density of many species is lower in logged forests (Caldecott 1987), and this is of concern to rural communities that rely on the forest for a substantial part of their food requirements.

The important investigations conducted by A. D. Johns suggest that virtually all species eventually return to logged forests, albeit in some cases at lower densities. There is evidence, then, at least in so far as a *single* logging operation is concerned, that many animal species can adapt to altered conditions. As Whitmore (1990, p. 178) has noted, however, ongoing short-cycle logging is very likely to eliminate the pockets of primary forest which provide the refuges at the first logging and to be too frequent to allow complete recovery of all the animal species. So, although Johns's research gives grounds for some optimism, in fact the long-term prognosis for species survival in repeatedly logged production forest is not good.

The fact that many species of animals can survive in logged forest has implications for conservation. It has been suggested that the long-term survival of many wide-ranging species is more likely in extensive tracts of logged forest than it is in undisturbed but small and isolated forest parks (A. D. Johns 1983a, 1985, 1987). However, the value of logged forests for con-servation of fauna will be much diminished if logging cycles are reduced, if old trees with nesting holes are not retained, if hunting pres-sure increases, and, of course, if shifting cultiva-tors are not kept out. Animal life in logged-over forests would benefit from retaining adjacent undisturbed forest as a source of seeds and animal colonizers, maintaining unlogged corrid-ors of mature forest along watercourses, and leaving uncut patches of forest within forest reserves (see Shelton 1985).

A number of mammals and birds can survive in rubber, oil-palm, and cocoa plantations, provided that patches of forest or scrub remain (Duckett 1976; Davies and Payne n.d., p. 190). More species tend to be found in rubber estates than in areas of oil palm because patches of forest are often left in swampy areas in the former and because several species that could survive in the latter are not tolerated because they eat the fruit. Rubber trees also provide better nesting sites for birds and squirrels. In Sabah, 55 species of birds were found to be breeding in a mixed-tree plantation and 33 species in a mature cocoa plantation, the latter also supporting the slow loris, the lesser tree shrew, the common palm civit, and the leopard cat (Davies and Payne n.d., p. 190).

Cumulative impact

In 1982 the Malayan Nature Society reported on the conservation status of Malaysian mammals and birds (B. H. Kiew 1982*a*; B. H. Kiew and Davison 1982). Species with populations of less than 1000 and less than 3000 were considered endangered and vulnerable, respectively. Sixty-one species of mammals and 16 species of birds were listed as endangered and 130 species of mammals and 148 species of birds as vulnerable. Two species of birds, the white-winged wood duck (*Cairina scutulata*) and the green peafowl (*Pavo muticus muticus*), were thought to be extinct (B. H. Kiew and Davidson 1982, p. 2). In 1985 the Malayan Nature Society listed what it considered to be Malaysia's ten most endangered species (Table 4.2). The precarious existence of seven of the 10 species was attributed to loss of habitat and to poaching. There is no reason to believe that the status of these species has improved since 1985.

Both in the Peninsula and in East Malaysia, the large mammals have generally steadily declined in numbers over the past two decades, the result of loss of habitat and hunting. The tiger population of the Peninsula dropped from an estimated 3000 in the early 1950s to some 500 in the mid-1970s, and then to about 250 in the 1980s (Locke 1954, p. 7; Malayan Nature Society 1975, p. 6; Mohd. Khan *et al.* 1983, p. 3). Estimates for the early 1980s indicated that there were some 480 seladang and 671 elephants in the Peninsula, elephant numbers having remained steady since the mid-1970s. A further 1000 elephants were thought to exist in Sabah in the early 1980s, but the banteng population was thought to have dropped to between 300 and 550 animals (Davies and Payne n.d., p. 87). Of all the large mammals, the Sumatran rhinoceros is undoubtedly the most endangered: few if any survive in Sarawak, only some 30 remain in Sabah, and only some 50–80 are left in the Peninsula (Davies and Payne n.d., p. 83; Mohd. Khan *et al.* 1983, p. 3; *New Straits Times*, 24 December 1986).

Of course, numbers are not the only measure of a species' chances of survival. A significant factor is the ability of an animal to adapt to altered forest environments, because animals that cannot readily do so will be put under increasing pressure in coming years. Another factor is the size of the area required by individual animals or groups of animals: some creatures are far more wide-ranging than others and, generally speaking, the larger the animal the bigger the area required. For example, in Sabah the density of the orang-utan ranges from 0.2 to 2.0 individuals per km^2, whereas there are from 0.7 to 4.25 groups of the grey leaf monkey (*Presbytis hosei*) per km^2, with each group comprising between 3 and 13 individuals (Davies and Payne n.d. pp. 100 and 110–13).

Yet another factor that can be significant is reproductive capacity. The tiger, for example, has a gestation period of between 105 and 113 days and a female may give birth to 30 young in her lifetime, whereas for the elephant the period is between 18 to 22 months and a female only gives birth every four to five years (Mohd. Khan 1977, p. 28). All of these factors need to be taken into account when conservation plans are being formulated.

Table 4.2 Malaysia's ten most endangered animals

Animal	Comments
Sumatran rhinoceros (*Dicerorhinus sumatrensis*)	Habitat threatened by agricultural development. Lowered reproductive rate. Poached for their horn.
Tiger (*Panthera tigris*)	Threatened by loss of habitat.
Dugong (*Dugong dugon*)	Rarely seen in Malaysian waters. Breeding disrupted by human activities.
Storm stork (*Ciconia stormi*)	Estimated population of less than 200. Threatened by forest clearance along lowland rivers.
Ghavial (*Tomistoma schlegelii*)	Endangered by hunting with firearms. Last stronghold is Tasek Bera, Pahang.
Clouded leopard (*Neofelis nebulosa*)	Threatened by loss of forest habitat and hunting.
Orang-utan (*Pongo pygmaeus*)	Threatened by habitat destruction and poaching.
Hawksbill turtle (*Eretmochelys imbricata*)	Threatened by drag- and trawl-net fishing.
Kelesa (*Scleropages formosus*)	The golden dragonfish; threatened by over-collection for aquaria.
Peripatus (*Peripatus* spp.)	The evolutionary link between arthropods and annelid worms. Known from only a few sites in Peninsular Malaysia.

Sources: B. H. Kiew *et al.* (1985, pp. 2–3); Anon. (1987, p. 43).

4.4 THE IMPACT ON HUMANS

4.4.1 Introduction

The indigenous[20] peoples who live in and around the tropical rain forest include small, scattered groups of hunter–gatherers and at least 140 million shifting cultivators. Increasingly, the lives of such peoples are being affected by deforestation and forest degradation and by the swelling numbers of new colonists, loggers, miners, road builders, government officials, and others with whom they come in contact.

Most of the growing pressure on forest resources arises from external forces—from governments, from corporations, from landed interests, from logging concessionaires—and there is much truth in the comment that 'the tropical forest and its human populations have rarely been treated with consideration by those desiring resources' (Unesco 1978, p. 447).

The great majority of outsiders (including government officials and decision-makers) know very little about the forest peoples, whose ways of life they view as primitive, whose lands they think of as largely unoccupied or 'empty', and whose

presence they view as an obstacle to modernization and economic development. Most government policies call for assimilation. Generally
marginalized and powerless, the indigenous
peoples increasingly regard themselves as colonized victims of 'progress', and around the world
they are becoming more active in defence of their
rights and interests (ICIHI 1987, pp. 31–40).

Among the many changes that usually accompany deforestation and forest degradation in the
homelands of the forest-dwellers are new roads
and enhanced accessibility; replacement of col-
lective rights to and custodial use of lands and
forests with state or private ownership of property; substitution of well-tried, sustainable
resource-use strategies with policies and practices that maximize short-term gain at the expense
of natural systems; labour migrations and new
planned or spontaneous settlements; higher population densities and greater human interaction;
new or exacerbated health hazards; diminishing
or degraded home ranges, depletion of game and
plant foods, economic hardship, and mounting
social tensions; and resettlement, or movement to
towns and camps or deeper into the forest. Above
all, loss of habitat is a major threat to the forest
peoples.

This section tackles two themes: firstly, it
explores the impact of forest conversion and
government policies on forest-dwelling peoples,
focusing mainly on the Orang Asli of the Peninsula and the Penan and other native groups of
Sarawak; and secondly, it describes some of the
relations between environmental change and
disease. Let it be clear, however, that:

(1) we are not suggesting that the benefits of
economic development should somehow be
denied to indigenous peoples, thus keeping
them, so to speak, in a state of 'arrested
primitivism';

(2) neither are we claiming that such peoples are
incapable of adapting to change; but rather

(3) we are declaring that there is a great and
pressing need to

acknowledge the right of the indigenous to be themselves, to have a voice and to pursue their aspirations,
whether these be the preservation of their culture and
traditions, the management of their lands or, indeed,
education and development as they perceive it (ICIHI
1987, p. xii).

4.4.2 Government policy, land, and the Orang Asli

The official term for the aborigines of Peninsular
Malaysia is 'Orang Asli'—the Malay words for
'original people'. Most scholars divide the Orang
Asli into three main groups: the Aslian-speaking
Semang (or Negritos) and Senoi, who are primarily opportunistic foragers and swidden farmers respectively, and the Austronesian-speaking
Proto-Malays, who are mainly horticulturalists
(see Rambo 1982; Benjamin 1985). Almost
entirely forest- or forest-fringe dwellers, there
were some 53 000 Orang Asli in the mid-1970s
(Carey 1976, p. 11; Malaysia 1976, p. 164).[21] The
Proto-Malays live in the upper reaches of river
valleys in the southern half of the Peninsula,
whereas the Semang and the Senoi are mainly
concentrated in the foothills and uplands of the
generally more rugged central-to-northern
region: the Semang below and the Senoi chiefly
above an elevation of about 500 metres (Rambo
1988b, p. 24 and map 2.1).

Although the ways of life of the Orang Asli vary
greatly, Rambo (1979b, pp. 55–8) has suggested
that the ecological relations of most groups reveal
certain common features, including the following:
generalized rather than specialized economic
adaptations; life-styles that are basically forest-
oriented; possession of remarkably detailed and
accurate environmental knowledge; generally
rather minor environmental disruption; absence
of a strong 'conservation ethic'; and, for the most
part, successful adaptation to rain-forest ecosystems (see also Dunn 1975; Endicott 1979;
Rambo 1979a). Compared with the forest-based
Orang Asli, the Malays and other more recent
immigrants know rather little about the rain
forest; lacking the detailed, comprehensive, and

systematized knowledge of the aborigines, they have tended to view the forest as an alien world, a world they are able to penetrate only by converting or destroying it (Rambo 1980*b*, p. 86).

During almost all of the colonial period, British officials regarded the Orang Asli with indifference, intervening in their lives only when they contravened some game or forest ordinance. The official policy—if that is the correct term—was one of benign neglect. It was only in the late 1940s, when the Orang Asli were suddenly thrust into strategic importance in the anti-Communist guerrilla war (the so-called Emergency that was officially ended in 1960), that the colonial regime began to show any real interest in the forest peoples, and it was only in the early 1950s, almost at the very end of the colonial period, that any firm policy was adopted.

A new Department for the Welfare of the Aborigines was created in 1950 and, initially with military strategy in mind, programmes were designed to solicit Orang Asli support for the government's struggle against the forest-based insurgents and to prevent the latter from gaining control and influence over the aborigines (Means 1985/86, p. 644). Some of the early activities of the department—its name was later changed to Department of Aborigines (Jabatan Orang Asli, or Jabatan Hal Ehwal Orang Asli)—included constructing a hospital at Gombak in Selangor, establishing health clinics in the rain forest, removing Orang Asli from Communist-controlled areas and resettling them elsewhere (generally with disastrous results, including many deaths from disease), and building a string of garrisoned jungle forts (A. Jones 1968, pp. 295–301).

By the early 1950s, the department had been reorganized and expanded and made responsible, as it still is, for the provision of medical care, education, and welfare facilities in Orang Asli areas. Following the end of the Emergency, the department was made a permanent governmental body in 1961, and since the late 1960s more attention has been given to rural land development and resettlement schemes (see Malaysia 1966, p. 186, 1971, p. 261, 1976, pp. 176, 424–

5, 1981, pp. 42 and 390, 1986, pp. 92, 534, and 538). The long-range goal of the government is to integrate the Orang Asli into Malay society, and, at least in a technical and statistical sense, this has already been achieved because the aborigines of the Peninsula (and indigenous peoples elsewhere in the country) were lumped together with Malays as *Bumiputra* (literally 'son(s) of the earth') in the 1980 census.

We now turn to the matter of land, an issue that is of vital importance to the Orang Asli. Endicott (1979, pp. 172–5) has noted that the Semang and the Temiar Senoi do not consider themselves to own the land of their home areas and that they view rights of ownership or possession of natural resources to extend only to certain fruit trees, to ipoh trees (from which dart poison is extracted), and to the collected or harvested products of the land. Such perceived rights—namely, in essence, the right to live in and use the resources of certain areas—have no force in Malaysian law. This requires some brief explanation.

The central goal of the colonial government was to promote rapid economic development. This, as we saw in Chapter 3, was achieved in part by adopting the Torrens system of land tenure, which provided for undisputed titles to land. Under this system 'all land is deemed to be vested in the crown . . . in the case of the Malay States, the Malay ruler was deemed to be the crown' (Means 1985/86, p. 639). Recognizing that this system threatened the customary land rights of the Malays, the colonial government passed the *Malay Reservation Enactment* of 1913, thereby providing for the creation of so-called 'Malay reservations' in which only Malays could lease or own land (see Emerson 1937, pp. 478–80; T. G. Lim 1977, pp. 108–16, 164–70; Winzeler 1988, map 5.2, p. 97). But no such protection was extended to the aborigines. Instead,

aboriginal lands were deemed to be the crown lands of the Malay rulers, and were treated as if they were unoccupied. . . . the aborigines, who were deemed to be without land title . . . were permitted to live on 'unoccupied lands' by sufferance, as dependents of the Malay rulers (Means 1985/86, p. 640).

Orang Asli rights to the lands they occupy are 'protected' only in the legally designated areas that were established under the *Aboriginal Peoples Ordinance* of 1954 (revised in 1974). But even in these places—they are called 'aboriginal areas' or 'aboriginal reserves'—the aborigines are essentially tenants at will because the enabling legislation made no provision to give them legal title to the land (Means 1985/86, p. 645). Most of the designated aboriginal areas are in the southern half of the Peninsula, so by no means all of the Orang Asli live in such areas; indeed, some groups—the Temiar for example—are mainly concentrated on state land.

Land development schemes, logging, road building, dam construction, government policies that have been implemented within aboriginal areas—all these processes have placed ever-greater pressure on the land and life of the Orang Asli. Here are a few specific examples of their impact (see Endicott 1979):

1. The amount of land available to local aboriginal groups is being reduced by land development schemes, and in some areas shrinking home ranges are in danger of being over-exploited. This is probably the case, for example, in lowland areas occupied by Senoi swidden farmers, because a combination of rapid population growth and a reduction in the extent of traditional territories is resulting in a cycle of shorter fallow periods. Indeed, such trends threaten the very survival of swiddening as a way of life (Rambo 1982, p. 281).

2. Several main roads have been constructed in the Peninsula in recent years, and others are projected (see above). Two that traverse areas where Negrito peoples live are the East–West Highway between Jeli in Kelantan and Gerik in Perak and the Kuala Krai (or Kerai)—Gua Musang Highway in Kelantan, both completed during the fourth national plan period, 1981–5 (Malaysia 1986, p. 431). Such roads mean loss of land, interference from outsiders,

greater exposure to the market economy, and easier access for loggers.

3. Logging (and the new roads, tracks, skid-paths, lorries, heavy machinery, and workers that inevitably accompany it) is increasingly encroaching on traditional Orang Asli territories in both lowland and upland areas. The results include loss of land, depletion of game and of forest products that enter into trade, soil erosion and nutrient loss, and, at least for some communities, probably a lowering of living standards.

4. Dam construction is another threat to some groups. An example is the building of the Temenggor Dam in Perak, which led to many Temiar (and a few Negritos) being moved to three new resettlement centres outside the reservoir area. This is only one of several dams that have been built in Temiar country (*New Straits Times*, 3 July 1984; Malaysia 1986, map 17-1, p. 458).

5. In some cases the implementation of government policies within Orang Asli areas has resulted in environmental disruption. For example, Negrito groups have been encouraged to move into new 'pattern settlements' (see Plate 13) and to adopt an agricultural way of life, but as Rambo (1982, p. 279) noted,

[t]he much higher concentration of population in the settlement areas has placed tremendously increased pressure on local resources, and consequently their depletion has been rapid. Wild game, rattan and fish have been virtually wiped out within a five-kilometer radius of these new settlement schemes.

The potential for further change in the land and life of the Orang Asli is great (but remains largely unexamined) because during the *Fourth Malaysian Plan* period, 1981–5

[a] comprehensive programme involving five regrouping projects was implemented for about 23,000 Orang Asli residents along the Central Range of Peninsular

Malaysia. Under this programme, 13,600 hectares of land were distributed, while another 2,000 hectares were planted with rubber and 1,300 hectares with fruits. In addition, a total of 270 units of houses was constructed (Malaysia 1986, p. 92).

Clearly, much of the emphasis here is on agriculture and resettlement.

We make four comments by way of conclusion. Firstly, the Orang Asli are becoming increasingly concerned about their land rights and this, among other things, is contributing to greater ethnic consciousness, a trend that is precisely the opposite of what the government claims to be promoting (see Gomes 1988, pp. 106–11). Secondly, among the Negritos and the Senoi-speaking Semai, for example, there appears to be a marked inclination to enter more fully into the market economy, and this can be expected to yield much greater interest in the issue of land rights. Thirdly, this emerging interest is occurring at a time of mounting competition for natural resources, and the possibility that the Orang Asli (especially those living outside the designated reserves or areas) will be largely dispossessed cannot be ignored. And fourthly, there is a growing desire on the part of the aborigines to define their own goals and aspirations. This is revealed in a remarkable document that was drafted in 1982, following the first conference of Orang Asli headmen (the text is in Means 1985/86, pp. 651–2). Addressed to the government and people of Malaysia, the document or memorandum declared, in part, that

We appeal that all Orang Asli land will be gazetted and a copy of the land title will be given to each village headmen [sic]. This will ensure that our land will not be taken away from us by any outsiders.

and

We appeal that the Government enact new laws to safeguard [against] the indiscriminate exploitation of our natural habitat by mining and timber companies. And that all present activities of such companies be ceased to safeguard the livelihood of our people in the villages.

4.4.3 Government policy, land, and the native peoples of Sarawak

Like their counterparts in the Peninsula, the native peoples of Sarawak have demonstrated increasing concern over the erosion of their customary land rights. But whereas the Orang Asli's concerns have gone largely unnoticed and unpublicized, the situation in Sarawak has made headline news in Malaysia and attracted widespread publicity overseas, in large part because of the plight of the Penan. In this section we discuss the impact of government land-use policies and logging operations on the native peoples. (The nature of the protest movement and reactions to it are described in Chapter 5.)

Land and forest laws

Much more than an economic resource, land to the native peoples of Sarawak also has great social significance and deep spiritual value (Hong 1987, pp. 37–8). Land generally belongs to the community: it is not a commodity that can be bought or sold, although individuals or families may have certain ownership rights with respect to some types of land and exclusive rights to fruit trees. The right to use the land and to take the products of the land are defined by traditional laws known as *adat*. In complete contrast are the concepts of individual landownership and the paramount rights of the state that are enshrined in current land and forest legislation. Native customary land rights have been steadily eroded and it is this issue that is central to the recent conflicts between native groups on the one hand and loggers and the state on the other.

The Brookes' position on customary rights was ambivalent: their legislation protected existing rights, but after the introduction of the 1863 Land Order, natives required permission to open up new land.[22] Subsequently, legislation was introduced to control shifting cultivation and to restrict the movement of native groups, and a clause in the 1933 Land Settlement Order gave the state the right to revoke native customary rights by notification in the government gazette.

During the period of British colonial administration (1946–63), land in Sarawak was divided
into five categories: mixed zone land, native area
land, native customary land, reserved land, and
interior land. This classification is still used. New
customary land can be created on interior land if a
permit is granted, or by gazette notification of
land as a native communal reserve. Communal
reserves can only be established with the approval
of the minister, who also has the power to revoke
them by an order in the government gazette.

The land laws are complemented by the provisions of the Forest Ordinance. The state's
forests are divided into two categories, Permanent Forests and Stateland Forests, the latter
being available if required for agricultural and
other types of development. In 1984, Permanent
Forests comprised 34 per cent of the total forest
area. There are three types of Permanent Forests:
Forest Reserves, Protected Forests, and Communal Forests. Natives are not allowed to enter
Forest Reserves, but if they have permission they
may hunt and gather forest products in Protected
Forests. Communal Forests are for the use of
native groups. In 1984, however, there were only
56 km^2 of such forests, representing a mere 0.17
per cent of the Permanent Forest area (Hong
1987, p. 75). Logging is permitted in Forest
Reserves and Protected Forests (where it opera-
tes, theoretically, on a sustained-yield basis) but
not in Communal Forests. The Minister for
Forestry can revoke Communal Forests by notification in the government gazette. Although interested parties have three months in which to object
to such a decision, the arrival of timber lorries is
often the first indication to native groups in the
deep interior that they have been deprived of their
customary rights.

Natives are also generally prohibited from
pursuing customary practices in national parks
and wildlife sanctuaries. There are currently eight
national parks and three wildlife sanctuaries in
Sarawak, and several additional reserves have
been proposed (see Chapter 5). The Penan of the
Gunung Mulu area were opposed to the creation of
the Gunung Mulu National Park[23] (it was estab-

lished in 1974), and the Berawan people of Long
Teru on the Tinjar River are concerned over the
proposal to create a national park at Loagan
Bunut (Hong 1987, p. 68; Colchester 1989,
p. 64).

Although the erosion of native customary
rights has a long history, the process probably did
not greatly affect most native peoples until quite
recently. During the 1980s, however, numerous
logging operations advanced deeper into the
interior, with results that included further losses
of customary rights, environmental degradation,
depletion of food resources, social disruption,
and alienation.

The impact of logging

We have already noted that most animal species
eventually return to logged forests, albeit in some
cases at lower population densities. What, then, is
the impact of logging on the food resources of
native groups? To answer this question we draw on
an important study by Caldecott (1987), who
quantitatively demonstrated the importance of
wild meat and fish to the peoples of the interior.

From a survey of 63 boarding schools in that
region, Caldecott found that some 203 tonnes of
fish and meat were consumed during 1984–5.
Bearded pig comprised 32 per cent of the total,
other wild meat 7 per cent, fish 18 per cent, and
domestic pork, beef, and chicken made up the
remainder. For a number of schools in the upper
Baram area, bearded pig accounted for 61 per cent
of the total. In the study area as a whole, wild meat
and fish accounted for approximately 60 per cent
of daily protein requirements. Where supplies
were not plentiful, they generally were not supplemented by more expensive domestic meat.

Wild meat and fish also provide a source of
income for the longhouses. During 1984, some
10 200 dressed bearded-pig carcases and 1400
deer with a total value of M $4 million were sent
downstream along the Rajang from Kapit to Sibu,
while 6400 pig and 2300 deer carcases were
purchased for local consumption. Trade in fish
was also found to be important. In a one-month
period, a Kapit trader exported 12 species of fish

weighing 3456 kg to Sibu, with empurau (*Tor dourensis*), mengalan (*Puntius bulu*), and tenggadak (*Puntius schwanenfeldii*) heading the list by weight and tenggadak being by far the most important by value. Caldecott (1987, p. vii) estimated that the amount of wild meat harvested in Sarawak each year was in the order of 20 000 tonnes, which would cost at least M $100 million to replace with domestic meat. The wild meat trade is clearly important, but it is doubtful whether it can be sustained as logging extends into the interior.

Information obtained by Caldecott from interviews with longhouse families indicated that there was a sharp decline in the amount of wild meat that they harvested from logged forests. This resulted from several factors, including habitat change, hunting by timber-company workers, greater availability of ammunition, and enhanced access to forests along logging roads, all of which combined to deplete resources. Caldecott's data showed that 3806 kg of meat per 10 families per year was harvested from unlogged forests, compared with 1240 kg during the first decade after logging, 534 kg during the second decade, declining to only 155 kg during the third decade. This is equivalent to a drop in meat consumption per head from 54 kg to 2 kg per year.

Logging also adversely affects fish populations. As described earlier in the chapter, present-day logging operations completely remove the vegetation cover from a substantial part of the logged-over area. As a result, large quantities of sediment are washed into streams and rivers. High turbidity levels and siltation, together with pollution from diesel oil, cause fish stocks to decline dramatically (Caldecott 1987, p. 97; Colchester 1989, p. 37). The magnitude of the problem can be judged by the fact that some 59 per cent of the state's rivers are considered polluted, and that a serious decline in fish stocks along a number of interior rivers was reported by 57 longhouses (Ngau *et al.* 1987, p. 178).

Logging also affects other food resources. For example, it destroys fruit trees and other food plants as well as trees that are favoured by bees.

The nomadic Penan are particularly affected by the loss of the sago palm (*Eugeissona utilis*), which is one of their staple foods (Jayl Langub 1988). In addition to food plants, logging also results in the destruction of plants that are harvested for domestic use (e.g. in house construction and basket-making) and for sale (e.g. rattans). Much resented is the loss of engkabang trees, from which illipe nuts are collected, and of ipoh trees, from which poison for blow-pipe darts is obtained.

Social impact

The loss of customary land and forest rights has given rise to a number of social problems. Many natives have drifted to the towns or moved to resettlement schemes, where adaptation for some has not been easy. Natives from the interior with little or no education or work experience have often found it difficult to obtain employment and many have ended up in squatter settlements; some women have turned to prostitution (Hong 1987, p. 209). It would appear that those who have joined land development schemes, where rubber and oil palm are the main cash crops, have generally found it difficult to adjust to the new way of life and to the cash economy; in addition, there have been complaints that government promises relating to compensation for lost customary rights have not always been fulfilled (Hong 1987, chapters 5 and 11; Colchester 1989, pp. 54–62). It is likely that more forest land will be cleared for cash-crop plantations (see Chapter 3 this volume) because both the state and the federal government appear to be convinced that land development programmes will help the economy, provide a means of rehabilitating 'idle' land, and bring better facilities to rural areas (Malaysia 1986, p. 196; Colchester 1989, pp. 54–5).

Those who cling to traditional ways are finding it increasingly difficult to subsist. In many areas wild meat harvests are declining and hill-rice production is falling, the latter resulting from less land being available for swiddening and the lower yields from swiddens that have been recultivated

after a fallow of only a few years. Reports suggest
that malnutrition was widespread in the interior in
the late 1970s and early 1980s (Hong 1987,
pp. 201–9). In one survey, 81 per cent of the
children examined were found to be suffering
from moderate-to-severe malnutrition. Although
medical services have undoubtedly improved,
malnutrition has almost certainly persisted and in
some areas the situation may have deteriorated.

The Penan in particular appear to be suffering
from declining food resources because loggers
have moved through some of their traditional
areas four or five times (Rainforest Information
Centre 1989a, p. 5, 1990, p. 10). At a meeting of
the Penan Association in January 1990, it was
reported by native elders that people were starv-
ing in the Long Kidah, Long Iman, and Ulu Baram
areas, in part because fruit trees had been
logged.[24] One elder lamented that '[w]e suffer
problems finding food and materials which we get
from the forest. It takes us days to find meat and
fish—sometimes we don't find anything'.

Native groups have attempted to save areas of
forest from logging by having them gazetted as
Communal Forest Reserves. Almost without
exception, however, their applications have been
unsuccessful. In Ulu Belaga, for instance, several
longhouses made more than 30 applications
during the early 1980s but only one was successful, whereas some 242 760 ha were allocated as
timber concessions. In the mid-1980s, 18 applications from longhouses in the Limbang and
Baram areas were all unsuccessful, although again
extensive areas were allocated to timber companies (Sarawak Study Group 1989, pp. 12–15;
Sahabat Alam Malaysia 1989, pp. 110–13). Similarly, applications by natives for logging licences
have been turned down.

In order to to survive, many natives have sought
jobs in the logging industry. The accident rate,
however, is very high (Hong 1987, pp. 146–52;
Jeyakumar Devaraj 1989). In 1977 there were 28
deaths and 726 non-fatal accidents; by 1984 the
figures had risen to 87 deaths and 1553 non-fatal
accidents. The logging-related death rate is some
20 times higher than it is in Canada. Minimal

safety standards, lack of training, poorly constructed logging roads, long working hours and
fatigue, a get-rich-quick mentality on the part of
some contractors—these are among the reasons
for the high accident rate.

It is not at all surprising that the native groups
feel that they are being unjustly treated or that
protests have taken place.[25] In addition to individual misfortunes, logging is threatening the very
existence of native cultures. Change is inevitable
and most natives are not opposed to it; what they
resent is the kind of change that is being foisted
upon them. The Penan, for example, would like
schools and clinics, but at the same time they want
to retain many of their traditional ways. They
would hardly agree with the Prime Minister's
assertion that their traditional way of life is a
'miserable life' (Rainforest Information Centre
1989b, p. 5, 1990, p. 10). There is, unhappily,
much thinly veiled contempt for the native
peoples. The issues are complex and, as Hong
(1987) has advocated, need to be urgently
addressed with the full participation of native
communities.

4.4.4 Forest change and health hazards

The closeness or intimacy of the human interaction with tropical rain forest varies between, let us
say, hunter–gatherers in largely undisturbed forest, shifting cultivators or loggers in altered forest,
forest-fringe or near-forest peoples who move
back and forth across the forest edge, and farmers
or plantation workers in extensive clearings.
These different associations have epidemiological
significance (Unesco 1978, chapter 17). The
many diseases that humans are heir to in the
humid tropics—and not a few of which they create
for themselves—is one reason why the rain forest
remained largely intact until quite recently.

Small communities of hunter–gatherers in
undisturbed forest usually experience relatively
good health, although disease is a constant threat
and mortality and morbidity among children are
generally high. Even for the well-adapted forest
peoples, the rain forest is by no means a paradise.

Among the Orang Asli, for example, filariasis, malaria, tuberculosis, and yaws are common diseases, as are several others.

In undisturbed forest, however, diseases tend to be less virulent and varied than they are in modified forest because, among other things, humans in such settings may not constitute the principal vertebrate host, and because certain vectors of disease are confined to the upper strata or canopy of the forest where they are normally isolated from potential human victims. Alteration or removal of the forest or elimination of preferred host animals may cause vectors to feed on humans and parasites to enter a purely human cycle, thereby greatly increasing the risk of sudden outbreaks of disease.

Forest alteration, deforestation, and new settlements create favourable conditions for the transmission of malaria, dengue fever, intestinal parasites, and other insect- and water-borne diseases. Most forest change is the result of the activities of people from outside the forest—people like slash-and-burn farmers, loggers, road builders, and organized colonists—most of whom have little or no immunity to the diseases of the forest ecosystem, while unsanitary living conditions in new settlements also pose a variety of disease-related problems. The forest-dwellers possess a certain degree of immunity to the diseases of the rain forest, but they too face new health hazards when they come in contact with outsiders in the forest or when they are brought together in resettlement centres. Two brief examples will serve to illustrate some of the health problems that result from changing human relations with the tropical rain forest.

Forest conversion and malaria

Malaria is probably the principal cause of sickness and death throughout the tropical underdeveloped world today. Plans to eradicate the disease have failed, and a large reservoir of endemic malaria persists in much of the tropics, especially in Africa. World-wide, there are some 5 million new cases annually, and some 1.2 billion people live in risk areas (ICIHI 1986, p. 22).

Control, let alone eradication, poses many problems because the 'mosquito vectors are resistant to several insecticides, the agent is resistant to major drugs, and the disease is resurgent in many countries where only a decade ago it was virtually eradicated' (Meade *et al.* 1988, p. 79).

Malaria is largely a human-made disease. Whereas there are generally few species of malaria vectors in the undisturbed forest and those that are present often prefer to bite animals, activities like shifting cultivation, logging, and land development create favourable light and water conditions for rapid increases in the populations of certain species and for transmission of the disease. New colonists in the rain forest are major carriers of malaria, and Moran (1988, p. 160) noted that at any one time more than 20 per cent of the settlers in his study area on the Transamazonian Highway frontier had malaria.

Malaria was a major killer on the Malayan plantation frontier, where it was associated in particular with the rapid adoption of rubber cultivation after the turn of the century (see Chapter 3). Certain human and environmental conditions proved ideal for the spread of the disease: non-immune Tamils from South India comprised the bulk of the labourers on the estates; living conditions on the labour lines and the floating nature of the labour force provided efficient mechanisms for the distribution of the malaria parasites; and rain-forest removal created highly favourable habitats for the breeding of certain vectors of the disease.

Several species of *Anopheles* mosquitos are responsible for the transmission of malaria in the Peninsula, and each has its own special habitat requirements (Hodgkin 1956; Sandosham 1970). Some of the early rubber plantations were established on low-lying land near the coast where the vector species included, for example, *A. sundaicus*, which breeds in brackish water (including mangrove swamps), and *A. letifer*, which breeds in fresh water (especially in the acid waters associated with peaty soils) and under shade. Control measures in coastal areas included improved drainage and removal of shade vegetation.

The spread of rubber cultivation into the interior foothills brought plantation workers into contact with a much more dangerous vector of malaria: this was *A. maculatus*, which found ideal breeding conditions in the countless clear, slow-moving streams that were exposed to sunlight by widespread forest removal. Whereas most other vectors of malaria in the Peninsula occur in relatively small numbers, usually prefer to bite animals, and are normally responsible for 'endemic, chronic malaria at low levels', *A. maculatus* readily bites humans, proliferates at a great rate, and 'as a vector of epidemic malaria is one of the most dangerous in the world' (Meade *et al*. 1988, p. 84). Prevention and control of the breeding of *A. maculatus* in rural areas was largely achieved by spraying streams and other breeding sites with mineral oil and by keeping water courses free of grass and other obstructions, while in most urban areas subsoil drainage by means of earthenware pipes proved satisfactory.

Malaria took a heavy toll of the lives of plantation workers, the great majority of whom were South Indians (see Sandhu 1969; Stenson 1980). In 1911, at the height of the rubber boom, the mortality rate among estate workers in the FMS was 62.9 per thousand, with the less-developed states of Negeri Sembilan and Pahang registering rates as high as 195.6 and 109.5 per thousand, respectively (Stenson 1980, p. 21). Some estates lost between a quarter and a half of their labourers (Ooi 1976, p. 242). There is no doubt that *A. maculatus* was mainly responsible for so much human misery.

Control and preventive measures reduced the mortality rate from malaria to 18 per thousand in 1921 and then to 9.9 per thousand by 1930, although it was not until the 1940s and 1950s, when residual insecticides like DDT and Dieldrin and more widespread use of prophylactic drugs were adopted, that the number of reported cases of the disease declined dramatically. Malaria is still a threat in newly cleared areas (see below) and remains a major danger to shifting cultivators like the Senoi.

Land development and health hazards

Land settlement schemes have been implemented in many developing countries, often with little regard for their ecological implications. Most schemes have been enormously expensive, some have resulted in social tensions between settlers and indigenous peoples, and not a few have failed to meet their social, economic, and other objectives (Oberai 1988). Housing, water, and sanitation conditions in many newly built settlements pose health risks, while settlers in modified or transformed rain-forest settings face hazards from diseases like malaria, dengue fever, scrub typhus, leishmaniasis, trypanosomiasis, and schistosomiasis.

Planned rural development has proceeded apace in Malaysia since the mid-1950s (see Chapter 3 this volume), much of it under the auspices of FELDA (Bahrin 1988). The main goals of the programme are to improve the socioeconomic status of hitherto poor and landless peasants and to produce agricultural products for export. This section, which is mainly based on the work of Meade (1976, 1977, 1978; see also Meade *et al*. 1988), looks at health conditions on the FELDA schemes (see Plate 5).

Land development in Malaysia has resulted in the replacement of complex, species-rich lowland rain forest with greatly simplified, species-poor agro-ecosystems featuring mainly rubber and oil palm, and new relations have been established between people, behaviour, and habitat that have important consequences for human health. Meade found that the FELDA schemes generally brought beneficial changes in public health. She showed, however, that health conditions were better on mature and accessible schemes than they were on new and remote ones, which suffered more from nutritional, infectious, and psychological hazards. Although improvements in the quality of drinking-water and the provision of latrines generally lead to reductions in enteric and helminthic infections, there were also five kinds of increased hazards that required more attention (see Meade *et al*. 1988, pp. 127–8).

Firstly, clearing the rain forest created breeding

habitats for *A. maculatus*, but the efficacy of drugs and insecticides had 'led to neglect of traditional concern for drainage, shading, oiling, and other techniques that had originally been developed in Malaysia' (Meade *et al.* 1988, p. 127). Secondly, cleared vacant land intended for commercial, institutional, and other uses was quickly invaded by *lalang*, an ideal habitat for rats and their mites—including the species of mite that transmits scrub typhus. (Oil-palm plantations also bring rats, mites, and humans into dangerous juxtaposition.) Thirdly, because settlers did little hunting or fishing, and vegetables were either not grown or were too expensive, nutritional hazards probably increased. Fourthly, remoteness and higher incomes resulted in many more motor cycles, and there was a great increase in accidents and mortality on the roads. And fifthly, altered mobility patterns had other implications for human health: for example, occasional journeys by entire families to places of origin facilitated the diffusion of infectious disease.

The FELDA schemes are generally healthy places, and no doubt they compare favourably with similar kinds of schemes in other tropical countries. They are not free of serious problems, however, and Meade's richly detailed studies suggest the need for vigilant attention to the subtle changes in health conditions that result from new relations between people, behaviour, and habitat. The now rapid pace of forest change in the humid tropics confronts many countries with serious health problems, problems that 'can only be resolved by sound forest management, settlement planning, adequate infrastructure and health care' (ICIHI 1986, p. 21, also pp. 71–3).

4.5 CONCLUSION

It appears unlikely that indigenous peoples caused any plant or animal species to become extinct in pre-colonial times. Since then, however, two large mammals and two birds have become extinct, and many species of both plants and animals are now endangered. The spread of agriculture, logging, and the widespread use of firearms are the major endangering processes.

The human impact on animal life has been a matter of some concern for almost a century, whereas it has been only in recent years that the precarious status of some of the flora has attracted any attention. Many species of trees that have not been reported for many years will almost certainly not be seen again. Hundreds of species of plants are endangered, and very little has been done to protect them.

Colonial officials were less concerned about wildlife than they were about the impact of tin mining and plantation agriculture on soils and water. Legislation and better management curtailed most of the worst excesses, but the fact that soil losses from agricultural land are higher than those from forested areas continues to have implications for drainage and floodplain management. The popular belief that forest removal leads to severe flooding is not well founded. Evidence suggests that the worst floods are caused by exceptional meteorological events and that such floods are little affected by land use.

There have been social costs to pay for the rapid pace of forest change, not the least of which have been a variety of health hazards. Peoples who rely on the forests for their livelihood are beginning to speak out against the erosion of their customary rights. This is most evident in Sarawak, where loss of land and destruction of forest resources have caused much resentment.

5 CONSERVATION: TOWARDS A SUSTAINABLE FUTURE

The Malay Archipelago has long been famed for its natural wealth. It was the lure of gold and tin and the riches of the forest and the sea that for some two millennia or more drew Arab, Indian, and Chinese traders to regions like the Malay Peninsula and the coasts of Borneo; and it was from the abundant harvest of natural resources that early maritime kingdoms and empires drew much of their wealth.

Western images of an opulent and plentiful East were conjured up by the aroma of the legendary spice islands, and the ineffable superabundance of plant and animal life in the luxuriant forests of the Malay world inspired some of the earliest scientific studies of tropical nature, including those of the great naturalist Alfred Russel Wallace. We have seen that the spread of British rule over the territories that now comprise Malaysia was motivated in large part by the desire to control and exploit the natural endowment of the region.

Today, as in the past, Malaysia's natural wealth is a focus of much attention: governments continue to exploit it, for a heavy reliance is still placed on primary production; vested interests vie for a larger share of it, for resources like timber mean great wealth for the select few; and conservationists struggle to save and protect it, for current trends threaten to expunge much of it for ever. Behind the growing number of land-use disputes are conflicting values and heightened competition for dwindling natural assets.

It was, as we have noted elsewhere, one of the many contradictions of colonialism that it fostered not only a speeding up of resource exploitation but also a certain interest in conservation. Several colonial officials—Stamford Raffles, Hugh Low, and Charles Hose are three names that spring immediately to mind—were also informed amateur naturalists, and it is generally not appreciated that the history of specific conservation measures spans a century or more in Malaysia.

Recent decades have witnessed a widening interest in environmental affairs and a growing realization that environment and development are two sides of the same coin. This relatively new departure—it is new just about everywhere—is reflected in several federal government initiatives on the environment, in the emergence of non-governmental organizations (NGOs) with various interests in the state of the Malaysian environment, and in growing public awareness of the need to protect the nation's remarkable natural heritage.

This chapter describes some of the measures that have been taken to conserve and manage Malaysia's natural wealth. The temporal coverage embraces both the colonial era and the national period, focusing mainly on the latter. We pay particular attention to legislation, parks and reserves, NGOs and their activities, land-use disputes, and recent conservation strategies. The chapter closes with a brief survey of the kinds of 'transitions' that will be required to put Malaysia on a path towards sustainable development. The next section outlines a few basic concepts.

5.1 CONCEPTS

5.1.1 Sustainable development and conservation strategies

The meaning of 'sustainable development' has proved elusive. The widely cited view is that it means 'development that meets the needs of the present without compromising the ability of future generations to meet their own needs' (WCED 1987, p. 43; see Chapter 1 this volume). It would appear that sustainable development is a way of addressing current and future stresses on natural systems and human well-being; it is a kind of critique or basic social guide. Sustainability is not a new idea (O'Riordan 1988), and the notion of sustainable development has an older relation in the concept of eco-development (see Redclift 1987, pp. 34–6).

Interest in sustainable development was given a major boost by the publication of the *World Conservation Strategy* (IUCN/UNEP/WWF 1980), which linked the concept to three principles: maintenance of essential ecological processes, preservation of biological diversity, and sustainable utilization of species and ecosystems. The strategy emphasized the mutual dependence between environment and development and noted, among other things, that poverty and inequality were major barriers to sustainable development.

Socio-economic and institutional considerations were highlighted in *Our Common Future*, the report of the World Commission on Environment and Development, which commented that

sustainable development is a process of change in which the exploitation of resources, the direction of investments, the orientation of technological development, and institutional change are all in harmony and enhance both current and future potential to meet human needs and aspirations (WCED 1987, p. 46).

The Commission underscored the links between poverty, inequality, and environmental degradation and called for a new era of socially and environmentally sustainable economic growth— growth that would fulfil fundamental human needs.

It has been argued, however, that sustainable development requires greater attention to such matters as the principles of social self-determination and the achievement of equity; that the needed economic growth in the developing countries ought to be offset by reductions in energy and materials consumption in the affluent societies (where various non-material needs have not been met); and that there are, in addition to conventional interpretations of sustainable development, a variety of alternative paradigms or prescriptions for a more sustainable future (see, e.g. Gardner and Roseland 1989*a,b*).

A conservation strategy is a guide or blueprint for the sustainable use of resources. It sets goals, outlines means of addressing problems and opportunities, and establishes criteria for assessing whether desired results have been achieved. The ideal strategy is holistic, integrative (as opposed to sectoral), multidisciplinary, co-operatively developed by all concerned parties, and derived from a society's fundamental values. A conservation strategy is an image or vision of the future.

Many countries have prepared national conservation strategies. These, as Strong (1986, p. 14) has commented, are intended to provide 'a means for integrating conservation measures into national development and resource management planning, policies and action'. To date, however, most such initiatives appear to have focused rather narrowly on natural systems and resources and the human impact on nature. There is a need to widen the scope of conservation strategies to include demographic, economic, health, equity, and other socio-economic considerations.

National parks and other protected areas have an important role to play in conservation strategies and sustainable development. Many such areas can be given over to multiple uses, thus providing a mix of protection and development. It is crucial that natural heritage areas be seen to contribute to sustainable development. Planning should aim to co-ordinate land uses within and around such protected areas (Nelson and Eidsvik 1990).

5.1.2 Environmental management: north and south

As practised in the North, environmental management is 'usually a responsive set of techniques rather than a framework for implementing policy' (Redclift 1987, p. 132); for the most part it is oriented to after-the-fact damage repair rather than to the anticipation and prevention of environmental ills. Economies tend to be viewed as perpetual growth machines that have little or no contact with the environment, whose supporting role is largely ignored and grossly undervalued. Management, which is 'technocentric' rather than holistic or 'ecocentric', is concerned more with inventories and 'more studies' than it is with formulating alternative development policies. When we turn our attention to the South, we should 'not exaggerate the importance of environmental planning and management in our own societies' (Redclift 1987, p. 136).

A good deal of development in the South has been in the form of large-scale, environmentally destructive projects, many of them financed by the multilateral development banks (e.g. Rich 1985; Fearnside 1987b). Vast sums have been loaned for capital-intensive agriculture, ranching, colonization, and transmigration schemes, all of which have resulted in massive deforestation. Generally speaking, the emphasis has been on project planning rather than on development policy, and environmental considerations have taken a back seat to such matters as commodity production for foreign exchange and the need to service massive debts. To the extent that it exists, most environmental management has been palliative at best.

In both the North and the South the term 'forest management' is something of a misnomer, because it is generally only timber that is being managed in the forested lands outside specially protected areas such as national parks and equivalent reserves. Timber management is a subset of forest management. The latter, as a Canadian professional forester has noted, is more holistic. Specifically,

[i]t involves a detailed knowledge and understanding of the many facets of ecology, wildlife management, and wilderness values. It requires a solid base of information about the plant and animal species present, their habitat needs, and what effect timber management and other forest activities will have on them in the short- and long-term. It makes allowances for these needs and designs timber management as a part—not the whole—of the use of forests (Dunster 1990, p. 44).

There is an urgent need to promote the idea that conservation and management must extend beyond the limits of parks and reserves, that it must, in short, embrace all lands and forests.

5.1.3 Conservation

To conserve something is to save it. Saving natural resources for future consumption is conservation according to Passmore (1980), who distinguished between conservation ('saving for') and preservation ('saving from'), which he defined as

the attempt to maintain in their present condition such areas of the earth's surface as do not yet bear the obvious marks of man's handiwork and to protect from the risk of extinction those species of living beings which man has not yet destroyed (p. 101).

While many resource managers and economists would probably agree with Passmore's definition of conservation ('saving for'), most biologists, ecologists, and conservationists would probably wish to give greater weight to evolutionary processes, as in the following definition by Frankel and Soulé (1981, p. 4), who noted that conservation means

policies and programmes for the long-term retention of natural communities under conditions which provide the potential for continuing evolution, as against 'preservation' which provides for the maintenance of individuals or groups but not for their evolutionary change. ... zoos and gardens may preserve, but only nature reserves can conserve.

In short, Passmore's 'preservation' ('saving from'), provided it included some statement dealing with the significance of evolution, is more or less synonymous with 'conservation' as viewed by Frankel and Soulé.

Conservation is a process whose purpose is to devise rational guidelines for human interaction with the natural systems that comprise the biosphere. (The conservation of cultural landscapes and their individual elements is not our concern here.) In the words of the *World Conservation Strategy*, conservation of living, renewable natural resources like forests is that aspect of management 'which ensures that utilization is sustainable and which safeguards the ecological processes and genetic diversity essential for the maintenance of the resources concerned' (IUCN/UNEP/WWF 1980, n.p.). Conservation guidelines are required not only for ecosystems that have already been modified but also for those that have, in varying degrees, escaped the imprint of humans (White and Bratton 1980). Broadly defined, then, conservation embraces preservation ('saving from'), as well as 'maintenance, sustainable utilization, restoration, and enhancement' (IUCN/UNEP/WWF 1980, n.p.).

Human well-being depends on conserving living resources and the diversity of life on earth, a point that has been emphasized by Ehrlich (1980), who noted that the public service functions of ecosystems include

helping to regulate climate, maintaining the quality of the atmosphere and of freshwater supplies, running nutrient cycles, disposing of wastes, producing and maintaining soil, pollinating crops, controlling . . . pests of crops and diseases of human beings, providing various aesthetic values and maintaining a 'library' of genetic information from which future crops, domestic animals, antibiotics, spices, drugs and tools for medical research can be drawn (Ehrlich 1980, p. 333; see also Ehrlich and Ehrlich 1983, pp. 91–120).

Degradation of ecosystems cancels many of these services, but as yet we know very little about how to make substitutions for those that are lost (Ehrlich and Mooney 1983). Conservation is therefore essential.

Especially relevant to the conservation of living resources is the new synthetic discipline of conservation biology, whose main goal is 'to provide principles and tools for preserving biological diversity' (Soulé 1985, p. 727). An important ingredient of the new discipline is the inclusion of ethical considerations. An expanded approach should recognize that environments and resources are socially constituted, mean different things cross-culturally, and are transformed not only by economic processes but also through experience.

5.1.4 Justifications for conservation

There are instrumental and intrinsic justifications for conservation. Both are related to the ways that we appraise something as valuable. Instrumental values are means values, that is, they are means to some valued end. For example, rain forest should be conserved for something other than itself—for, let us say, its scientific value. On the other hand, something that is intrinsically valuable is good or valuable in itself; consequently, rain forest should be conserved for its own sake, regardless of its usefulness to humans.

Instrumental or utilitarian justifications for conservation claim that we 'should look after nature because nature looks after us'. In other words, self-interest is served by the realization of the economic, scientific, genetic, and other values that, in this case, rain forests carry. Where, as is often the case, such values conflict (Tribe *et al.* 1976), resolution is usually sought through cost–benefit analysis. This suggests that all values can be measured in monetary terms, which, in turn, assumes that all values are commensurable. For some, however, such economic considerations are simply out of place because the values at issue are not economic in the first place (Ehrenfeld 1976; Godfrey-Smith 1979). By 'basing all arguments on enlightened self-interest the environmentalists', as Evernden (1985, p. 10) has noted, 'have ensured their own failure whenever self-interest can be perceived as lying elsewhere'.

The prevailing Western, primarily Cartesian conception of nature reduces everything other than human consciousness to the laws of mechanics; views nature as something separate and apart and over which humans can exert mastery and control; establishes a dualism between ourselves and non-human species, which are regarded as 'things' or objects; denudes the world of subjects; separates knowledge from value; and tends to accept what science tells us about the world in preference to our own experience of value in the world. What is required is a less mechanistic, more ecological philosophy of life, a philosophy that, in the view of what appears to be a growing number of scientists and philosophers, would abandon the sharp distinction between ourselves and everything else and would recognize that ecosystems and non-human species have their own intrinsic value (Ehrenfeld 1978; Godfrey-Smith 1979; Skolimowski 1981; Webb 1982; Devall and Sessions 1985; Rolston 1985, 1988; Naess 1986).

While there is much dispute in philosophy between those who argue for a new 'environmental ethic' and others who claim that we need to change our behaviour (Passmore 1980), most agree that we are badly in need of a more caring, sensuous, or passionate involvement with nature. Greater compassion for and empathy with non-human species would go a long way. The so-called real world is not simply an economic world.

5.2 FROM ABUNDANCE TO DECLINE: CONSERVATION IN THE COLONIAL PERIOD

5.2.1 The Peninsula

A considerable body of conservation legislation was introduced during the colonial period, some of which forms the basis of current legislation. In addition, several game reserves and a large national park were established. Conservation efforts during this period focused on three main areas of concern: forest protection, wildlife protection, and the control of tin-mining activities. The first two are described below.[1]

Forest protection

The earliest forest regulations were introduced in the 1880s to control the widespread cutting of timber in the tin-mining areas (see Chapter 3). Furnaces that required high-quality hardwoods were banned in Negeri Sembilan, Perak, and Selangor between 1887 and 1898, and between 1888 and 1915 regulations were introduced in Pahang, Perak, and Selangor prohibiting the cutting of certain species (Aiken *et al.* 1982, p. 113). Meanwhile, the rate of agricultural expansion, especially during the rubber boom, caused concern among foresters, who pressed for a permanent forest estate to safeguard future timber supplies. Although no such estate materialized, extensive tracts of forest were reserved.

As noted in Chapter 3, forest reservation commenced in the Straits Settlements (SS) in the 1880s and in the Federated Malay States (FMS) in the 1890s, and some 492 km^2 had been reserved by the turn of the century (Wyatt-Smith and Vincent 1962, p. 210). A single forest department for the SS and the FMS was established in 1901, and thereafter the forest reservation programme expanded rapidly in the FMS; and following the introduction of legislation and the creation of forest departments, forest areas were also reserved in the unfederated states in the 1920s and 1930s. By 1940, some 25 900 km^2 of forest had been reserved: approximately two-thirds as productive forest, the remainder as protective forest.[2] Mainly located in mountainous areas, the latter provided a safeguard against soil erosion. Although they were not created with the protection of wildlife in mind, the forest reserves did serve this function to a certain extent. Timber cutting in the productive forests was still unmechanized, and its impact on wildlife was probably not very great.

During the 1930s, however, foresters became increasingly concerned that as more forest was cleared and as logging operations extended into new areas, relatively little unaltered forest would remain. In order to preserve representative areas of untouched forest it was decided that a network of Virgin Jungle Reserves should be established (Wyatt-Smith 1950, 1961 *c*). These, it was hoped, would:

(1) preserve natural habitats for long-term, basic scientific study;

(2) provide accessible areas throughout the country in which botanical, phenological, and ecological studies of natural forest could be carried out;

(3) retain specimens of all types of natural vegetation as reference areas for managed forest;

(4) provide natural arboreta for the benefit of foresters and interested members of the

general public by setting aside permanent sample plots in which trees would be numbered and identified;

(5) preserve outstanding forest or individual trees as a national heritage;

(6) provide wildlife sanctuaries within large tracts of managed forest (see Rahman-Ali and Wong 1968, p. 364).

A total of 54 reserves incorporating examples of most major forest and vegetation types had been established by 1960. Most of these were between 40 ha and 80 ha, so they were too small to function as wildlife sanctuaries (Rahman-Ali and Wong 1968). Unfortunately, many of the reserves appear to have been degazetted to make way for agricultural development.

In addition to placing erosion-prone steepland areas in forest reserves, legislation was introduced in the SS in 1937 that required a permit to cut vegetation in areas that had been gazetted as 'hill lands', and between 1922 and 1940 legislation designed to prevent erosion in steepland areas was introduced in the FMS as well as in Kedah and Kelantan (Leigh 1973; Aiken *et al.* 1982, pp. 122–3 and 260). It would appear, however, that few areas were gazetted and that enforcement was minimal (Hartley 1949).

Wildlife conservation

Pre-1930 Between 1884 and 1930 legislation to protect wildlife was introduced in every state except Kedah and Perlis, and a number of game reserves were established (Table 5.1). Birds with bright plumage, whose feathers were much sought after by the milliners of the day, were protected by an ordinance that came into force in the SS in 1884.[3] Similar legislation was introduced in Selangor and Perak in 1889. The first legislation to protect game animals was introduced in Pahang in 1896, when it became necessary to obtain a licence (free of charge until 1911) to kill seladang, elephant, and rhinoceros. Similar legislation that was introduced in Selangor and Perak

in 1902 and in Negeri Sembilan in 1903 was amended in all three states in 1904 to include fees for big-game licences and to prohibit the taking of immature animals and female sambar. The enactments provided for the creation of game reserves, and Malaya's first reserve was established at Chior in Perak in 1903 (Table 5.1).

In 1911, a single enactment applicable to all four federated states was introduced, but it differed little from the earlier state laws; it made no real provisions for enforcement, and licences continued to be issued indiscriminately. Consequently, the slaughter of big game proceeded, an example being the elimination of a large herd of seladang in Pahang (Hubback 1923). In an attempt to redress the situation a new enactment was introduced in 1921 that provided for the appointment of game wardens. But no funds were forthcoming, with the result that only honorary part-time wardens were appointed (Wild Life Commission of Malaya 1932, II, p. 17).

Although conservation legislation in the SS was mainly directed towards the protection of birds, the killing or taking of sambar and barking deer on Singapore Island was prohibited in 1923, and in the following year the killing or taking of elephant, tapir, or rhinoceros in the Dindings was banned. Legislation was introduced in Johor in 1912 that attempted to protect wildlife by making licences very expensive. Legislation that was modelled on the 1921 FMS enactment was introduced in Terengganu in 1923, but the only animals that it covered were tapir, seladang, elephant, and rhinoceros. Licences were also required to kill these animals in Kelantan after 1921, and the export of doves and other birds from that state was banned in 1930.[4] Legislation to protect wildlife was not introduced in Kedah and Perlis until 1956.

Growing concern for wildlife resulted in the creation of several game reserves during the 1920s (Table 5.1), these including the large Kerau (552 km²) and Gunung Tahan (1425 km²) reserves in Pahang. But the reserves had no security of tenure and, in response to growing demands for more agricultural land, a number were degazetted, including the Serting Reserve in

Table 5.1 Wildlife reserves and parks in the Peninsula, *c.*1964

State	Reserve or park	Approximate area (ha)	Date established
Perak	Chior	4330	1903
	Sungkai	2428	1928
	Batu Gajah	5	1952
Pahang	Kerau	55 200	1923
	Fraser's Hill	?	1957
	Cameron Highlands	68 600	1962
	Pahang Tua	1335	1954
Johor	Segamat	31 100	1937
	Endau–Kluang	75 900	1933
	Endau–Kota Tinggi (W)‡	80 600	1933 (1941)
	Endau–Kota Tinggi (E) †	18 700	1933 (1941)
	Four Bird Islands	2	1954
Negeri Sembilan	Port Dickson Islands	0.4	1926
Selangor	Fraser's Hill	2 978	1922
	Kuala Selangor	44	1922
	Bukit Kutu	1943	1922
	Klang Gates	130	1936
	Bukit Sungai Puteh	40	1932
	K. L. Golf Course‡	403	1923
	Bukit Nanas	16	1906
	Templer Park	1214	1956
	Sungai Dusun	4232	1964
National Park	Taman Negara	434 300	1938–9

† The Endau–Kota Tinggi Reserve was divided into two portions (western and eastern) in 1941.
‡ The original reserve, which was established in 1923, is now occupied by the Royal Selangor Golf Club. Stevens (1968, p. 39) remarked that the 'golf courses of most cities are wildlife sanctuaries and almost everyone agrees that they should be because hunting and golf, though both use directed projectiles, do not readily mix within the same piece of real estate'.
Source: Stevens (1968, table 1, p. 18).

Negeri Sembilan (Comyn-Platt 1937, p. 49), the Sungai Lui Reserve in Pahang, and all of Johor's reserves. There was pressure to degazette the Kerau Game Reserve, but fortunately it survived. Apparently there was no similar pressure on the Gunung Tahan Reserve, presumably because it was 'too remote to fall under the avaricious eyes of commerce' (Burkill 1971, p. 206).

The Wild Life Commission The appointment of the Wild Life Commission of Malaya in December 1930/January 1931 resulted from the efforts of Theodore Hubback, sometime Honorary Game Warden of Pahang and well-known big-game hunter. Hubback opposed the continuing slaughter of Malaya's wildlife, deplored the general ineffectiveness of the prevailing legislation, and had little time for the avarice of the planters and their commercial backers. He expressed his concerns in a letter to the Chairman of the Society for the Preservation of the Fauna of the Empire, and in 1929 he visited the Society and the Colonial Office (Hubback 1924; Page 1934, p. 36). Hubback was rewarded by being appointed to head the Commission, whose terms of reference were as follows (Wild Life Commission 1933, p. 38):

1. To report on the existing regulations for the Protection of Wild Life throughout Malaya.

2. To report on the steps necessary to constitute a National Park or Wild Life Refuge in the vicinity of Gunong Tahan, including portions of the State of Pahang, and, if the Governments of those States shall signify their concurrence, of the States of Trengganu and Kelantan.

3. To report on the existing Game Reserves and their value or otherwise for the permanent preservation of the fauna of Malaya.

4. To inquire into the allegations of damage done to agriculture by wild life; to call for and reveal evidence bearing on such matters and to suggest methods of dealing with any situations which may be revealed.

5. To report on the organization required to administer the regulations for the preservation of wild life, including the administration of National Parks, Wild Life Refuges, or Game Reserves.

The Commission held 64 sessions throughout Malaya and heard evidence from Orang Asli, Malay villagers, Chinese and European planters, and colonial administrators. As discussed in Chapter 4, the evidence presented to the Commission clearly indicated that many species of birds, mammals, and fish were in urgent need of greater protection and that some species were on the verge of extinction.[5] The Commission also found that there was general support for greater protection, except, predictably, from the planting community. The Assessor G. Hawkins commented that in the view of the Rubber Growers' Association the choice was between the ruin of agriculture and the extermination of wildlife (Wild Life Commission of Malaya 1932, II, p. 249).

The Commission, which clearly did not sympathize with the views of the agricultural lobby, recommended that:

1. A commissioner should be appointed who would be responsible for the execution of policy for wildlife preservation throughout Malaya.

2. A national parks enactment and a new preservation of wildlife enactment should be introduced.

3. A wildlife fund should be established.

4. Wildlife sanctuaries should be established under secure tenure.

5. The Krau (now Kerau) Game Reserve should be elevated to a national park and a Gunong (or Gunung) Tahan National Park should be established.

6. Open seasons for the shooting or killing of game should be declared.

7. Commercialization of wildlife should be totally prohibited.

8. Wild bird refuges should be created.

9. The conservation of 'riverine fish' should come under the control of the commissioner for wildlife.

10. More honorary deputy game wardens should be appointed (Wild Life Commission of Malaya 1932, II, pp. 287–8).

Implementation of the Wild Life Commission's recommendations A number of the Commission's recommendations were implemented before the end of the colonial era. Perhaps the most significant of these was the creation of an extensive national park in the Gunung Tahan area, where the three states of Kelantan, Pahang, and Terengganu adjoin.[6] The King George V National Park, now Taman Negara, remains the Peninsula's only national park (Fig. 5.1). Owing mainly to administrative decentralization in the FMS and competition from other land uses, very few additional wildlife sanctuaries were established in the years following the release of the Commission's report (Hubback 1938a, pp. 18–19; 1938b, p. 107).

Game conservation in Malaya was centralized under a Chief Game Warden in 1937, but this arrangement lapsed after the war when the Game Department was once again decentralized (Stevens 1968, p. 16). Co-ordination of legislation, however, was achieved just before the end of the colonial period. The federal Wild Animals and Birds Protection Ordinance of 1955 provided total protection for seven mammals (Javan rhinoceros, Sumatran rhinoceros, pangolin, gibbon, binturong, slow loris, and tapir) and protection for eight others and a number of bird species. The ordinance was adopted by all of the states, thereby giving uniformity of legislation to the Peninsula some 60 years after the introduction of the first wildlife conservation ordinance (Mohd. Khan 1988, p. 254).

5.2.2 North Borneo and Sarawak

Forest protection

The colonial forest departments of North Borneo and Sarawak also embarked on forest reservation programmes, although progress was slow in the pre-World War II period. In North Borneo, a Forest Department was established in 1914 and a reservation programme was initiated in 1920 (Fyfe 1964, p. 85; John 1974, p. 76). By 1930, 19 separate reservations covering some 30 065 ha had been created, although this represented only 0.37 per cent of the land area of the state. At the end of 1939 some 2.1 per cent of the land area had been reserved, far short of the target of 10 per cent for that year. Reservation proceeded more rapidly after the war and by the end of 1963 some

25 618 km^2, or 34 per cent of the state, had been reserved (Fyfe 1964, p. 85). The reservations afforded some wildlife protection because a permit was required to hunt in them. Protection and Amenity forest reserves afforded the greatest protection, and at the end of 1963 these covered 2 308 km^2 and 311 km^2, respectively.[7] In addition, 40 Virgin Jungle Reserves covering some 5606 ha had been created by the early 1960s (de Silva 1968b, p. 146).

In Sarawak, regulations prohibiting the felling or damaging of jelutong, 'ketio', and engkabang trees were introduced in 1912, although they do not appear to have been very effective because there was a lack of staff to enforce them (Smythies 1963, pp. 236 and 240). Reservation commenced following the establishment of a Forest Department in 1919. By 1929, some 551 km^2 had been reserved, and by 1934 this had risen to 1388 km^2, or 1.2 per cent of the state (Mead 1925, p. 98; Smythies 1963, p. 239). The lack of suitable base maps was the main reason for the relatively slow rate of progress. By 1940, some 6474 km^2 (or about five per cent of the state) had been reserved, and by 1960 this had risen to 29 785 km^2, representing 24 per cent of the state. Hunting and collecting within Permanent Forest Reserves, other than Communal Forest Reserves, was not allowed without permission (Hong 1987, pp. 73–6).

Wildlife conservation

The main legislation protecting mammals and birds in North Borneo was introduced in 1936, although legislation protecting some game animals dates from 1897. The 1936 Wild Animals and Birds Protection Ordinance made it an offence to kill or capture banteng, elephant, rhinoceros, tarsier, orang-utan, gibbon, and proboscis monkey without a licence. Shooting deer with firearms also required a licence, and several birds were protected. No more licences to kill rhinoceros were issued after early 1947 (Burgess 1961, p. 144). The Ordinance provided for the establishment of bird and game sanctuaries and five of the former were established before the end of colonial rule.[8]

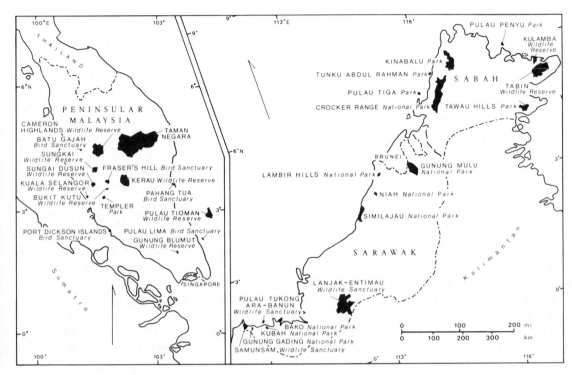

Fig. 5.1 Malaysia's national parks, wildlife reserves, and sanctuaries, *c.*1990. (From Davies and Payne n.d., p. 211; Sarawak n.d.*a*; WWFM 1985, p. 9 and map 1.1; Mohd. Khan 1988, p. 253; Anon. 1990, pp. 34–6.)

NB There are two conservation areas in the Kuala Lumpur region that cannot conveniently be shown on this figure: they are Kuala Lumpur Golf Course and Klang Gates Wildlife Reserve (see Tables 5.1 and 5.2). Quartz ridges and their biota are protected in the latter, but degradation has resulted from burning and farming.

Although the game sanctuaries were not established, a large national park encompassing Mt. Kinabalu was constituted in May 1963, the National Parks Ordinance of 1962 providing the framework for its management (Carson 1968, p. 492).[9]

A more comprehensive piece of legislation, the Fauna Conservation Ordinance, was prepared near the end of the colonial period. It was passed in 1963 but did not become effective until July 1964. The new ordinance gave total protection to the Sumatran rhinoceros, tarsier, orang-utan, gibbon, and proboscis monkey and required a licence to hunt elephant (male only), clouded leopard, ban-

teng, and bear. In addition, 55 bird species were placed on a protected list (see de Silva 1968*b*, annex 1).

There appears to have been little formal protection of wildlife in Sarawak during the period of Brooke family rule. The British colonial administration (1946–63) introduced a Game Ordinance and a Rhinoceros Protection Ordinance in 1947, and these were replaced in 1958 by a Wild Life Protection Ordinance. The latter totally protected the rhinoceros but, as Caldecott (1987, p. 97) has noted, the animal was virtually extinct by this time. As in North Borneo, wildlife sanctuaries were not created prior to independence, although a national park was established at Bako in 1957 under the provisions of the National Parks Ordinance of 1956, and during the early 1960s two potential park sites were investigated—one at Niah Caves, the other in the Gunung Mulu area (Sarawak n.d. *b*, p. 117).

5.3 CONSERVATION IN THE MALAYSIAN PERIOD

A vast area of tropical rain forest has been eliminated or variously degraded in recent decades. Indeed, more forest has been lost in the brief national period (since 1963) than during all of the long colonial era. Anthropogenic forest change has been especially rapid in the Peninsula, where land development and selective logging have taken a very heavy toll of the species-rich lowland and hill forests. Outright elimination of forests has proceeded more slowly in Sabah and Sarawak, although here too the pace of change has accelerated rapidly since *c*.1970. Only some 5–6 million ha of largely undisturbed forest remain in the whole country. Whether most, or indeed any, of this can be protected is a moot point.

In addition to the destruction caused by the rapid removal and alteration of the forests (see Chapter 4), Malaysia now faces a wide range of other environmental problems: effluent from rubber-processing factories and palm-oil mills has polluted many of the Peninsula's rivers and streams; plants, animals, and humans are increasingly threatened by ever-larger applications of agricultural chemicals; delicate coral reefs have been degraded by water pollution and by blast fishing, dredging, and overcollection of corals and shells; pollution of estuarine and coastal waters has disrupted subsistence and commercial fisheries; and the growing concentration of people, industries, and vehicles in the Peninsula's west coast cities—most notably in the Kelang Valley conurbation—has resulted in serious air and water pollution, massive soil erosion from construction sites, and a variety of environment-related health hazards (see Aiken

and Leigh 1975, 1978, 1988; A. S. H. Ong *et al.* 1987). Although these various problems lie outside the scope of this work, we mention them *en passant* because their all-too-apparent impact has given rise to growing environmental concern over the state of the Malaysian environment, including what is happening to the country's forests.

There have been several responses to the widening circle of environmental problems: the federal government has become more active in environmental affairs; several urban-based non-governmental organizations (NGOs) have struggled to keep environmental issues on the government's policy agenda; support for such groups has grown among the sizeable middle class; several land-use disputes have attracted widespread public interest and a good deal of media coverage; and certain indigenous peoples have revealed increasing frustration over loss of land, erosion of customary land tenure, and destruction of their ancestral lands and forests (see Chapter 4). It is to a discussion of these various responses, in so far as they pertain to the tropical rain forest, that we now turn.

5.3.1 Forests, conservation, and the federal government

The central government has played a more active role in environmental affairs since the early 1970s. Since then it has, among other things, established a Department of Environment (sometimes referred to as a Division) within the Ministry of Science, Technology and the Environment, devoted greater attention to the environment in certain of its national plans, introduced several important pieces of environment-related legislation, and co-operated with other ASEAN nations (i.e. with Indonesia, the Philippines, Thailand, Singapore, and Brunei) on common regional environmental concerns (see Sharp 1983, 1984; McDowell 1989). Overall, however, the government has adopted a shilly-shallying stance on environmental problems. This section focuses on three themes: forests and conservation-related matters in the national five-year plans, legislation, and the government's vacillation.

National plans

The five-year development plans outline national aspirations and policies, discuss intended actions, set targets, and review achievements. Few countries showed much interest in the environment prior to the late 1960s, and Malaysia was no exception. The new nation inherited what appeared to be an unlimited supply of lands and forests, and resource exploitation proceeded apace, with a heavy emphasis on land development. Economic growth was the overriding priority, and environmental damage was probably viewed as 'unfortunate' or an unavoidable by-product of 'progress'. It is therefore hardly surprising to find virtually no mention of environmental concerns in the early national plans (Malaya 1956, 1961; Malaysia 1966, 1969, 1971, 1973). By the early 1970s, however, the highly destructive impact of resource exploitation could no longer be ignored.

The *Third Malaysia Plan 1976–1980* not only broke with tradition but also offered great hope to environmentalists (Malaysia 1976). Unlike all earlier plans, this one devoted a whole chapter to the theme of environmental protection and management, and it declared that the objectives of development ought not to be 'negated by the costs of environmental damage' (p. 218; see B. H. Kiew 1982*b*). Recognizing that there was a need to protect a representative sample of the country's natural systems, the plan called for a total of 23 new national parks, nature reserves, nature monuments, and wildlife sanctuaries. This was indeed encouraging, although forests and forestry were given rather short shrift.

The fourth plan came as something of a shock to nature lovers, concerned citizens, and NGOs, because it devoted only three short paragraphs to environmental matters and allocated the paltry sum of M $13 million (spread over five years) to wildlife management and conservation activities (Malaysia 1981, 1984). The plan promised two new national parks for the Peninsula (neither of which has been created). Again, forests and forestry received only passing mention (see Environmental Protection Society Malaysia 1981).

The fifth plan (for the period 1986–90) reinstated a chapter on the environment, in which it was noted that environmental standards will be made consistent with the 'development goals of the country rather than [with] the high environmental quality standards of the industrialized countries' (Malaysia 1986, p. 289). Does this reflect a waning of federal interest in environmental matters? What does it imply in the light of the government's own policy of fostering rapid industrialization? It would appear to mean that certain standards will be 'relaxed'.

The fifth plan makes no mention of new parks and reserves, no doubt because most earlier promises have gone by the board. There are, however, certain new things in the plan; for example: it recognizes the need for greater cooperation with other ASEAN nations on common environmental concerns; it places greater stress on preventive rather than on curative environmental measures; it gives some attention to environmental impact assessment procedures; and it acknowledges the role of public awareness in environmental affairs. On the other hand, less than one of the massive plan's 549 pages is devoted to forests and forest conservation and forests are barely mentioned in the chapter on the environment.

None of the national plans makes any attempt to set development within an environmental context, although some attention is given to spatial considerations. The plans give the distinct impression that development and environment are two discrete entities, whereas in reality they are inextricably connected. The two plans that include chapters on the environment tend to reinforce rather than to dilute the impression of separateness, because they look for all the world like afterthoughts.

5.3.1.2 Legislation

Like several other ASEAN countries, Malaysia has sought solutions to environmental degradation through legislation and environmental law (see McDowell 1989). While this is by no means unimportant, it should be kept in mind that most of the causes of environmental ills are essentially economic and structural. We focus here on three pieces of legislation that bear on forests and conservation: the 1972 *Protection of Wild Life Act* (Act No. 76), the 1980 *National Parks Act* (Act No. 226), and the 1984 *National Forestry Act* (Act No. 313). We mention in passing another important piece of legislation, namely the 1974 *Environmental Quality Act* (Act No. 127). This Act was introduced to control various types of pollution (see note 12). It is not directly concerned with habitat protection (see Aiken *et al*. 1982, pp. 265–6).

The human impact on the country's fauna increased dramatically during the early years of independence (see Chapter 4). The Malayan Nature Society drew attention to the plight of a growing number of species, but to no avail (Wyatt-Smith and Wycherley 1961). Eventually the government recognized that the earlier *Wild Animals and Birds Protection Ordinance* (Ordinance No. 2 of 1955) had outlived its usefulness, and new legislation was introduced. Whereas the 1955 legislation provided protection for only 15 mammal species, the new 1972 law placed 34 mammal species on the 'totally protected' list and a further 35 on the 'protected' list; in addition, 465 totally protected and 60 protected bird species were listed (see Aiken *et al*. 1982, appendix B).

While admirable in its intention, the 1972 Act has proved difficult to enforce. Illegal hunting is widespread, and that there is insufficient manpower to bring poachers to heel is suggested by reports of protected species turning up in food-stalls, restaurants, and pet-shops (e.g. *New Straits Times*, 21 August 1986). We suggest that 'companion legislation' is urgently needed to protect a rapidly growing list of wild plants.

The 1980 *National Parks Act* established a legislative framework for the creation and management of this kind of conservation area. The Act empowers the appropriate federal minis-

ter to *request* the reservation of any state land (including marine areas) for the purpose of a national park. Such reservation cannot subsequently be revoked by the state authority without the concurrence of the federal minister. It should be noted that the Act does not apply to Taman Negara, the Peninsula's only national park, nor to Sabah and Sarawak (East Malaysia). So far, not a single park has been established under the Act, for the simple reason that the states are reluctant to surrender their jurisdiction over land to the federal government. It is essentially for this reason that the proposed Endau–Rompin national park has not been established (see below).

The long-awaited *National Forestry Act* need not detain us here, for we have discussed it elsewhere (see Chapter 3). We reiterate that the Act is not enforceable, because control over forests rests with the states.

Vacillation

A vacillating and somewhat waning concern appears to have characterized the federal government's stand on environmental affairs since the late 1970s. Much inconsistency is apparent: for example, the need for conservation and nature protection was clearly expressed in the third national plan, whereas these matters were virtually ignored in the fourth and fifth plans; logging of the core area of the proposed Endau–Rompin national park was openly criticized in the late 1970s, whereas to date, there appears to have been no attempt to halt the ongoing rape of Sarawak's forests or to intercede on behalf of the beleaguered Penan (see Sarawak Study Group 1989); and legislative measures have been adopted to control environmental degradation, whereas the main causes of such degradation are the very economic policies—e.g. rapid land development—that the government continues to pursue.

The federal government's half-hearted response to environmental deterioration has obviously been dictated by its economic goals, including the need to attract foreign investment. Economic growth is the overriding priority, and, as in most countries, only lip service has been paid to the needs of conservation. Most initiatives, to date, have been merely palliative, and it is clear that environmental considerations will not be allowed to stand in the way of the real task at hand, which is more growth. More emphasis in recent years has been placed on heavy industry and on privatization, and it is most unlikely that this trend will leave much scope for environmental considerations.

In addition to its half-heartedness, there are three reasons why the federal government has been reluctant to intervene in the environmental affairs of the states: firstly, because lands, forests, and other resources come under state, not federal, jurisdiction; secondly, because federal authorities realize considerable revenue from taxes, duties, and other charges on the export of raw materials (e.g. timber, oil, gas); and thirdly, because the central government and virtually all of the state governments are dominated by the *same* political party. Although it is of considerable importance, this third factor has been largely ignored in the environmental literature on Malaysia. Open federal intervention in the Endau–Rompin dispute (see below) was a rare case of washing dirty political linen in public (see Shafruddin 1987, chapter 9).

5.3.2 NGOs and environmental awareness

Third World NGOs with environmental interests have proliferated in recent decades (Myers and Myers 1983). Notable examples of such organizations in Malaysia (see Woon and Lim 1990) include the Consumers' Association of Penang, the Environmental Protection Society Malaysia, Sahabat Alam Malaysia (Friends of the Earth Malaysia, generally known as SAM), and the Malayan Nature Society (MNS).

The MNS is the country's premier conservation organization. Formed in 1940 by a small group of nature lovers (see Edgar and Shebbeare 1961), it now has close to 3000 members. The MNS publishes the *Malayan Nature Journal*, in which it has, among other things, called for rational conservation policies, including the creation of a comprehensive network of parks and reserves

(Malayan Nature Society 1971, 1974). The Society's other activities have included extending membership to schools and universities, releasing press statements on environmental issues, and mobilizing public opposition to further forest destruction. The MNS possesses two very important resources: knowledge and the strength of its ideas and beliefs. The emergence of other NGOs since the late 1960s has reflected the growing environmental awareness of Malaysia's expanding middle class. SAM in particular has won considerable popular support.

All of the previously mentioned NGOs are urban-based and all have originated in the most dynamic and developed region of the nation, namely the west coast lowlands of the Peninsula. This is by no means a coincidence because it is here to a greater extent than elsewhere that the human impact on nature is most apparent, it is here that ways of life are more modern, complex, and functionally specialized, it is here that frustration with government policies (or the lack of them) on the environment is most intense, and it is here that demands for greater public participation in decisions affecting the environment have been voiced the loudest.

The federal government will attempt to promote greater environmental awareness on the part of the public, public interest-groups, and the private sector according to the fifth national plan, which also noted that 'citizen groups increasingly played a role in the process of creating greater environmental awareness' (Malaysia 1986, pp. 279 and 281). Yet there is abundant evidence that the government is highly suspicious of the activities of such groups. Here are two cases in point. Firstly, in 1981 the government introduced amendments to the Societies Act, one of which, had it been fully implemented, would have effectively forbidden criticism of its policies by environmental organizations. Although certain of the amendments were subsequently modified, NGOs had clearly been warned against 'interfering' in government policies (see especially Gurmit Singh 1984; also Das 1982a,b; Segal 1983; Barraclough 1984). Secondly, in 1987 Harrison Ngau, a member of

SAM and an activist in the Baram–Limbang logging dispute (see below), was imprisoned without trial under the Internal Security Act and subsequently placed under house arrest (Hanbury-Tenison 1990).

Among other factors that have conspired to restrict the influence of NGOs are weak opposition to the ruling Barisan Nasional, restrictions on the activities of the media, trade unions, and student bodies, absence of elected local government, and the subordinate role of women and women's organizations. Two experts on the Malaysian political scene have noted that NGOs play a minimal role in the process of passing legislation and that they

have few opportunities to state their views on proposed laws through consultation. It is only after a law has been passed that the groups can agitate for amendments, or at least for favorable implementation (Milne and Mauzy 1986, p. 98; see also Suhaini Aznam 1987*a,b*).

Among the 93 persons detained in late 1987 under the Internal Security Act were members of SAM and the vice-president of the Environmental Protection Society Malaysia (Suhaini Aznam 1987*c,d*; McDowell 1989, p. 321). Environmental groups are probably viewed with even greater suspicion by most state governments, and in this regard Sarawak would appear to lead the way (see *Borneo Post*, 17 July 1990).

Although variously hampered and restricted, Malaysia's NGOs have managed to play a vitally important role in environmental affairs, and it is greatly to their credit that they have been able to win considerable popular support. Now well-organized and sophisticated groups, their achievements include creating public awareness of environmental degradation, fostering at least some political support for greater action on the environment, assembling, interpreting, and distributing environmental information (including SAM's important 'State of the Malaysian Environment' reports), and playing a valuable watch-dog role in the environmental arena, which has entailed, among other things, pressing the government to enforce its own environmental regulations.

5.3.3   Land-use conflicts: forests, logging, and dams

Several land-use disputes have flared up in Malaysia since the late 1970s. Four in particular—two involving logging, two relating to proposed dam construction—have attracted much public interest and media coverage. We discuss the four disputes in this section, where they are referred to, in shorthand, as the Endau–Rompin, Baram–Limbang, Sungai Tembeling, and Bakun controversies. Although distinct in character and widely separated from one another, the locales of the four disputes share certain characteristics: all are largely wilderness areas, all are valued by the peoples living in or near them, and all are in less developed parts of the nation—two in the eastern region of the Peninsula, two in still largely undeveloped Sarawak.

Endau–Rompin

There are only four extensive tracts of lowland rain forest remaining in the Peninsula: in northern Kelantan, in part of Taman Negara, in the Kerau Wildlife Reserve, and in the Endau–Rompin region. Situated astride the boundary between Pahang and Johor and long recognized as an important conservation area, Endau–Rompin is one of the last refuges of several endangered species of animals, including the very rare two-horned Sumatran rhinoceros (Stevens 1968; Flynn and Mohd. Tajuddin Abdullah 1984). The wilderness region possesses great potential for recreation and tourism, provides scientists with a vast outdoor laboratory of inexhaustible interest, and sustains several communities of Orang Asli.

An agreement reached in 1972–3 between the federal government and the state governments of Pahang and Johor called for the creation of a 202 343 ha national park in the Endau–Rompin region. Within the proposed park area, logging would be restricted to an extensive buffer zone, leaving an inviolate core area (some of which had been logged under licences issued in the 1960s) of some   91 365 ha—52 610 ha   in   Johor   and 39 355 ha in Pahang (Furtado 1978). Although the third national plan called for the federal-state agreement to be implemented, the park was not gazetted (Malaysia 1976, p. 225).

In March 1977, the Pahang state government approved the setting aside of 12 140 ha of the core area of the proposed park as a logging concession, claiming that it had not made any formal commitment to the idea of a national park (see Wain 1978). The result of this action was the first major environmental controversy in Malaysian history.

On 7 May 1977 the MNS (see above) and five other societies placed a half-page advertisement in the *New Straits Times* outlining the nature of the Endau–Rompin region and why it should be protected. There then followed a rash of letters to the editor, feature articles on the issue, and editorials in favour of saving the region. When logging failed to cease, the MNS, acting as an umbrella organization, launched a campaign to save Endau–Rompin, adopting the threatened Sumatran rhinoceros as its symbol (see Plate 14). The tactics adopted by the MNS and its supporters included letter-writing campaigns, organizing petitions, attracting media coverage, holding public meetings, and reporting on the values of the wilderness region (see Aiken and Leigh 1984, 1986a).

The Pahang government (Johor was not involved in the dispute) argued that economic development must take precedence over conservation, pointing out that revenue from logging licences would aid state development. Although the federal government was openly opposed to the logging, it was unwilling to invoke certain of its constitutional powers to take possession of the disputed area (see Aiken and Leigh 1986a, pp. 173–4 for details). There is evidence, however, that it did intervene at the level of intra-UMNO politics (both the central government and most state governments, including that of Pahang, have long been dominated by the same political party, the United Malays National Organization) and that considerable pressure was exerted on the state government to bring logging to a halt (see Shafruddin 1987, chapter 9). Eventually, the Pahang government agreed to stop all logging in the disputed area after August 1978, and it duly kept its word. A preliminary management plan for the proposed park was submitted to the (federal) Department of Wildlife and National Parks in 1980 (Flynn 1980), and in the same year the long-awaited *National Parks Act* (see above) was proclaimed. But the national park was not gazetted.

The next significant event in the continuing saga of the proposed park was the 'Malaysian Heritage and Scientific Expedition' to the forests of the Endau–Rompin region. Organized and planned by the MNS in co-operation with *The Star* newspaper, the expedition had three objectives: to document all encountered species of plants and animals, to provide government agencies, scientists, and other interested parties with readily available information about Endau–Rompin, and to involve young persons in field studies. More than 80 projects were conducted on the physical and biological characteristics of the wilderness region, while extensive media coverage brought a previously little-known part of the country to the attention of many Malaysians (see Barlow 1985; Tho and Lim 1985; Rubeli 1986a; B. H. Kiew 1986a,b, 1987; B. H. Kiew *et al.* 1987). The expedition was a popular success and a very considerable co-operative scientific achievement.

In mid-1987 the state governments of Pahang and Johor declared that they would not surrender jurisdiction over any of the wilderness region to the federal government, hence there would be no Endau–Rompin national park (*New Straits Times*, 11 and 22 June 1987). Instead, it was announced that a total of some 90 000 ha would be gazetted as two adjoining *state* parks (*New Straits Times*, 17 and 18 July 1987; Seaward 1987a). This is far from encouraging, because several state parks and reserves have been wholly or partially excised in the past, whereas a park gazetted under the *National Parks Act* cannot be revoked except with the concurrence of the federal minister. It was reported in mid-1990 that Pahang had gazetted part of the Endau–Rompin wilderness as a permanent forest reserve from which logging would be excluded and that it planned to gazette 40 200 ha of the rain forest as a state park (*New Straits Times*, 14 June 1990; see also Rubeli 1989). We suspect that the controversy is far from over.

Baram–Limbang

The main focus of logging activity in Sarawak has shifted from the peat-swamp forests of the lowlands to the dipterocarp-dominated hill forests of the interior, where it has met with considerable resistance from several indigenous peoples since the early 1980s. International attention was drawn to Sarawak in 1987, when the Penan and several other ethnic groups erected barricades across the logging roads in the Baram and Limbang districts in the northern part of the state. The story of the timber blockades has been told in considerable detail elsewhere, hence we focus on only a few main points (see Apin 1987; Hong 1987, chapter 7; Seaward 1987*b*; Colchester 1989, pp. 40–4).

Access to logging concessions in Sarawak is largely determined by political power and family connections. Evidence suggests that those who benefit most from the timber industry are the Malay–Melanu ruling élite and the mostly Chinese contractors who actually carry out the work, whereas the 'communities whose land is affected gain little, if anything, from logging, though they have to live with its often disastrous human and environmental consequences' (Jomo 1989, p. x; also *New Sunday Times*, 22 February 1987; Colchester 1989, pp. 34–6; Sarawak Study Group 1989).

The rapid expansion of logging has been facilitated by land and forest regulations that have progressively restricted the realm in which customary land rights can be exercised. Although the law recognizes such rights, it also includes provisions that make it relatively easy to dispossess the natives and to extinguish native customary tenure. While native groups have been greatly concerned over, and have protested against, the environmental destruction caused by advancing logging companies, the real fight has been for control of land, a matter that is of vital importance to indigenous peoples everywhere. It has also been

an ethnic and cultural fight, a struggle for recognition and a greater share of power (Scott 1988).

Native groups responded to logging in a variety of ways. For example, they sought monetary compensation from logging companies for destruction of fruit and nut trees, depletion of fish and game stocks, water and noise pollution, and other forms of environmental degradation; they made numerous applications for Communal Forest Reserves to meet their domestic needs (there were only 56 km^2 of such reserves in Sarawak in 1984); some communities applied for their own logging licences; and they attempted to garner a share of the profits of the logging companies (Sarawak Study Group 1989, pp. 10–16). It is important to note, then, that native groups have not been totally opposed to logging; what they want, in part, is a greater share of the profits of the industry and compensation for depleted or depreciated natural resources. It was largely because their various responses to changing circumstances were rebuffed by state authorities that they turned to direct action.

The timber blockades were initiated by the Penan, most of whom are concentrated in the Baram district. There are some 9000 Penan in all. Estimates suggest that about five per cent are still fully nomadic, that about 60 per cent are seminomadic, and that the rest are settled. Most of these people are heavily dependent on the rain forest for food, trade items, building materials, and many other needs. They have a profound affinity with the forest (Jayl Langub 1988). Among other mainly forest groups who quickly joined the Penan were representatives of the Kayan, Kenyah, and Kelabit peoples.

The blockades commenced in March 1987 in the Sungai Tutoh area of the upper Baram River, and the idea soon diffused to numerous other areas, mainly in the Baram and Limbang districts. It is believed that a sympathetic Swiss citizen, Bruno Manser, who had been living with the Penan since 1984, may have been involved in organizing the protests. Manser's presence in the rain forest focused a great deal of national and international media coverage on the blockades, and in Europe and Australia, where he has been something of a *cause célèbre*, his supporters roused interest in the plight of the Penan and led demonstrations aimed at persuading governments to ban timber imports from Sarawak. The federal and Sarawak governments countered by portraying Manser—he secretly departed from Sarawak in early 1990 (*Age*, 14 April 1990)—as an extremist for whom there was little local support.

The protectors won widespread public support both at home and abroad. Among their allies were SAM, which attempted to forge an alliance between environmentalists and native groups, various social and religious groups in the Peninsula, and, at the international level, organizations like the World Wildlife Fund, Greenpeace, the Rainforest Information Centre Australia, and Survival International. A person who played a key role in the protest campaign was Harrison Ngau, a local Kayan, who ran SAM's office in Marudi and whose other activities included organizing an attention-catching visit to Kuala Lumpur by a delegation of native leaders (*New Straits Times*, 13 and 16 June 1987; Seaward 1987*b*). He was, as we have already noted, subsequently imprisoned.

Blockades or other kinds of protests have persisted, and it is impossible to present all the details here. Instead, we close with a few comments on the kinds of responses that the dispute has elicited. Probably because it has been much embarrassed by so much media coverage, the main response of the Sarawak government has been one of defiance: it has pulled down the barricades and arrested numerous protesters. Logging will continue. As for the federal government, it has been unwilling to intervene in what it considers a state matter. All three state coalition parties are members of the ruling federal coalition, the Barisan Nasional, and this has undoubtedly influenced the central government's decision not to intervene in the dispute (Suhaini Aznam 1987*e*).[10] The international response included rallies and protests in many cities around the world and a continuing campaign, most notably in Australia and in several Euro-

pean countries, to ban imports of Malaysian (and other) tropical hardwoods (*New Straits Times*, 22 March 1987; *New Sunday Times*, 25 October 1987; *Age*, 13 September 1989).

The federal government's main response to adverse international publicity was to invite the International Tropical Timber Council to undertake a study of forest management in Sarawak. As a result, the International Tropical Timber Organization (ITTO) established a mission headed by the Earl of Cranbrook

to assess the sustainable utilization and conservation of tropical forests and their genetic resources as well as the maintenance of the ecological balance in Sarawak ... and to make recommendations for the further strengthening of sustainable forest management policies and practices, including areas of international cooperation and assistance (ITTO 1990, p. vii).

The mission visited Sarawak in late 1989 and again in early 1990. The question of native land rights was considered an 'awkward difficulty' by the mission and, strictly speaking, none of its business. However, it was accepted that this matter could very well be germane to any assessment of sustainable forest utilization in the state.

That ITTO was chosen to undertake such a study has been greeted with scepticism in some quarters. Essentially a trade 'club' of nations that produce and consume tropical timber, the Yokohama-based organization has the dual role of regulating the timber trade and protecting the forests. It is the view of many conservationists that such duties are antithetical (Cross 1990; Pearce 1990, p. 24). Critics have claimed that the mission's brief ought to have included the question of native land rights, and the fact that no social scientist was appointed to the mission has raised concerns (Pearce 1990).

In the mission's report, which was submitted in May 1990, it was noted that sustainable forest management has been partially achieved and that certain 'admirable features' were aready in place, including

an established Permanent Forest Estate, a Forest Policy which sets watershed protection as a primary objective, the reservation of areas for the protection of biological diversity, an effective system for tracing and controlling the movement of logs, comprehensive management planning for the production forest and a data bank of research information unequalled in the tropics (ITTO 1990, p. vii).

But on the other hand the mission found, among other things, that the hill dipterocarp forests were being overcut, that inadequate catchment management was resulting in gross siltation of many rivers, that there was insufficient staff in the Forest Department to supervise and monitor logging operations, that the training of logging operators left much to be desired, that the existing set of totally protected areas, though impressive, was 'still insufficient to protect the full range of habitats and biological diversity in the State' (p. vii), and that, in order to facilitate sustainable management, an assessment of forest resources should be made in relation to long-term prospects for trade in forest products. The mission also found that logging practices result in widespread environmental disruption and excessive damage to residual trees, with the result that recent levels of production (see Table 3.2) cannot be maintained.

The mission made recommendations (pp. 71–3), including the following: the staff of the Forest Department should be comprehensively strengthened, slopes greater than 60 per cent should not be logged, catchment protection must be improved, and the rate of logging should be reduced to a sustainable yield of 9.2 million m^3 per year. As for native land rights, the mission opined that a simple solution would be to suspend timber harvesting in disputed areas until legal issues had been resolved (pp. 8–9).

Conservationists think that the mission should have given greater attention to the issue of native land rights, and in their view a sustainable yield of 9.2 million m^3 per year is unattainable because the ideal conditions required for such a level of production are unlikely to materialize; they would, in short, prefer a much lower figure (Cross 1990). Meanwhile logging continues unabated and there is no indication that the Sarawak government is

prepared to reduce the annual harvest by any significant amount.

Protests and blockades also continue, and it is possible that native resistance will be further galvanized by Harrison Ngau's election, in October 1990, to the federal parliament. Ngau successfully campaigned against the Barisan Nasional's timber company-supported candidate, the deputy Minister of Public Works, on the issue of indigenous peoples' rights (Hanbury-Tenison 1990, p. 13).

Sungai Tembeling

Taman Negara is the only national park in the Peninsula. Covering an area of 4343 km^2 and originally called King George V National Park, it was created in 1938–9 from parts of the three adjoining states of Pahang, Terengganu, and Kelantan. The outstanding biological, scientific, recreational, and other values of the conservation area have been widely recognized, yet parts of the park have periodically been threatened by proposed logging, road building, and dam construction schemes (Rubeli 1976, 1986*b*; Rachagan 1983; Anon. 1986*a*). In 1982 the National Electricity Board (NEB) revived a scheme (first announced in 1971 but shelved in 1978) to build a power-generating and flood-mitigation dam on the Sungai (River) Tembeling, which forms the south-eastern boundary of Taman Negara. This decision led to widespread public protest.

The Tembeling is an upper tributary of the powerful Sungai Pahang, the Peninsula's longest river. It is the only means of travel for the 12 000 or so inhabitants of the 50 or more kampungs (villages) in its vicinity, and it provides the only access to the national park. The river valley is rich in archaeological sites, and the forests of the region have long been home to groups of Orang Asli. As might be expected, most of Taman Negara's species are concentrated in the lowlands, parts of which are traversed by the Tembeling.

Building on the experience they had gained during the Endau–Rompin dispute (see above), conservation groups mounted a vigorous campaign against the NEB's project, which they considered unwarranted on several scores: parts of the species-rich lowlands of Taman Negara would be flooded, thereby threatening scores of plants and animals (including the already endangered seladang and the tiger); projected logging of the reservoir area would entail constructing roads into and within the area, so damage would not be confined to the immediate site of the dam and the reservoir; the park headquarters would have to be moved, as too would some 3000–5000 kampung-dwellers and several groups of Orang Asli; and archaeological sites would be destroyed.

Following an announcement in May 1982 that the government definitely planned to go ahead with the dam, a number of societies and associations mounted various types of protest campaigns, including the distribution of posters, T-shirts, and car-stickers, a postcard campaign aimed at the Prime Minister, and the organization of a nation-wide petition. At the instigation of the MNS, the Tembeling Co-ordinating Committee was formed and it was this body that drafted the Tembeling Declaration:

We oppose the proposed Tembeling hydroelectric dam as it will undermine the viability of Taman Negara and will have adverse environmental impacts both on the people and the wildlife in its immediate vicinity. We call upon the Federal Government to permanently abandon this proposal and urge all Malaysians to join our campaign to prevent Taman Negara from becoming Taman Tenaga ['Power Park'] (Anon. 1982, p. 2).

The declaration was immediately signed by the three leading conservation groups in the country—the MNS, SAM, and the Environmental Protection Society Malaysia—as well as by seven other groups, including the Malaysian Trade Union Congress. The protest campaign continued during the latter half of 1982, and by the end of the year there were some 25 000 signatures on a petition against the dam (Aiken and Leigh 1983, 1986b).

It was announced in January 1983 that the government had decided to abandon the project completely, and it was conceded—probably largely as a consequence of the strength of public opposition—that environmental considerations had played a part in this decision. It should be noted, however, and this is not meant to detract from the sterling efforts of NGOs, that none of the states was involved in the dispute and that power generation in the Peninsula is a federal responsibility. In short, this was not a particularly complex dispute. Since 1983 a number of other dams have been built in rain-forest areas, including the Kenyir Dam in Terengganu and the Batang Ai Dam in Sarawak.

Bakun

A second controversy, also involving a dam, erupted in the mid-1980s. This one was over the Bakun Hydroelectric Project, which called for the construction of a power-generating dam at the Bakun rapids on the Sungai Balui, in the upper Sungai Rajang Basin, some 37 km upstream from the town of Belaga in Sarawak's Seventh Division.

Virtually everything about the scheme was on a massive scale, including its 204-m high dam (almost twice the height of the Aswan Dam), its reservoir covering an area of 695 km² (larger than Singapore Island), its maximum generating capacity of 2400 MW (only a little less than the Peninsula's required generating capacity in 1985), its two 650-km long undersea cables running between Sarawak and the Peninsula (longer than any in the world), and its cost of some M $8–10 billion. It would be the largest dam in South-east Asia. Less impressive was the dam's projected life-span of only 30–50 years (Fernandez 1985; Mohun and Omar Sattaur 1987). Like other dams of its type, Bakun was part of the government's energy diversification programme, which aims to reduce the nation's dependence on oil for power generation (Malaysia 1981, p. 339, 1986, pp. 453–4). With its many fast-flowing rivers, Sarawak was viewed as having great potential for hydropower development. Although long-term experience with large dams is still limited, the available evidence suggests that most have been fraught with numerous social and environmental problems, perhaps none more so than those that have been built in areas of tropical rain forest (Petr 1978; Goldman 1979; Caufield 1983; Monosowski 1983).

On the basis of nine feasibility studies (none of which was made available in full to the public), the government declared the giant project safe, technically and economically viable, necessary, and financially beneficial to the whole nation; furthermore, in addition to producing energy for industrial development and domestic consumption, it was claimed that the scheme would yield other benefits, including a reservoir fishery, opportunities for tourism, employment, technological transfer, improved upriver navigation, and flood control (*New Straits Times*, 5, 19, 20 March and 4, 5 April 1986; *Borneo Post*, 11 February 1987; Mohun and Omar Sattaur 1987, p. 38). On the other hand, opponents of the dam argued that these largely hypothetical benefits would be more than offset by the project's numerous deleterious social and environmental consequences and that public opinion had been entirely ignored by the decision-makers.

Opponents of the giant project included the Environmental Protection Society Malaysia, SAM, the Consumers' Association of Sarawak, certain Sarawakian politicians, and many local people; in addition, the federal Minister of Science, Technology and the Environment expressed concern over the impact of the project on local communities, noting that it 'will be difficult to resettle them. We need to heed their wishes' (cited in Colchester 1989, p. 59; Anon. 1986*b*; Gurmit Singh 1986; Nicholas 1986; Sim 1986). For the most part, however, the central government showed rather little sympathy for the views of those in the vicinity of the project site (see Hong 1987, chapter 11).

The Bakun Dam would have displaced more than 4000 people and inundated a vast area of mainly primary rain forest along with ancestral lands, crops, fruit trees, and burial grounds. Press reports revealed that many Kayan, Kenyah, and persons from several other ethnic groups were opposed to the dam, and a memorandum bearing the signatures of 2000 opponents of the project was dispatched to the Prime Minister in early 1986. Meanwhile, various public-interest groups continued to pressure the government to abandon the scheme, environmentalists noting, among other things, that the planned accelerated logging of the reservoir site would cause soil erosion and river sedimentation and that flooding of the site would have a disastrous impact on wildlife. Government promises of resettlement and compensation were greeted with scepticism because many of the locals were aware of the hardships that Iban communities had encountered when they were displaced by the Batang Ai Dam (completed in 1985, the dam dislodged some 2800 people from 26 longhouses; see *New Straits Times*, 21 June 1984; Hong 1987, pp. 170–9).

It was announced in May 1990 that the federal cabinet had decided not to go ahead with the Bakun Dam, opting instead for a series of smaller dams along the Sungai Rajang. Environmentalists disputed the government's claim that the new scheme would have little impact on the environment: it was, according to the leader of one NGO, merely 'a clever way of camouflaging the original Bakun proposal . . . [and] does not seem to change much from the old plan in terms of its possible impact on the natives' (*Borneo Post*, 29 May 1990; *New Straits Times*, 7 July 1990).

The disputes: a concluding note

Wilderness areas are fast disappearing, competition for natural resources is increasing, and awareness of the benefits that societies derive from nature protection is growing. Under these circumstances, it is perhaps not surprising that land-use conflicts have flared up. State and federal politicians and government authorities now know that public opinion and the tactics of pressure groups are forces to be reckoned with. It is clear that concerned citizens and interest groups now wish to play a more active role in environmental decision-making. Presently, however, authoritative institutions are a good deal stronger than participative ones. Recent disputes have served to heighten ethnic consciousness, a trend that runs counter to what the government wants.

There is a need to view dam projects not only in narrow technical–economic terms but also in relation to such matters as environmental quality, social well-being, citizen participation in decision-making, consideration of alternative schemes, and interlocking planning for soil, vegetation, and water—preferably at the scale of drainage basins. Protection of the remaining rain forest is greatly hampered by the heavy reliance that certain states place upon timber as a source of revenue and by the constitutional division of powers between the centre and the states. On these matters two Malaysian geographers have made the following astute observation:

Sound management of the forests may be possible only if alternative sources of revenue for the states is found and if these sources of revenue are made conditional on sound forest management. Alternatively, the Federal Government will have to rely on the Constitutional provisions which allow it to acquire state land for federal purposes. This will call for conservation to be accepted by the courts as a legitimate federal purpose (Rachagan and Bahrin 1983, p. 502).

5.4 RECENT DEVELOPMENTS: TOO LITTLE, TOO LATE?

Much has been done in recent years to protect the environment and to conserve the country's flora and fauna. Whether this has been too little, too late, remains something of a moot question because, as we note in the concluding section, several 'transitions' will be required to put Malaysia on a path towards a sustainable future. In this section we describe what has been done during the national period to protect areas of rain forest in parks and other kinds of reserves and we take a brief look at certain other kinds of approaches to conservation.

5.4.1 Parks, reserves, and sanctuaries

Peninsular Malaysia

The parks and wildlife reserves that were present in the Peninsula shortly after the end of the colonial period are listed in Table 5.1 and those in place as of c.1990 are shown in Fig. 5.1 and listed in Table 5.2.

The lack of progress is disappointing. Only two wildlife reserves have been created on the mainland since independence (1957), namely Cameron Highlands (1962) and Sungai Dusun (1964), the latter to protect the Sumatran rhinos that were known to be in the vicinity (Mohd. Khan 1988, p. 268); in addition, Pulau Tioman was established in 1972 (see Fig. 5.1) and there are to be two state parks in the Endau–Rompin region of Pahang (see above). But there have also been losses, because the wildlife reserves at Chior in Perak and at Segamat in Johor (see Table 5.1) no longer exist (Stevens 1968, p. 42; Musa bin Nordin 1983, p. 113). In short, as far as the area devoted to parks and reserves is concerned, the net increase to date has been rather modest. Indeed, the situation is probably worse than before, because the Peninsula's protected areas increasingly resemble islands in an extensive sea of development.

There have been several calls for the creation of more parks and reserves. In 1968 a visiting American, B. E. Weber, recommended the adoption of a parks system that would comprise national parks, national nature monuments, and national historic sites, and to this list Stevens, a visiting Canadian, added wildlife reserves and wildlife sanctuaries (Stevens 1968, pp. 80–1). In his 1968 report to the Game Department, Stevens recommended the creation of 13 new wildlife reserves and sanctuaries and, as suggested by Weber, six national nature monuments. This proposal was subsequently developed by the Malayan Nature Society (1974), whose 'Blueprint for Conservation' identified 64 sites in the Peninsula (including some already protected) that were considered worthy of some form of protection.

Stevens's list was included with only minor modifications in the *Third Malaysia Plan* (Malaysia 1976, table 11-3, p. 225), which called for two new national parks, Endau–Rompin (Pahang/Johor) and Gunung Blumut (Johor), and for large wildlife reserves at Belum and Gerik (Perak), Tasek Bera and Tasek Cini (Pahang), Ulu Muda (Kedah), Sungai Nenggiri (Kelantan), and in Ulu Terengganu. Two new national parks, Endau–Rompin and Kuala Koh

(Kelantan), were promised in the *Fourth Malaysia Plan* (Malaysia 1981, p. 391). No promises of new parks appeared in the *Fifth Malaysia Plan* (Malaysia 1986).

At the time of writing, only part of the Endau–

Rompin wilderness and some of the smaller proposed sites had been afforded protection (e.g. Batu Caves in Selangor). There are probably two reasons why the larger areas have not been protected: firstly, because the states are loath to

Table 5.2 National parks, wildlife reserves, and sanctuaries in Malaysia, *c*.1990

Region/state	Name	Area (ha)[1]
Peninsular Malaysia		
Perak	Sungkai Wildlife Reserve	2428
	Batu Gajah Bird Sanctuary	5
Pahang	Kerau Wildlife Reserve	53 095
	Cameron Highlands Wildlife Reserve	64 953
	Pahang Tua Bird Sanctuary	1336
	Pulau Tioman Wildlife Reserve	7160
Johor	Endau-Kluang Wildlife Reserve	101 174
	Endau-Kota Tinggi Wildlife Reserve (West)[2] [3]	61 959
	Endau-Kota Tinggi Wildlife Reserve (East)	7413
	Pulau Lima Bird Sanctuary	5
Negeri Sembilan	Port Dickson Islands Bird Sanctuary[4]	0.5
Selangor[5]	Fraser's Hill Bird Sanctuary[6]	2979
	Kuala Selangor Wildlife Reserve[7]	44
	Bukit Kutu Wildlife Reserve	1943
	Klang Gates Wildlife Reserve	130
	Kuala Lumpur Golf Course[8]	403
	Templer Park[9]	1011
	Sungai Dusun Wildlife Reserve	4330
Pahang/Kelantan/Terengganu	Taman Negara	434 351
Regional total		744 719.5
East Malaysia		
Sabah[10]	Kinabalu Park	75 370
	Tunku Abdul Rahman Park	1289
	Pulau Penyu (or Turtle Islands) Park	15 [11]
	Pulau Tiga Park	607
	Tawau Hills Park	27 792
	Crocker Range National Park[12]	139 919
	Tabin Wildlife Reserve	120 521
	Kulamba Wildlife Reserve	20 682
Total for Sabah		386 375

Region/state	Name	Area (ha)[1]
Sarawak	Bako National Park	2730
	Similajau National Park	7070
	Niah National Park	3140
	Lambir Hills National Park	6950
	Gunung Mulu National Park	52 890
	Gunung Gading National Park	5430
	Kubah National Park	2316
	Batang Ai National Park[13]	27 060
	Samunsam Wildlife Sanctuary	6040
	Lanjak-Entimau Wildlife Sanctuary	168 750
	Pulau Tukong Ara-Banun Wildlife Sanctuary	1.4
Total for Sarawak		282 377.4
Regional total		688 752.4
Grand total		1 413 471.9

1. Reported figures vary somewhat; a few have probably been rounded.
2. Includes Gunung Blumut (see Fig. 5.1).
3. Available evidence suggests that these are primarily forest reserves and that their value as conservation areas has been severely reduced, hence they are not shown in Fig. 5.1.
4. There are three islands: Pulau Burong, Pulau Babi, and Pulau Perjudi.
5. Other areas that receive some protection include Bukit Sungai Puteh Forest Reserve (about 40 ha), Bukut Nanas Forest Reserve (about 9 ha), and Batu Caves.
6. Variously called a Bird Sanctuary, a Wildlife Reserve, or a Wildlife Sanctuary. The reported area excludes the small portion of the reserve that extends into the neighbouring state of Pahang.
7. Also called a Hill Reserve.
8. See Table 5.1.
9. Primarily used for forest recreation; reported area varies widely.
10. An enactment of 1984 reconstituted the then existing 'national' parks as state parks.
11. Land areas only; the sea areas (in ha) are 3640, 1725, and 15 257, respectively.
12. Whether the word 'National' should be retained in the official name of this park is not clear from available evidence.
13. Created after this book went to press, hence it is not shown in Fig. 5.1.
NB Even leading authorities appear to differ over such matters as the size, status, and official names of some of the conservation areas; consequently, this table (and Fig. 5.1) should be read with caution.
Sources: Anon. (1990, appendix V, pp. 37–9) is the main source; see also the sources for Fig. 5.1, above.

'tie up' large tracts of forest that are likely to prove a future source of timber revenue; and secondly, because the federal government has been wary of antagonizing the states by relying on certain of its constitutional powers in order to acquire state land for federal purposes. A priority of the federal and state governments should surely be to protect the eight large areas that have been identified by the Director-General of Wildlife and National Parks as desirable for new parks and wildlife reserves (see Table 5.3).

A positive development has been the creation of more Virgin Jungle Reserves (VJRs; see pp. 112–13). In 1990, there were 71 VJRs in the Peninsula, compared with 54 in 1960. For the most part, however, the reserves are relatively small: many are less than 200 ha, with some only 40–80 ha (Anon. 1990, appendix VI).

Sabah and Sarawak
Much more has been achieved in Sarawak. The two areas that were identified as potential parks by

Table 5.3 Peninsular Malaysia: proposed parks and wildlife reserves

Site	Approximate area (ha)	Comments
Ulu Muda, Kedah	115 200	To conserve large mammals that concentrate at salt licks in the Muda River area.
Gerik, Perak	54 400	Probably the richest wildlife area in the Peninsula. Suffers from poaching.
Kuala Gula, Perak	890	A bird sanctuary in an area of mangrove.
Belum, Perak	214 900	Rich wildlife and an area of scenic beauty. A potential national park.
Ulu Selama, Perak	23 300	Sanctuary for Sumatran rhinoceros and other large mammals.
Tasek Bera and Tasek Cini, Pahang	33 184 4856	Natural freshwater lakes rich in fish and birdlife.
Ulu Terengganu	104 400	Reserve for tiger, elephant, and seladang. Includes the Kenyir Dam.
Nenggiri, Kelantan	36 800	Would protect seladang and smaller animals.

Source: Mohd. Khan (1988, pp. 269–71).

the colonial authorities, Niah Caves and Gunung Mulu, were gazetted in 1974 and since then five other national parks have been established: Lambir Hills, Similajau, Gunung Gading, Kubah, and Batang Ai; in addition, three wildlife sanctuaries have been established: Lanjak–Entimau (mainly for orang-utan protection), Samunsam (mainly for the protection of the proboscis monkey), and Pulau Tukong Ara-Banun (see Fig. 5.1 and Table 5.2 this volume; Aken and Kavanagh n.d., p. 18). The World Wide Fund for Nature Malaysia has recommended that four additional parks and an additional sanctuary should be established and that two of the existing sanctuaries should be extended (Table 5.4; see Colchester 1989, p. 63), and it was announced in June 1990 that there are plans to establish a 500 000 ha 'trans-frontier' park along the Sarawak–Kalimantan border in the Kapit–Sri Aman–Sarikei area (*New Straits Times*, 16 June 1990). If these additional parks and sanctuaries are established, Sarawak will have an

excellent array of protected areas. So far no VJRs have been established in Sarawak.

The process of establishing parks and reserves has moved more slowly in Sabah. Since 1963, when the state's first national park was established at Kinabalu, several other conservation areas, including the extensive Crocker Range National Park and two wildlife reserves, Tabin and Kulamba, have been created (Fig. 5.1 and Table 5.2). However, certain of the parks comprise groups of small offshore islands (they serve to protect birds and turtles), the Klias Park, which was established to protect mangroves, was degazetted in 1981, and 40 per cent of the Tawau Hills Park was logged immediately prior to its establishment in 1979 (Davies and Payne n.d., pp. 207–15).

But the scene is less bleak than it might at first appear because Sabah possesses several other important conservation areas. Notable among these are Virgin Jungle Reserves (VJRs). In 1990, Malaysia's 119 VJRs covered 109 571 ha, of

Table 5.4 Sarawak: proposed national parks and wildlife sanctuaries

Site	Area (ha)
National parks	
Santubong, First Division	2363
Tanjong Datu, First Division	1214
Loagan Bunut, Fourth Division	10 740
Pulong Tau, Fourth and Fifth Divisions	164 500
Wildlife sanctuaries	
Samunsam extension, First Division	14 810
Lanjak–Entimau extensions,	
Third and Seventh Divisions	18 414
Sibuti, Fourth Division	1212

Source: Colchester (1989, table 4, p. 63, after WWFM 1985).

which 48, accounting for 88 299 ha, were in Sabah (Anon. 1990, appendix VI). The VJRs embrace a variety of forest types and habitats and some are quite extensive, including the following: Maligan (9240 ha in montane forest), Pin-Supu (4696 ha in freshwater-swamp and lowland dipterocarp forest), Kabili Sepilok (4294 ha in lowland dipterocarp forest), Lungmanis (6735 ha in lowland dipterocarp forest), Sungai Imbak (18 113 ha in lowland ditperocarp to montane forest), and Sepilok (1235 ha in mangrove forest). In addition, the Danum Valley (214 500 ha), a former Sabah Foundation timber concession area, is now jointly managed for conservation purposes by the Foundation, the Forest Department, and the Parks Authority (*New Straits Times*, 9 October 1986). Davies and Payne (n.d., pp. 231–9) have recommended the establishment of four additional conservation areas: Silabukan, an area of lowland rain forest in eastern Sabah, to protect elephant and rhinoceros; Tanjung Linsang, a deltaic area on the east coast, to protect mangroves and freshwater-swamp forest; Kumambu in eastern Sabah to protect banteng; and Gunung Lotung to protect diverse montane and upper-hill dipterocarp forest and to include the mountain's spectacular sandstone ridge.

An obvious factor that must be borne in mind when evaluating the effectiveness of parks and reserves is their size: generally speaking, the bigger they are the better, especially if their purpose is to maintain a diverse range of animals and plants. Small parks and reserves, particularly those that are isolated or lack buffer zones, are often unable to sustain many species, although they can fulfil specific functions, such as protection of specimens or rare plant species. Davies and Payne (n.d., p. 218) have estimated that the minimum area of continuous forest required for the conservation of 200 adult animals is in the order of 50–400 km^2 for various species of hornbill, 150–200 km^2 for the orang-utan, 800 km^2 for the bear and the clouded leopard, and 6000 km^2 for the elephant and the Sumatran rhinoceros. Clearly, few of the conservation areas are of these dimensions. As far as trees are concerned, there is still no definitive answer to the question of how many individuals are required to maintain the genetic integrity of a species in the long run (Ashton 1976*b*; Franklin 1980). A '*very rough rule is 50 breeding individuals short-term, 500 long-term*' (Whitmore 1990, p. 193).

5.4.2 Planning and management

State conservation strategies

Conservation of Malaysia's rain-forest flora and fauna requires a holistic approach to policy formulation, planning, and management; ideally, this would transcend departmental boundaries and, where necessary, state boundaries. Such an approach is advocated by the World Wide Fund for Nature Malaysia (WWFNM), a non-governmental organization that over the past decade has prepared conservation strategies for most of the states. The WWFNM's approach is based on the three main principles of the *World Conservation Strategy*: maintenance of essential ecological processes, preservation of biological diversity, and sustainable human use of species and ecosystems (WWFNM 1990, p. 2). To date, strategies have been prepared for Negeri Sembilan (1982), Melaka (1983), Terengganu (1983), Kedah (1984), Perlis (1984), Sarawak (1985), Selangor (1988), and the Federal Territory of Kuala Lumpur (1989); one is being prepared for Kelantan. The strategies[11] contain recommendations relating to policy, priorities, species conservation, area conservation, and management. Prior to 1986, implementation of the strategies appears to have progressed very slowly; since then the WWFNM has worked closely with Perlis on state planning matters and Selangor and Kuala Lumpur 'have made practical use of their strategies' (WWFNM 1990, p. 8).

Integrated forest management

The conservation of Malaysia's flora and fauna requires two complementary approaches: creation of more parks and reserves (and greater security of tenure for those that exist) and multi-purpose management of the large areas of productive forest reserves that still exist. In light of the limited success of the former, conservationists are calling for greater emphasis on the latter.

As noted in Chapter 4, A. D. Johns (1983*a, b,* 1987) has suggested that the long-term survival of many wide-ranging species is more likely in extensive logged forests than it is in relatively small and (often) isolated parks and reserves (see also Mohd. Khan 1988, p. 259), and Davies and Payne (n.d., pp. 219–20) have suggested that areas of logged forest in excess of 80 000 ha that contain tracts of primary forest (each at least 2000 ha) should be reserved for the conservation of plants and animals that are especially sensitive to logging. The adoption of this kind of approach will require co-operation between government departments and logging companies and the careful preparation of management plans. Since April 1988, the federal Environmental Quality Act has required the preparation of an Environmental Impact Assessment (EIA) for certain prescribed activities, including logging that covers an area of 500 ha or more and drainage of wetland, wildlife habitat, or virgin forest covering an area of 100 ha or more (see Anon. 1990, pp. 7–8).[12] At the present time, however, the EIA requirement is not being applied to logging operations in Sarawak (ITTO 1990, p. 19), and we have no evidence that it is being applied to similar operations in other states or, indeed, that the preparation of EIAs is markedly affecting either the rate or the nature of forest clearance and logging activities.

There is now a great danger that some species will be lost before they can be protected in any kind of reserve or by any other means. There are, however, some signs of hope, including:

(1) the total protection from felling of four species of trees in Sarawak: *Casuarina equisetifolia* (to prevent coastal erosion), *Dipterocarpus oblongifolius* (for control of river bank erosion), and *Shorea macrophylla* and *S. splendida* (needed to safeguard illipe nut production);

(2) the proposed protection of an additional 42 species of plants in Sarawak;

(3) a planned programme to conserve the genetic resources of six of Peninsular Malaysia's endangered tree species: *Neobalanocarpus heimii, Dryobalanops aromatica, Dyera costulata, Shorea gratissima, Parkia speciosa,* and *Calamus manan* (Anon. 1990, pp. 14 and 20).

There is also a need for greater attention to *ex situ* conservation, including new or expanded programmes to collect seeds of endangered species for cultivation in gardens and arboreta. Through the use of *ex situ* plantings, some species may eventually be re-established in their original habitats. At best, however, such off-site conservation can save only a tiny fraction of the great species richness of tropical rain forests (Ng 1983, p. 374; R. Kiew 1989*b*, pp. 9–12; Whitmore 1990, pp. 186–8).

Regional and international co-operation

The ASEAN nations have collaborated on several common environmental concerns (see McDowell 1989; also *New Straits Times*, 19 June 1990). As far as forestry is concerned, they agreed in 1981 to co-operate in matters of policy development, technical and institutional concerns, and intra-regional timber trade, and also to take a common stand on international forestry-related issues. Two ASEAN projects are based in Malaysia: the Canadian-supported ASEAN Institute of Forest Management and the European Community-supported ASEAN Timber Technology Centre. Another co-operative venture is the France–SEA-MEO (Southeast Asian Ministers of Education Organization) programme, whose objectives include the provision of training in the collection and analysis of satellite data that can be used to detect deforestation and to monitor reforestation (see Anon. 1990, pp. 24–6). The proposed Malaysian–Indonesian trans-frontier park (see above) suggests that there is scope for further regional co-operation.

Malaysia is a signatory of the World Heritage Convention, but, to date, not a single site has been nominated for inclusion in the World Heritage List. The federal government has the constitutional power to submit such nominations (and there are many sites that would qualify), but has probably chosen not to do so because of predictable opposition from the states.[13] Malaysia has several other international commitments: it is a signatory of CITES (Convention on International Trade in Endangered Species of Wild Fauna and Flora); it is working with ITTO, of which it is a member, to develop appropriate forest management systems; it is involved in a project with the United Nations Development Programme and the Food and Agriculture Organization of the United Nations to develop, among other things, an effective and efficient continuous forest inventory system; and there is collaboration between the Forest Research Institute of Malaysia and the Dutch government on a project to document scientific information on the plant resources of South-east Asia (Anon. 1990, pp. 25–7).

5.5 TRANSITIONS: TOWARDS A MORE SUSTAINABLE FUTURE

This section outlines several 'transitions' that Malaysia will have to undergo in order to arrive at a more sustainable relationship between human activities and natural systems. Although much remains unknown about how tropical rain forests function and change, what is most urgently required is not new knowledge or technical breakthroughs, but rather, fundamental changes in institutions and values, focusing mainly on human needs and the maintenance of life-support systems. We recognize that Malaysia is part of an international economic system that exerts various pressures on the natural world, but an analysis of this matter remains beyond the scope of this work.

5.5.1 An information transition

Although Malaysia's forests have been very well served in the scholarly literature, the same cannot be said for the compilation of reliable, up-to-date data on various aspects of the resource; indeed, some quite basic information is still lacking. Thus, for example, only estimates are available for the following: the total area of forest in the country, the area of forest in each of the three major regions, the rates of forest removal and alteration, the amount of primary forest remaining, the volume of merchantable timber, and the area of forest logged each year. As Chin (1989, pp. 6–7) has noted, a fundamental shortcoming is the lack of detailed knowledge about the extent of the country's forest resources. This information can be put together using coarse and fine resolution satellite imagery: the acquisition, processing and analysis of this data for the whole country needs to be given top priority.

Management of the forests requires reliable baseline data. Without such information—and it is often not available even in the rich countries—'it is impossible to determine what intensity and mix of activities the forest can be expected to support sustainably, or to carry out effective monitoring of the changes wrought by these activities' (Dunster 1990, p. 48).

Even less appears to be known about Malaysia's non-timber, or so-called minor forest resources, although they remain of vital importance to many indigenous communities both within and on the fringes of the forests. What, for example, is their economic value *vis-à-vis* that of timber? Some estimates are available, but more 'hard facts' are needed. The same is true of land use; what is the extent of idle land in Malaysia and does it exceed the total amount of land that has been cleared and settled by FELDA and other agencies? How much forest do shifting cultivators clear each year and what kind of forest is it? It is not even clear how many shifting cultivators there are. Rational land-use planning requires more accurate, up-to-date data, preferably on a continuous basis. A forest inventory covering the whole country would be a useful place to begin.

5.5.2 A demographic transition

Population pressure is often singled out as the major cause of tropical deforestation (see Chapter 1). This is a rather simplistic notion, for the role of population in environmental change is invariably interrelated in complex ways with numerous other socio-economic, cultural, and environmental factors. That having been said, however, the importance of human numbers and growth rates should not be underestimated because they place increasing pressure on natural systems and greatly exacerbate the problem of raising the living standards of

the global underclass. A more stable or sustainable future requires striking a balance between population, resources, natural systems, and development processes (Salas 1987). Population pressure is only one of several factors that account for rapid forest change in Malaysia. Other, generally more important, processes include persistent poverty, government-sponsored land development, extensive selective logging of hardwoods, and the demands of the rich nations for timber, tin, plantation crops, and other commodities.

Factors that account for poverty include population growth, fragmentation and uneconomic size of many landholdings, both official and employer resistance to raising wages on rubber and oil-palm estates, inefficient use of existing agricultural land, the fact that many of the poorest rural dwellers do not benefit from development policies (Shamsul 1983; Hing 1984), and, especially in East Malaysia, erosion of customary land and forest rights. *In situ* solutions to these various problems would greatly reduce the perceived need for more land development, thereby reducing the pressure on the remaining forests. Both the incidence of poverty and the rate of population growth are high in Sabah and Sarawak; when combined with the unresolved issue of customary land tenure, these conditions add up to a prescription for land degradation.

Although fertility rates have declined substantially, Malaysia's population continues to grow at an average rate of about 2.5 per cent per annum (Kamarrudin Sharif 1982; Hirschman 1986). It is noted in the fifth national plan (for 1986–90), albeit rather vaguely, that the 'impact of the growing population and human activities...on the environment will be measured, where possible, and accounted for' (Malaysia 1986, p. 289). On the other hand, we have no evidence that the government has seriously considered the environmental implications of its anachronistic population policy, which calls for a total population of 70 million, or about four times the size of the current population, by the year 2100 (Anon. 1983; Jones and Tan 1985; Malaysia 1986, p. 133). An increase of this magnitude—apparently the main goal of the policy is to boost domestic demand for the products of the government's industrial programme—would greatly increase pressure on natural systems; even the remaining forests in parks and reserves would suffer from such an increase because there is a growing demand for recreation in wilderness areas.

Conservation and sustainable development require a transition to a stable population. This will require, among other things, giving greater attention to poverty alleviation (especially in East Malaysia), reducing fertility and mortality rates, and enhancing the status of women.

5.5.3 An institutional transition

Natural and economic systems are interconnected and the consequences of human agency in nature are interrelated, whereas bureaucracies and institutions are invariably fragmented. Herein lies a problem, because natural systems cannot be adequately managed by separate, often competing entities such as departments of forestry, agriculture, energy, and the like. Neither can we reasonably expect departments or ministries of environment adequately to protect natural systems when, as is frequently the case, their activities conflict with those of more powerful ministries and agencies. Referring to an example of such conflict in Malaysia, Lowry and Carpenter (1985, p. 246) drew attention to the unsuccessful attempt by the Department of Environment 'to bring soil erosion controls into the creation of oil palm plantations by the Federal Land Development Authority (FELDA)'. In Malaysia, as in virtually all other countries, there is a need to give environmental agencies

more capacity and more power to cope with the effects of unsustainable development policies. More important, governments must make their central economic, trade and sectoral agencies directly responsible and accountable for formulating policies and budgets to encourage development that is sustainable (MacNeill 1989, p. 163).

What has actually happened in Malaysia is that the federal government's 'top-down', largely sectoral approach to development planning has spawned a proliferation of statutory bodies, special public enterprises, and other institutions whose multifarious activities tend to be fragmented, overlapping, competitive, and uncoordinated (Gullick 1981, pp. 151–4; Majid and Majid 1983; Mehmet 1986, p. 49). This has, among other things, resulted in a good deal of waste and inefficiency as well as an overconcentration of power and decision-making in the hands of the ruling élite. The compartmentalized nature of institutions, the government's less than explicit commitment to conservation and sustainable development, the broad extent and rapid pace of resource exploitation (especially in relation to land development and logging), and the fact that much remains unknown about tropical rain-forest ecosystems—all of these add up to a prescription for environmental disorder.

One of the main values of developing conservation strategies (see above) is that they bring various interested parties together. As explained by Manning (1990, p. 24), instead

of having economic development strategies, land-use planning strategies or social development strategies developed in isolation, conservation strategies ideally integrate all of these to form a blueprint for concerted action.

Arriving at a consensus on the optimal use of natural resources in order to promote sustainable patterns of development will require greater cooperation between government institutions and bureaucracies, which tend to favour compromise and stability, and NGOs like SAM, which generally favour change and the adoption of new perspectives on conservation and development. As we have seen, however, authoritative institutions are stronger than participative ones, hence the scope for influencing decision-making is still rather limited. We hope that this will change.

5.5.4 A social and participatory transition

Conservation and sustainable development are first and foremost about people. Conservation is a tool or technique, not an end in itself, and it must be seen to contribute to the ongoing process of sustainable development, whose essential elements include maintenance of natural life-support systems, satisfaction of basic human needs, freedom from unwanted dependence, and social equity. True equity requires not only the sharing of wealth but also the sharing of power, and this in turn requires, among other things, effective citizen participation in decision-making processes. Indeed, interpretations of the meaning of sustainable development increasingly recognize that people and their communities should be encouraged to define and develop their own strategies for sustainable development and eco-system integrity; that many indigenous peoples have devised resource-management techniques that conserve moisture, nutrients, and species, suppress pests, yield balanced diets, and, in general, support sustainable livelihoods (see Chapter 1); and that from the sharing of power under conditions of social self-determination stems 'the potential for community self-reliance, cultural integrity, enhanced creative and problem-solving capabilities, and individual development and fulfillment [sic] ...' (Gardner and Roseland 1989a, p. 29; also B. W. Walker 1987).

Decision-making powers in Malaysia are very heavily concentrated in the hands of ruling federal and state élites, leaving relatively little scope for any kind of 'bottom-up', participatory approach to conservation and sustainable development. Neither is there much evidence that such an approach will be encouraged. On the contrary, a great welter of formal institutions and an expanding bureaucracy have largely disempowered communities and citizens, with the result that they have very little say in decisions that affect their lives. Dissatisfaction with this state of affairs is clearly evident in East Malaysia, where unresolved issues of native customary land tenure have given rise to mounting social tensions and contributed to heightened ethnic consciousness. Development in this region is largely élitist and, as we have noted, there is much thinly veiled contempt for some of the 'yet-to-be-modernized' forest peoples.

In short, the government's approach to development planning does very little to foster community initiative and self-reliance, fails to build on indigenous environmental knowledge, and acts to retard, rather than to enhance, the ability of local people to adapt to changing social, economic, and environmental circumstances. There is an obvious need for greater power-sharing between governments, NGOs, and communities, focusing mainly on human fulfilment and environmental integrity.

5.5.5 An economic transition

Many environmental problems have resulted from the reckless pace and increasing scale of forest exploitation. The time has come to slow down and to give greater attention to smaller-scale, carefully planned sustainable patterns of land use. It does not make much sense to clear more and more forest for land development when a vast area of agricultural land sits idle or remains underutilized. Instead of clearing more forest, greater importance should be attached to land rehabilitation and reforestation.

Rapid and destructive logging of Malaysia's forests has been abetted by state government policies that generally concede a highly disproportionate amount of rents to concessionaires, thereby robbing treasuries of much-needed development funds—funds that could be channelled into sustainable agriculture and forestry. Vast profits have been available to licensees and their favoured contractors, so it is hardly surprising that there has been a post-war timber boom. Certain concession terms (such as short-term leases) and the form of certain government charges for use of public forests (such as basing charges on the volume of timber extracted rather than on the volume of merchantable timber in the tract) have combined to encourage overexploitation of timber resources (see Gillis 1988).

The imposition of much heftier royalties on timber would generate revenue that could be used to promote sustainable development projects. Logging concessions should be granted for longer terms, thus encouraging sustainable forestry practices. A more intractable problem is how to deal with corruption and the important role that concessions play in the system of patronage politics. One solution would be to auction concessions competitively. The area of remaining primary forests is dwindling at a very rapid rate, and there is now an urgent need for *all* timber production to come from secondary forests and plantations. The compensatory plantation programmes must be speeded up.

Some aspects of forest exploitation simply do not make any economic sense. Consider the case of Sarawak's mangrove forests, which are rapidly being felled for wood chips or converted to other uses. In the mid-1980s the annual revenue from mangrove-dependent fisheries was in the order of US $38 million, whereas the comparable revenue from timber products was 'no more than about US $5 million, two-thirds coming from the one-time harvesting of wood-chips and cordwood for export' (Bennett and Caldecott 1986, p. 81). Fortunately, there is a growing realization in Sarawak that intact mangroves are money-spinners, and this may serve to protect them (*Borneo Post*, 20 June 1986).

Similarly, the economic and use value of non-timber resources has been greatly underestimated, this being especially the case when the value of wild game and other food products is included in this category. 'In no year', according to Gillis (1988, p. 152), 'has the value of processed wood exports from Sarawak exceeded the value of meat harvested in the forest' (see Chapter 4 this volume). Non-timber resources—so-called minor forest products—are of great subsistence and commercial value to the peoples who live in or close to the forests, and this is one of the major reasons why there has been much opposition to the destructive practices of advancing logging companies.

That there is now some formal recognition of the importance of non-timber resources is suggested by the formation, in 1982, of the Rattan Information Centre and the creation, in 1987, of a national committee on plant genetic resources. In addition, however, we suggest that three other initiatives are required. Firstly, the conflict between loggers and indigenous peoples in East Malaysia must be solved. Secondly, more forests from which logging is excluded should be set aside for the use of local communities. And thirdly, encouragement should be given to decentralized, small-scale agroforestry projects and to forest-based local industries that rely not only on timber resources but also on a wide range of non-timber products—e.g. rattans, orchids, fruits, essential oils, exudates, dyes, and spices. In addition to the benefits of being labour-intensive and potentially conserving of resources, such forest-oriented activities can usually be adapted to a wide range of socio-economic and environmental conditions. Experience in several countries suggests that the participation of local groups in the planning and development of such activities is a prerequisite for their success (Winterbottom and Hazlewood 1987). The pre-eminence of a single forest product, namely timber, is a very recent phenomenon, whereas the exploitation of non-timber forest resources has a history that spans more than 2000 years; the survival of the remaining primary forests demands that Malaysia get back on course.

The fate of the forests may also depend, in part, on whether a new fiscal relationship can be worked out between the central government and the state governments. Although the individual states that comprise the federation enjoy control over lands and forests, their powers and duties tend to exceed what their revenue sources can sustain and they are barred from raising loans without central government consent (see Rachagan and Bahrin 1983, p. 501). Partly as a result, the states have attempted to surmount their financial constraints by overexploiting their forests. In the case of East Malaysia, for example, a greater share of oil revenue (under the present arrangement 95 per cent goes into the federal coffers) might be returned to the two states, and this might help to take some pressure off the remaining primary forests (Suhaini Aznam 1990).

5.6 CONCLUSION

Conservation in Malaysia has a long history and a good deal has been achieved: the record of environment-related legislation spans more than a century, the first wildlife reserve was established as long ago as 1903, a large and magnificent area of rain forest was set aside as a national park in 1938–9 (now Taman Negara), a national forest policy has been prepared, and some new parks and reserves have been established. For the most part, however, conservation has been accorded rather low priority, and much remains to be done. What is needed above all else is the formulation and adoption of a rational, nation-wide land-use policy. More parks and reserves are needed, but that will not be enough. Conservation practices must be adopted everywhere.

As in most other countries, conservation in Malaysia has been a rearguard activity, seeking for the most part to protect the remnants of the natural world. Reaction, not anticipation, has been the order of the day; thus, for example, the rubber boom was virtually over before much thought was given to its impact on wildlife, and the vigorous land development programme of more recent times was well under way before much official attention was given to conservation and the need for a permanent forest estate.

The complex administrative structure of the Peninsula during the colonial period proved to be a major obstacle to the adoption of a uniform forest policy, and yielded wildlife legislation that varied considerably from state to state. A major problem today, at least as far as a co-ordinated approach to conservation is concerned, is the constitutional division of powers between the central government and the state governments. States that derive much of their revenue from timber have been reluctant to 'lock away' large areas of forest in parks and reserves or to curtail selective logging operations. Some pressure might be taken off the remaining forests by adopting a new revenue-sharing arrangement. Another alternative, but one that is unlikely to be adopted, is for the federal government to rely on its constitutional powers to acquire land for federal purposes.

Conservation is not an end in itself and it must be seen to contribute to human needs and aspirations. A more stable or sustainable future will require several new departures or 'transitions'. Whether these can be made remains a moot point. What kind of future do Malaysians want? Will there be scope for NGOs and governments to co-operate on environmental and social issues? Will ordinary citizens be empowered? We have no answers to these questions. Meanwhile the rain forest continues to vanish.

6 EPILOGUE

6.1 AN ESSAY BY WAY OF CONCLUSION

Deforestation has a long and very complex history, most of which remains to be written. Systematic interest in the subject spans little more than a century, beginning with G. P. Marsh's seminal *Man and Nature* (1864/1965). Palynological and archaeological evidence suggests that anthropogenic forest change can be traced to at least the early Holocene, while literary and other sources reveal many examples of deforestation since ancient times.[1] Consider a few greatly simplified examples.

The causes and consequences of forest loss in the Mediterranean region during pre-classical and classical antiquity have been examined by geographers and historians, who have emphasized the role of such factors as expansion of arable land, overgrazing, population growth, urbanization, mining, metal smelting, shipbuilding, and the ancient timber trade. One result was much land degradation. Mount Lebanon was famous for its valuable cedars for more than 2000 years, but today much of it is rocky and barren.[2]

There had been much forest loss in Attica by the fifth century BC, and in the *Critias* Plato observed that what remained was 'like the skeleton of a sick man, all the fat and soft earth having washed away, and only the bare framework of the land being left'.[3] It would appear, however, that most deforestation in the northern and southern rim of the Mediterranean occurred in the last century.

Other examples of the human impact on the world's forests are legion, including forest loss in Western and Central Europe beginning in the Middle Ages, deforestation of the eastern seaboard of North America following European settlement, destruction of the great pine forests of Wisconsin and Michigan after the mid-1800s, and forest removal in China, both ancient and modern.[4]

Clearly, the recent assault on the tropical rain forest is only the most recent episode in a very long saga of forest clearance. But in at least three respects it is an episode that differs from past experience: firstly, the overall pace of change is much more rapid, because *each year* an area of forest equivalent to the size of most of Wisconsin and Michigan combined is cleared or variously degraded; secondly, tropical rain forests are less resilient and more sensitive to human impact than are temperate or most other forests; and thirdly, the consequences of forest loss are likely to be far more profound, because current trends threaten to eliminate much of the earth's biodiversity

Tropical rain forest is the characteristic natural vegetation of Malaysia, which belongs to the far-flung floristic region called Malesia. A variety of forest formations are present, depending on edaphic, geological, hydrological, and other conditions. Greatest and most extensive of the formations is the dipterocarp-dominated lowland forest, home of an astonishing wealth of species: few other forests are so wonderfully grand; few other terrestrial ecosystems can boast a more bounteous cornucopia of life; few other wild landscapes are more awe-inspiring.

Humans have lived in or alongside the forests since at least the early Holocene, perhaps for a good deal longer. Prior to the nineteenth century, however, the Peninsula was very thinly populated, and even today most of Sabah and Sarawak remain lightly settled. In 1786, when the English East India Company acquired Pulau Pinang (Penang), the population of the Malay Peninsula probably did not exceed about a quarter of a million. Making up most of that number were the *kampung*-dwelling Malays, the majority of whom were concentrated in parts of the river valleys where they tended orchards and kitchen gardens and cultivated rice in forest clearings.

Living in or on the edges of the Peninsula's forests were scattered groups of opportunistic foragers, shifting cultivators, and horticulturalists whose descendants are now called the Orang Asli. Although small in numbers and possessing a rather weak technology, thousands of years of human agency on the part of these aboriginal peoples had selectively altered the genetic resources of their green and complex world. Roaming the coastal waters of the Peninsula and the northern coast of Borneo were the boat people, the Orang Laut, whose livelihood was oriented towards the sea.

A bird's-eye view of the Peninsula in the early nineteenth century would have revealed a few extensive clearings in the rain forest—in parts of Kedah and Negeri Sembilan for example—but these were infinitesimal compared with the extent of the forest cover, for virtually everywhere the great green mantle clothed the land, stretching away to the horizon, extending down to the coastal waters of the straits and the sea.

Across the South China Sea, in north-western Borneo, the cultural scene was even more complex, but at least in the early nineteenth century certain conditions there were similar to those in the Peninsula: virtually everywhere the forests prevailed, clothing the plains, rolling over the rugged hill country, and climbing into the mountains; population numbers were very low and most people lived in the river valleys; long tendrils of trade had linked the great island and its peoples to the outside world since ancient times (thereby giving the lie to the still widely held view of Borneo as an enduring and remote bastion of primitiveness); and there too the forests must have long been subjected to selective pressures from specialized collectors of jungle products and from swidden farmers—most notably from the Iban, whose great migration into Sarawak began in the mid-sixteenth century.

Forests continued to prevail virtually everywhere during the nineteenth century. Only in the Straits Settlements and in parts of the neighbouring Malay states was there any appreciable increase in the pace of anthropogenic forest change. Although Britain continued to pursue a policy of non-intervention in the affairs of the peninsular states until the 1870s, her presence in the Straits acted as an economic magnet to a swelling tide of Chinese immigrants, many of whom began flocking to the Malay states in the early decades of the century. The activities of the Chinese included tin mining, which devastated parts of Perak and Selangor, and a resource-depleting form of shifting plantation agriculture that left behind an expanding hollow frontier of degraded soils and forests.

European commercial enterprises met with little success during the nineteenth century: a variety of commercial crops were introduced but most of these proved disappointing and eventually failed, and it was not until the turn of the century that attempts to break into the tin-mining industry met with any real success. In short, the Chinese dominated the nineteenth-century economy, and it was largely as a result of the application of their capital and labour (with some assistance from European sugar and coffee planters) that forests in some areas began to retreat.

The colonial presence was also felt in Sarawak, where the paternalistic rule of the Brooke family was established in 1841, but no encouragement was given to either European or Chinese commercial enterprises. Attempts were made to introduce a variety of commercial crops but all were dogged by failure prior to 1905 when successful commercial planting of rubber began. The remarkable territorial expansion of the Iban continued apace and the Chinese presence increased in importance from the 1880s, but these events had only a modest impact on the forests.

Britain adopted a new forward policy in the 1870s, and between 1874 and 1914 colonial control was extended over the entire Peninsula, whose landscape and life were thereby eventually transformed. Imperial expansion was not limited to the Peninsula, for in 1881 the British government granted Alfred Dent a royal charter to form the British North Borneo Company. For some time, however, the Company struggled merely to survive in the thinly populated, largely inaccessible, and remote northern end of the huge island. As for Sarawak, territorial expansion proceeded apace there until 1905, but at that time the total population of Charles Brooke's sprawling dominion was still only about 400 000. Both Borneo territories remained undeveloped and heavily forested.

The spread of British rule in the Peninsula set the stage for the first major assault on the region's forests. After 1874 the colonial government began to establish the framework for a profitable export economy based on tin and plantation agriculture: an efficient administration was established, a modern system of communications was laid down, liberal land and taxation policies were implemented, a trigonometrical land survey was started, cheap labour was made available to planters and miners, and law and order were upheld. In short, the colonial government began to erect an edifice in which private enterprise could flourish.

Tin-mining was already well established in the western lowlands and it was here, especially in the tin-rich states of Perak and Selangor, that economic development proceeded most rapidly. Until about 1914 most of the government's revenue was derived from the export duty on tin, and the bulk of this was ploughed back into building railways and roads both within and between the main mining areas, thereby stimulating existing mining enterprises and also serving to attract an increasing inflow of European capital, management, and technology.

Conditions were also becoming more favourable for the development of commercial agriculture: peace and security, better communications, support for botanic gardens, generous loans, a light tax burden, improvements in public health, favourable land laws—all of these combined to place agriculture on a firmer foundation. Planters and officials were now poised to take full advantage of any new and profitable enterprise that might come along.

The story of commercial agriculture after about 1905 is largely the story of rubber, whose adoption spread like wildfire in the first two decades of the new century. It was planted first in the core area or heartland of the country, the western lowlands, where it soon proved adaptable to a wide range of environmental conditions, eventually spilling over

into the region east of the Main Range. The new boom crop spread at the expense of the rain forest, which was often recklessly swept away in the pell-mell rush to bring more land under cultivation.

The two great prongs of the colonial economy, tin and rubber, transformed much of the western lowlands into an extensive humanized landscape. Over much of the rest of the Peninsula, however, great forests continued to hold sway and in 1957, when the Federation of Malaya became an independent country, some 75 per cent of the Peninsula remained forested.

Economic development proceeded at a much slower place in the two Borneo territories. In North Borneo, where tobacco flourished for a while, timber and rubber became the mainstays of the economy, the former concentrated in the east, the latter in the west. The colonial economy of Sarawak in this century featured mainly oil, rubber, pepper (especially during the 1950s), and timber, the last becoming increasingly important after World War II. Economic development in the two territories always lagged far behind the rate of progress in the western Malay states, and in 1963, when they joined the Federation of Malaysia, forests still covered about 80 per cent or more of their respective areas. Both were, as they still are, heavily dependent on primary production.

The rapid pace of change in the Peninsula in the period down to the 1920s raised some concern over the plight of a growing list of birds and mammals, consequently legislation was passed to protect wildlife and several game reserves were established. A Wild Life Commission was formed in 1930–1 and one of its recommendations, namely that a large national park should be established in the centre of the Peninsula, was acted on in 1938–9 when the King George V National Park (now Taman Negara) came into existence. It remains the Peninsula's only national park. Some attention was also given to forest reservation and management, but a uniform policy for the Peninsula was bedevilled by the complex administrative subdivision of the region and by decentralization in the FMS during the 1930s. Extensive areas of forest were also reserved in North Borneo and

Sarawak, but it was not until the very end of the colonial period that national parks appeared on the scene. Only two were established: Bako (1957) in Sarawak and Kinabalu (1963) in North Borneo.

Most deforestation in Malaysia has occurred in the relatively brief post-colonial period. The new nation inherited an export-oriented economy based mainly on rubber and tin (with timber of growing importance), and primary-producing activities remain of great importance, especially in East Malaysia. Most of lowland Peninsular Malaysia has been transformed into an extensive humanized landscape, and anthropogenic forest change has greatly picked up speed in Sabah and Sarawak.

Vast areas of forest have been cleared to make way for land development schemes or variously degraded by selective logging of hardwoods. Road-building and dam construction have also played a part. Shifting cultivation in East Malaysia (it is no longer widely practised in the Peninsula) is another factor of some importance, although its role has been exaggerated by certain politicians and government officials.[5]

Most land development schemes feature sprawling monocultures of rubber and oil palm. While it is true that these tree-crop replacement systems can provide considerable protection for soils and nutrients, they are not without their problems. Soil erosion is generally higher under plantation crops than it is under forest cover, and often very much higher where cover crops have not been immediately established. Many of the Peninsula's rivers were or still are severely polluted with effluent from palm-oil mills, while the increasing use of fertilizers and pesticides contributes to water pollution. Plantations are subject to pests and diseases, and they support only a very impoverished array of wild plants and animals.

Many Malays in the Peninsula have benefited from land development schemes, but the ecological cost has been the near elimination of the species-rich lowland forest. The same land development model has been implemented in East Malaysia, but there, under very different environmental, economic, and socio-cultural conditions,

it has met with much less success. East Malaysia remains a largely underdeveloped periphery, given over mainly to the extraction and overexploitation of natural resources. Lands and forests are being degraded at an alarming rate.

Logging and forest-based industries have taken a heavy toll of Malaysia's forests, and in all three major regions log production has increased dramatically since c.1960. Mounting overseas demand, improvements in technology, and not a little greed and corruption are among the main reasons for this trend. Having depleted much of the available lowland dipterocarp-dominated forest of the Peninsula, logging in that region has moved progressively upward into the hill-dipterocarp forest. It is expected that the Peninsula will become a net importer of timber by the turn of the century.

The forests of Sabah and Sarawak are being 'mined' for their valuable timber, and taken together the two states now account for some two-thirds of all log exports from developing countries. As of c.1980, just about all of Sabah's productive forest had already been either logged over or assigned to concessionaires. The peat-swamp forests of Sarawak continue to be an important source of timber, but some prized species have been severely depleted and logging activities have been pushing into the extensive hill-dipterocarp forests since c.1970. Various reports suggest that Sarawak's forests are being raped by a coterie of local politicians and their favoured clients. Social tensions are mounting as native peoples struggle to protect their lands and strive to participate in decisions that affect their lives.

Some of the generally baleful consequences of selective logging include extensive damage to residual stands, soil erosion and soil compaction, damage to or elimination of seedlings and saplings, water pollution from sawdust and diesel oil, lost habitat and declining populations of certain species, depletion of minor forest products, and great wastage of timber.

Shifting cultivation is still widely practised in Sabah and Sarawak, but most data concerning the activity appear to be unreliable. That shifting cultivation results in at least temporary forest loss, and that it can, under certain circumstances, cause severe land degradation, cannot be denied. Compared with logging, however, it affects a much smaller area of primary forest. Officials have long tended to view shifting cultivation as a uniform, unchanging, and prodigial form of land use, especially now that timber is so valuable. Studies of the Iban and other shifting cultivators of Sarawak, however, suggest that the system is variable, dynamic, generally resource-conserving, and responsive to changing circumstances. This last attribute, that of responsiveness, should serve the shifting cultivators very well, because competition for lands and forests is increasing rapidly and it cannot be expected that the widespread practice of swidden agriculture will be permitted to survive for much longer.

The great evergreen lowland mantle now shows signs of much use and abuse, and soon only a few protected remnants of it will remain. The majority of the country's plants and animals are confined to the lowland forest, but without suitable habitat they cannot survive.[6] Some species that we will never know anything about probably have already vanished, and a growing number of others teeter on the brink of extinction. Human ways of life are also threatened by loss of habitat.

The lowland forests in dryland areas have borne the brunt of human activities, but other forest ecosystems have also been variously modified or transformed: for example, freshwater-swamp forests have been cleared for padi cultivation; mangrove forests, which for long have been an important source of fuel and other products, have been 'reclaimed' for building sites and agriculture, converted to pond aquaculture, and used for woodchip production; and montane forests are increasingly threatened by new or planned road construction.

The federal government's response to the widening circle of environmental problems has included establishing a Department of Environment, introducing several pieces of environment-related legislation, promising to establish several new national parks (some are long overdue), including policy statements on the environment in

certain of its recent five-year national plans, demonstrating support for nature conservation in certain land-use disputes (but not in others), and co-operating with other ASEAN nations on regional environmental issues and concerns.

But, on the other hand, legislation has been very slow to appear, the Department of Environment's activities often conflict with those of more powerful ministries and agencies, forest conservation and management were virtually ignored in both the fourth and fifth national plans, little or nothing has been done to halt the rape of Sarawak's forests, most emphasis to date has been on after-the-fact repair, and the prevailing sectoral approach to planning has meant that interrelated problems of air, water, soil, and forest degradation have usually been viewed as discrete and unrelated problems.

While it is true that the federal government could do more to protect the Malaysian environment, it must be emphasized that jurisdiction over lands, agriculture, water, and forests rests with the individual states of which the federation is composed. In other words, the constitutional division of powers between the states and the federal government greatly restricts the latter's ability to legislate in several key areas of environmental concern, and this in turn has hampered the formulation and implementation of nation-wide policies on resource conservation and management.

Several NGOs are active in the environmental arena, and it has been largely thanks to their attention-seeking activities that environmental issues have been kept on the federal government's agenda. Organizations like Sahabat Alam Malaysia and the Malayan Nature Society have played important roles in several environmental disputes since the 1970s, and they have been instrumental in fostering greater public awareness of environmental issues and problems. In Malaysia, however, authoritative institutions are still much stronger than participative ones, and the NGOs are viewed with considerable suspicion by the federal government. We suspect that any environmental group is probably anathema to the generally more conservative state governments.

Local and overseas defenders of the rain forest would do well to give greater financial and moral support to Malaysia's NGOs because, among other things, such grassroots organizations have a vitally important watch-dog role to play in environmental affairs.

Malaysia's forests have been overexploited but underutilized. Put rather bluntly, the pace of forest removal for land development has been too rapid, and logging has proceeded as though the forests were a non-renewable resource to be 'mined' as quickly as possible. Nowhere are the destructive consequences of these trends more apparent than they are in Sarawak, where both forests and peoples are being exploited for the benefit of a select few.

Only in the past few decades has timber taken precedence over all other forest products (always excepting Sabah, where the industry has long been important). This represents a major break in the history of forest-product exploitation, which for many centuries featured a wide range of plant and animal products. Pressure could be removed from some of the remaining primary forests by exploit-ing secondary forests more fully; in particular, there is a need to give much greater attention to decentralized, small-scale, sustainable, people-oriented, forest-based local industries and agro-forestry projects, featuring not only timber but also a variety of so-called minor forest products.

Officials since early colonial times have considered the removal of the lowland forest and its replacement with commercial agriculture (or mining) to be the 'most productive' use of such areas. While this may be true from a strictly neo-classical economic perspective, and while it would be unrealistic to insist that most of the low-land forest should remain intact, the now long-standing perception of what forests are worth has tended to ignore the many ecological and environmental services that they perform, to downgrade their scientific, genetic, aesthetic, recreational, educational, and intrinsic value, to overlook their vital importance to forest-dwelling peoples, and, ironically, to underestimate the value of minor forest products. Money alone cannot measure the worth of a global treasure.

NOTES TO THE TEXT

PREFACE

1. A few notes on the meaning of certain terms may prove useful to the general reader:

   Malaysia: As Turnbull (1981, p. 1) observed, this term 'acquired its current precise political connotation only in 1963. The word was apparently first coined in the 1830s and by the end of the nineteenth century was in general usage to describe a geographic–zoological–botanical region comprising the Malay peninsula, Singapore, Borneo, Sumatra and Java'. In Emerson's (1937) classic study it was used to cover Malaya and the Netherlands Indies.

   Malay Archipelago: This refers to the great chain of islands (together with the Malay Peninsula) extending from Sumatra in the west to beyond New Guinea in the east (see Wallace 1869).

   Malay Peninsula: The term refers to the narrow extension of the Asian mainland south of the Kra Isthmus.

   Malaya/British Malaya: The major part of the Malay Peninsula (see above) came under British rule between 1786 and 1914. This area became known as Malaya or British Malaya (see the next note).

   Sabah, Sarawak, Peninsular Malaysia: These are official names. The last was formerly called West Malaysia. Sabah and Sarawak together are invariably called East Malaysia.
2. Not to be confused with 'Malaysia' (see above). Prior to the ill-fated Malayan Union of 1946–48, British Malaya had a complex political structure comprising the Straits Settlements (Singapore, Melaka, and Pulau Pinang), the Federated Malay States (Perak, Selangor, Negeri Sembilan, and Pahang), and the Unfederated Malay States (Perlis, Kedah, Johor, Terengganu, and Kelantan). The Union was replaced by the Federation of Malaya, which became an independent state in 1957. Although it remained a colony, Singapore enjoyed self-government after June 1959.
3. Sarawak was ruled for a century (1841–1941) by the 'three white rajahs' (the Brooke family). It became a Crown Colony in 1946. North Borneo was administered under a royal charter granted in 1881. It likewise became a Crown Colony in 1946.
4. Kuala Lumpur was part of the state of Selangor until 1974, when it became a federal territory. A new state capital for Selangor was subsequently established at Shah Alam. Formerly part of Sabah, the island of Labuan has been federal territory since 1984.
5. All data for 1985 are estimates; all are from Malaysia (1986, chapter 4).
6. All data for this year were taken from Population Reference Bureau (1989).
7. In order to avoid confusion with (political) Malaysia, this Latin transliteration, which apparently was first introduced in the mid-nineteenth century, was reinstated in the 1960s to describe the botanical region (see Steenis 1981).
8. This estimate is based on Aiken and Leigh (1988, fig. 2, p. 294); see also Anon. (1990, table 1, p. 9).
9. See Bennett (1976).
10. See Indonesia (n.d.); also Morgan (1984).

CHAPTER 1 WHITHER TROPICAL RAIN FOREST?

1. Fig. 1.1 represents the kind of 'potential' vegetation that 'might be expected in the absence of man's interference' (Longman and Jeník 1974, p. 17). Such absence is increasingly rare. The scale of the figure inevitably disguises the great heterogeneity of most tropical rain forest, depending on such factors as seasonal distribution of rainfall, soil and geological conditions, drainage, and altitude. There is a brief but cogent description of the distribution in Ayensu (1980, pp. 10–19).
2. Seminal research on shifting cultivation was conducted by Conklin (e.g. 1957, 1961). There is a useful discussion with many references in Unesco (1978, pp. 436–51). See Padoch (1982) and Hong (1987) for judicious assessments of the impact of shifting cultivators on the forests of Sarawak (also Brookfield 1988, pp. 218–23).
3. Many writers do not define what they mean by

'deforestation'. The term's meaning differs in several important reports (Allen and Barnes 1985, p. 167), and it has been used in connection with so many different human activities—including some that patently do not result in deforestation—that it is virtually meaningless when not accompanied by a modifying phrase or clause (see Lanly 1982, pp. 74–7; L. S. Hamilton 1987a). Different ways of classifying the extent of human influence on vegetation have been proposed (see Goudie 1986, pp. 26–7).

4. Only about 4.4 per cent of the area of closed tropical forests is under any kind of management plan, and over three-quarters of this is in India (Lanly 1982, p. 43). Plantations are not without problems: they are, for example, susceptible to attack by insects and disease, and fertilizers are usually required to maintain productivity.

5. In many countries, however, much greater destruction has resulted from government-sponsored schemes like land colonization and road building. And it should be kept in mind that many encroaching settlers occupy secondary forest or logged forest, 'because undisturbed rain forest is seldom accessible except where it has been opened up by logging roads' (Guppy 1984, p. 938).

CHAPTER 2 THE RAIN FORESTS OF MALAYSIA

1. See especially Whitmore (1984a), including the extensive bibliography.

2. Many botanists accepted the land-bridge theory. Because the seeds of many rain-forest species are heavy and their viability is of limited duration, long-distance dispersion by sea or air generally was considered a less attractive explanation.

3. The concept originated with Wegener (1924). Lam (1934), later supported by Airy Shaw (1943), argued that the origin of the sub-Antarctic flora of the mountains of New Guinea and Australia could best be explained by continental drift, but no plausible mechanism was apparent. The reality of plate tectonics and sea-floor spreading in the South-east Asian region was established during the 1970s (e.g. Fitch 1970; Audley-Charles et al. 1972; W. Hamilton 1973, 1979), and published reports linked these mechanisms to the evolution of the Australian and Malesian floras (e.g. Raven and Axel-

rod 1972; Whitmore 1973; Gill 1975; Raven 1979).

4. The nature of the migration has been disputed. One theory is that it occurred along now denuded mountain chains (Muller 1972; Steenis 1972; cf. Flenley 1979; Smith 1982). A second theory is that the mountain peaks from which the forest cover had retreated during glacial periods were at those times used as 'stepping-stones' for seed transport by birds (Smith 1982).

5. *Kerangas* 'is an Iban term for land carrying forest which, when cleared, will not grow hill rice' (Whitmore 1984a, p. 161), while 'heath forest' is a transliteration of *Heidewald*, the term introduced by Winkler (1914, as cited in Whitmore 1984a, p. 327); see also Richards (1936a,b) and Browne (1952).

6. It may have been a cyclone that flattened extensive areas of forest in Kelantan in 1883 (Browne 1949; Wyatt-Smith 1954). Some 70 years later the regrowth 'Kelantan Storm Forest' (now largely cleared) could be readily distinguished from adjoining unaffected forest.

7. A build-up of combustible materials follows abnormal droughts. Lightning strikes may start large fires. Evidence from the Amazon Basin suggests that the forests of that region have long been disturbed by fire (Sanford et al. 1985).

8. Ho et al. (1987) postulated that the forest on soils of the Segamat series was inundated by the exceptional flood of 1926–7, causing the death of many large trees. The resultant gaps were occupied by *Elateriospermum tapos* and other non-dipterocarp species, which apparently thrive on free-draining sites.

9. Species richness varies between the three rain-forest blocks, with the African forest being the least rich. There are variations in species richness between formations, as noted in the text. Certain tree species grow in almost pure stands—e.g. the *Shorea curtisii* stands on the ridges in the hill forests of Peninsular Malaysia and those of *Shorea albida* in the peat-swamp forests of Sarawak (Anderson 1963; Burgess 1968, 1975). Species-rich plant communities occur outside the tropical rain-forest biome—e.g. in the South African heathlands in a Mediterranean-type climate (Whitmore 1984a, p. 238).

10. It is beyond the scope of this work to present species lists or descriptions of various species. See,

NOTES TO CHAPTER THREE

for example, on mammals: Medway (1977), Cran-brook (1983, 1984), Payne *et al.* (1985), Davies and Payne (n.d.); on birds: Medway and Wells (1976), Smythies (1981); on snakes: Haile (1958), Grandison (1978); and on butterflies and moths: Corbet and Pendlebury (1978), Barlow (1988), and, as cited in the latter, various works by Holloway.

11. Quantitative information for the Malesian region has come from two locations in Papua New Guinea (e.g. Turvey 1974; Edwards 1982), from four sites at Gunung Mulu in Sarawak (Proctor *et al.* 1983*b*), and from sites within the Pasoh Forest Reserve in Negeri Sembilan and at Gombak in Selangor, both in Peninsular Malaysia (e.g. Kenworthy 1971; Manokaran 1977; Lim 1978).

12. For example, the seeds of *Shorea curtisii* lose their viability if they do not receive a soaking within six to seven days. This, together with the fact that dipterocarp seedlings will not establish readily on slopes in excess of 45°, has serious implications for silvicultural management of many hill forests.

13. In the absence of nearby seed trees, recolonization by dipterocarps is slow. At Kepong in Selangor, for example, Kochummen (1978) noted that it took 23 years for a patch of regrowth forest to be recolonized, even though a 'mother' seed tree was only 180 m distant.

14. Traditionally expressed by foresters as annual production of wood in volume per unit area (m^3 ha^{-1} yr^{-1}). More meaningful measures are those of biomass and net primary productivity. Although these are notoriously difficult to determine, they provide some indication of the relative efficiency of competing land uses (see Whitmore 1984*a*, pp. 111–19).

15. A similar spread of growth rates has been estimated for a number of species in Sabah by Nicholson (1965) and F. O. Wong (1973).

CHAPTER 3 RESOURCE UTILIZATION AND FOREST CONVERSION: PROCESSES AND POLICIES

1. It was not until World War II, when the Japanese overran the region, that the term 'South-east Asia' began to come into general use to designate a part of Asia that 'is more than a mere indeterminate borderland between India and China' (Fisher 1964, p. 9; see Osborne 1979, chapter 1). It is now recognized that South-east Asia has long had a cultural identity of its own and that its Indic and Sinic cultural elements represent 'more or less deliberate imports by Southeast Asians who . . .modified and reshaped these elements according to their own needs and interests' (Hutterer 1982, p. 559, also 1988, p. 72).

2. That is, central Java and Nusatenggara. This is a region of seasonally dry, more open, partially deciduous, and readily combustible forests where there is less leaching and soils have been enriched by volcanic materials. Rice cultivation in permanent irrigated fields supports high population densities (see Bellwood 1985, pp. 11–13).

3. On the exceptionally complex matter of early settlement and the eventual introduction of domesticated plants and animals, see Hutterer (1983) and Bellwood (1978, 1985).

4. There is a very fine discussion of the origin and development of South-east Asian agriculture in Hutterer (1983).

5. Brockway (1979, p. 7) observed that the British Royal Botanic Gardens at Kew 'served as a control center which regulated the flow of botanical information from the metropolis to the colonial satellites, and disseminated information emanating from them'. In addition to plantation crops, there were many transfers of flowers, shrubs, and ornamental trees (see Cox 1961; Coats 1970).

6. Settlers from Minangkabau in Sumatra had been arriving in substantial numbers since the mid-seventeenth century. Earlier movements can be traced to the fifteenth century. Most migrants settled in the hilly areas of Negeri Sembilan, where skilled farmers 'had recreated their traditional mixed farming, based upon rice, fruit and animal husbandry, and introduced irrigation techniques' (Turnbull 1981, p. 63).

7. Strictly speaking, padi is unhusked rice, but the term is often used to describe the method of cultivation. A locality or region dominated by padi cultivation is sometimes referred to as a 'padi landscape'.

8. The following selected works include good descriptions of mining techniques: Doyle (1879), Wray (1894), Warnford-Lock (1907), Scrivenor (1928), Ooi (1955).

9. In Negeri Sembilan, for example, tapioca planters

were responsible for the complete deforestation of the Gemencheh area by 1889 and for the extensive tracts of *lalang* that covered much of the Rembau area by 1890 and parts of the Linggi area by the turn of the century (see Jackson 1968*b*, p. 78). Turnbull (1972, p. 159) noted that tapioca 'growing required little labour, the plants did not need fertile soil, processing was easy and a return on capital could be realized within two years . . . but it exhausted the soil and the tapioca factories used up large quantities of wood'.

10. 'In 1888 there were twenty-one Chinese sugar estates in Krian, occupying 16,414 acres [6 631 ha]; of this area, half had been taken up between 1877 and 1880. By 1894, 31,000 acres [12 524 ha] had been alienated for sugar planting, 25,000 acres [10 100 ha] to Chinese' (Sadka 1968, p. 346).

11. See the following for details concerning the introduction of Brazilian species: Anon. (1903*a*), Anon. (1903*b*), Anon. (1910), Drabble (1973, pp. 3–8).

12. Since about 1910 the Peninsula's rubber industry has been based on the Para rubber tree—*Hevea brasiliensis*. Prior to that time other species were tried, including the Brazilian *Manihot glaziovii* (Ceara rubber) and *Castilloa elastica* as well as the indigenous *Ficus elastica* (rambong). Rubber was also collected in the forest, particularly from jelutong (*Dyera costulata*) and gutta percha (*Palaquium gutta*) trees, the trees often being destroyed in the process (Anon. 1900; Anon. 1902; Anon. 1904).

13. The modern, 'industrial' plantation originated in the Malay Peninsula in the late nineteenth century. It differed from the plantations that had earlier come into existence in the Americas. Whereas the latter provided a way of life for a privileged class of owners, there was no scope for plantation societies to emerge in Malaya (or elsewhere in South-east Asia) because ownership soon passed from individuals to limited liability companies—a trend that was increasingly followed in the Americas as well (see Graham 1984).

14. According to Ooi (1976, p 268), it was introduced in the 1850s but 'was cultivated only as an ornamental plant until 1917 when the first plantation was started in Kuala Selangor' (cf. Jackson 1967, p. 319).

15. Straits Settlements, Forest Ordinance, Ordinance No. 22 of 1908. In the FMS, four separate but identical state enactments introduced in 1907 were combined into a federal forest enactment in 1914 (revised in 1918).

16. Johore 1920 (Enactment No. 6 of 1921); Kedah 1923 (Enactment No. 9 of 1926–7); Kelantan 1933 (Enactment No. 9 of 1934); Trengganu 1936 (Forest Rules, No. 8 of 1935–6).

17. They were an important source of firewood and of poles for fishing stakes and house construction. The bark was used in the tanning industry (Furnivall 1903; Cubitt 1920, p. 6). Some 330 000 tonnes (the figure has been rounded) of mangrove wood were cut in the FMS in 1919 (Cubitt 1920, p. 10).

18. There is a very useful discussion of the decentralization issue in Andaya and Andaya (1982, pp. 240–5).

19. A 70-year rotation period was envisaged for light and medium hardwoods, extending to a double-rotation period (140 years) if heavy hardwoods were present. Previously, the emphasis had been on a 130-year rotation period for heavy hardwoods (Wyatt-Smith and Vincent 1962, p. 206).

20. The chain-saw was being used in the Peninsula by *c.*1962; it was introduced into North Borneo in 1961 (Sabah 1967; Burgess 1973, p. 131).

21. In Sabah, post-felling silvicultural treatment was abandoned after 1977 because it was found that logging operations destroyed most of the unwanted timber (see Chai and Udarbe 1977).

22. H. S. Lee (1982, p. 8) suggested that the length of cycle chosen was based more on the period of amortization for forest industries than it was on silvicultural principles.

23. Since 1985 there has been an almost total ban on the export of logs from the Peninsula (Kumar 1986, p. 45), but substantial quantities of sawn timber, plywood, veneer, and mouldings are exported, mostly to Singapore and Western Europe (for up-to-date data, see *Maskayu*, The Monthly Bulletin of the Malaysian Timber Industry Board).

24. It is interesting to note that forest reservation has tended to be a prolonged procedure, whereas forest reserves have been degazetted with comparative ease (Salleh 1988, p. 131).

25. According to the Director-General of Forestry, Peninsular Malaysia, the objectives of the National Forest Policy are to

1. Establish a strategically located [Permanent] Forest Estate for production of timber and other commodities, and for conservation of soil, water and environmental quality.
2. Ensure security of Forest Estate against destructive agents.
3. Practise sound forest management.
4. Encourage multiple use of forest.
5. Promote integrated timber industries and efficient utilization.
6. Employ modern scientific principles and appropriate technology.
7. Upgrade forestry research, education and training.
8. Intensify Bumiputra participation in forest industries.
9. Promote sound development of trade and commerce of forest products (Muhammad Jabil 1980, p. 1).

26. The Act stipulates, in part, that

The Director, with the approval of the State Authority, shall by notification in the *Gazette*, classify every permanent reserved forest under one or more of the following classifications which shall be descriptive of the purpose or purposes for which the land is being or intended to be used: (a) timber production forest under sustained yield; (b) soil protection forest; (c) soil reclamation forest; (d) flood control forest; (e) water catchment forest; (f) forest sanctuary for wild life; (g) virgin jungle reserved forest; (h) amenity forest; (i) education forest; (j) research forest; (k) forest for federal purposes.

27. The figures in this paragraph were taken from Spears (1985), Myers (1989), and Anon. (1990); see also Table 3.3 this volume.

CHAPTER FOUR THE HUMAN IMPACT

1. The account that follows is a condensed version of Aiken *et al.* 1982, pp. 122–44.
2. Clean-weeding was a standard arable practice in Europe. The technique was utilized by coffee and tea planters in Ceylon and from there it was introduced into Malayan coffee and rubber estates (Carruthers 1908a,b).
3. For details of mining techniques and environmental problems associated with mining, see Warnford-Lock (1907), Fermor (1939), L. K. Wong (1965), Yip (1969), Aiken *et al.* (1982, pp. 109–19); see also note 8, Chapter 3.
4. Calls for miners to rehabilitate their mining leases date from the last century, but so far no regulations have been introduced (Ridley 1903; Mitchell 1957).
5. According to R. Kiew (1983a) and the Malayan Nature Society, plants are 'endangered' because they are on the verge of extinction, 'vulnerable' because their habitats are being destroyed, and 'rare' because they occur in only one place or are thinly distributed.
6. The term 'genetic erosion' has been used to describe this process; genetic erosion combined with species loss has been referred to as 'genetic degradation' (Ewel and Conde 1980, p. 21).
7. Damage of a similar magnitude has been recorded in Kalimantan (see Kartawinata *et al.* 1981b, p. 116; also Uk Tinal and Palenewen 1978).
8. A ban on the export of rattan from Peninsular Malaysia came into force on 1 December 1989 (Pearce 1989a, p. 76). This decision may have been motivated as much by the need to protect the furniture industry as it was by the need to protect the rattans.
9. In Sarawak, some 85 of the 213 recorded species are utilized (Pearce 1989a, table 1).
10. The collection and export of *Nepenthes rajah* is illegal because this plant is listed in Appendix I of CITES, to which Malaysia is a signatory.
11. This species, *Melicope suberosa*, was not known until 1982, when it was discovered during the construction of a new road in the Genting Highlands. Sadly, several plants were destroyed by a bulldozer in 1983, leaving only a sole survivor (B. H. Kiew *et al.* 1985, p. 3).
12. 'Portraits' of threatened plants appear from time to time in the *Malayan Naturalist*; 19 such plants were featured during 1983–9.
13. Rhinoceros horn was (and still is) much sought after by the Chinese for its alleged aphrodisiacal and medicinal properties. Both Malays and Chinese utilized almost every part of the tiger: skin, meat, bones, organs, and grease were made into compounds that were thought to cure numerous ills, while claws, bones, and whiskers were prized as powerful charms (Ranee of Sarawak 1913, p. 256; Locke 1954, pp. 176–84).

14. But in the case of wild cattle (including the Malayan seladang and the Bornean banteng) the reverse may have occurred because these species thrive in clearings vacated by swidden farmers (see Rambo 1978, p. 215, 1979a, p. 62).

15. Big-game hunting was a dangerous sport, and at least one prominent colonial official met an untimely end while indulging in the activity (see Gullick 1983, p. 87). There are graphic descriptions of the sport in Hubback (1905), Maxwell (1907/1960), and Foenander (1952).

16. The Commission's three-volume report is a truly remarkable document. Based on evidence presented by Malay villagers, Chinese and European planters, colonial administrators, and Orang Asli, it assembled much interesting and valuable information on the status of Malaya's wildlife.

17. There is a more detailed account of the human impact on the Peninsula's fauna during the post-colonial period in Aiken and Leigh (1985).

18. Referring to the late 1960s, Stevens (1968, pp. 22–3) commented that the game departments were equipped with an arsenal of 51 large-bore rifles capable of killing elephants and that they had more than 140 shot-guns. He suggested that the time was ripe for a re-allocation of funds!

19. It was reported that five rhinos were killed by poachers on the east coast of Sabah in early 1987, and a dead female rhino, from which only the horn had been taken, was found in the Sungai Dusun wildlife reserve in Peninsular Malaysia in 1986 (*Sunday Star*, 24 May 1987; Mohd. Khan 1988, p. 255).

20. According to the ICIHI (1987, p. xiv, see also pp. 3–13), '[t]here is no universally accepted description or definition of indigenous peoples. But whether they are called indigenous, autochthonous, tribal, Fourth World or First Nations, there is a growing consensus about them and their continuing plight'.

21. In the 1980 census the Orang Asli 'were included in the Malay category in the tables, so that more recent census data on their numbers and demographic trends are not available to the public' (Means 1985/86, p 638).

22. There is a detailed account of land and forest legislation for the period 1842–1979 in Hong (1987, chapters 4 and 6).

23. But note that they are permitted to hunt, fish, and collect forest products in the park (see Hanbury-Tenison 1980, chapter 12; Anderson *et al.* 1982, pp. 117–21 and 162–8; ITTO 1990, p. 116).

24. The Sarawak Penan Association was formed in September 1987 at a meeting in Marudi organized by Sahabat Alam Malaysia (SAM) to discuss problems facing various native groups (Sahabat Alam Malaysia 1989, p. 37).

25. The views of several natives are quoted at length in Hong (1987) and in Sahabat Alam Malaysia (1989).

CHAPTER 5 CONSERVATION: TOWARDS A SUSTAINABLE FUTURE

1. For a description of the environmental problems associated with tin-mining and the controls that were introduced, see Aiken *et al.* (1982, pp. 110–16).

2. A list of forest reserves established during the colonial period can be found in Rahman-Ali (1968, pp. 117–24).

3. The account of the early wildlife legislation is mainly based on Wild Life Commission of Malaya (1932, II, pp. 15–27 and III, appendix VII).

4. In 1929, 5500 ground doves were exported live from Tumpat to Singapore (Wild Life Commission of Malaya 1932, II, p. 23).

5. The three-volume *Report of the Wild Life Commission* is not easy to obtain but a detailed précis can be found in the *Journal of the Society for the Preservation of the Fauna of the Empire* for 1933.

6. The creation of the park was made possible by the co-operation of the three states, each of which introduced enabling legislation.

7. 'Protection Forest Reserves' are 'for safe-guarding water supplies, soil fertility and environmental quality; and the minimization of damage by floods and erosion to rivers and agricultural land'; 'Amenity Forest Reserves' are 'for amenity and arboretum use' (Davies and Payne n.d., pp. 203 and 206).

8. Namely, Pulau Bohay-dulang, Pulau Sipadan, Kota Belud, Labuan, and Pulau Mantanani–Pulau Lun-sungari.

9. A proposal made in 1933 to establish a reserve for rhinoceros in the headwaters of the Segama and Tengkayu rivers was not pursued because of opposition from timber interests (Burgess 1961, p. 145).

10. On the other hand, in January 1991 federal authorities arrested the Chief Minister of Sabah Datuk Seri Joseph Pairin Kitingan for alleged corruption, one of the charges being that he had given relatives a timber concession of 2000 ha. Kitingan's Parti Bersatu Sabah broke away from the Barisan Nasional in the run up to the October 1990 federal election (Suhaini Aznam 1991).

11. With the exception of the Selangor strategy, these are confidential documents that can only be obtained with state-government approval.

12. Legislation enabling the preparation of EIAs was introduced (as an amendment to the Environmental Quality Act 1974) in 1986 (Act A636, 1985). The Environmental Quality (Prescribed Activities) (Environmental Impact Assessment) Order 1987 came into effect on 1 April 1988.

13. The Australian experience has shown that such opposition is not unique (see Aiken and Leigh 1986a, 1987).

CHAPTER 6 EPILOGUE

1. Two useful overviews are provided by J. F. Richards (1986) and Williams (1989).

2. There is some very fine literature on deforestation in this region; see, for example, Semple (1919), Thirgood (1981), and Meiggs (1982). Land degradation is discussed in Murphey (1951) and Mikesell (1969).

3. As cited in Glacken (1967, p. 121).

4. See, for example, Darby (1956), Tuan (1968), Smil (1983), and Williams (1989).

5. See ITTO (1990, p. 106).

6. Suitable habitat includes logged forest, provided that settlers and hunters are kept out. Logging practices that favour animals include retaining enclaves of intact forest and leaving behind hollow trees to serve as nesting sites (Whitmore 1990, pp. 188–90).

REFERENCES

Absy, M. C. (1985). Palynology of Amazonia: the history of the forests as revealed by the palynological record. In *Amazonia* (ed. G. T. Prance and T. E. Lovejoy), pp. 72–82. Pergamon Press, Oxford.

Age, The Age (Melbourne), 13 September 1989; 14 April 1990.

Aiken, S. R. (1987). Early Penang hill station. *Geographical Review*, **77**, 421–39.

Aiken, S. R. and Leigh, C. H. (1975). Malaysia's emerging conurbation. *Annals of the Association of American Geographers*, **65**, 546–63.

Aiken, S. R. and Leigh, C. H. (1978). Dengue haemorrhagic fever in South-east Asia. *Transactions, Institute of British Geographers*, New Series, **3**, 476–97.

Aiken, S. R. and Leigh, C. H. (1983). Ending the threat to Taman Negara: a Malaysian Gordon-Franklin won. *Habitat*, **11** (2), 11–14.

Aiken, S. R. and Leigh, C. H. (1984). A second national park for Peninsular Malaysia? The Endau–Rompin controversy. *Biological Conservation*, **29**, 253–76.

Aiken, S. R. and Leigh, C. H. (1985). On the declining fauna of Peninsular Malaysia in the post-colonial period. *Ambio*, **14**, 15–22.

Aiken, S. R. and Leigh, C. H. (1986a). Land use conflicts and rain forest conservation in Malaysia and Australia: the Endau–Rompin and Gordon-Franklin controversies. *Land Use Policy*, **3**, 161–79.

Aiken, S. R. and Leigh, C. H. (1986b). Hydro-electric power and wilderness protection. *Impact of Science on Society*, No. **141**, 85–96.

Aiken, S. R. and Leigh, C. H. (1987). Queensland's Daintree rain forest at risk. *Ambio*, **16**, 134–41.

Aiken, S. R. and Leigh, C. H. (1988). Environment and the federal government in Malaysia. *Applied Geography*, **8**, 291–314.

Aiken, S. R. and Moss, M. (1975). Man's impact on the tropical rainforest of Peninsular Malaysia: a review. *Biological Conservation*, **8**, 213–29.

Aiken, S. R., Leigh, C. H., Leinbach, T. R., and Moss, M. R. (1982). *Development and environment in Peninsular Malaysia*. McGraw-Hill, Singapore.

Airy Shaw, H. K. (1943). The biogeographic division of the Indo-Australian archipelago. 5: some general considerations from the botanical standpoint. *Proceedings of the Linnean Society of London*, **154**, 148–54.

Aken, K. M. and Kavanagh, M. (n.d.). Species conservation priorities in the tropical forests of Sarawak. In *Species conservation priorities in the tropical forests of Southeast Asia*, Occasional Papers of the IUCN Species Survival Commission (SSC), No. 1 (ed. R. A. Mittermeier and W. R. Konstant), pp. 17–22. IUCN, Gland, Switzerland.

Allen, G. C. and Donnithorne, A. G. (1957). *Western enterprise in Indonesia and Malaya*. Allen and Unwin, London.

Allen, J. C. and Barnes, D. F. (1985). The causes of deforestation in developing countries. *Annals of the Association of American Geographers*, **75**, 163–84.

Allen, R. (1981). *How to save the world: strategy for world conservation*. Littlefield, Adams, and Company, Totowa, New Jersey.

Allen, R. and Prescott-Allen, C. (1982). *What's wildlife worth? economic contributions of wild plants and animals to developing countries*. International Institute for Environment and Development, London.

Altieri, M. A., Letourneau, D. K., and Davis, J. R. (1983). Developing sustainable agroecosystems. *BioScience*, **33**, 45–9.

Andaya, B. W. and Andaya, L. Y. (1982). *A history of Malaysia*. Macmillan, London.

Anderson, J. A. R. (1961). The destruction of *Shorea albida* forest by an unidentified insect. *Empire Forestry Review*, **40**, 19–29.

Anderson, J. A. R. (1963). The flora of the peat swamp forests of Sarawak and Brunei including a catalogue of all recorded species of flowering plants, ferns and fern allies. *Gardens' Bulletin Singapore*, **20**, 131–228.

Anderson, J. A. R. (1964a). The structure and development of the peat swamps of Sarawak and Brunei. *Journal of Tropical Geography*, **18**, 7–16.

Anderson, J. A. R. (1964b). Observations on climatic damage in peat swamp forests in Sarawak. *Commonwealth Forestry Review*, **43**, 145–58.

Anderson, J. A. R. (1966). A note on two tree fires caused by lightning in Sarawak. *Malayan Forester*, **29**, 19–20.

Anderson, J. A. R., Jermy, A. C., and Cranbrook, Earl of (1982). *Gunung Mulu National Park: a management and development plan.* Royal Geographical Society, London.

Anderson, J. M., Proctor, J., and Vallack, H. (1983). Ecological studies in four contrasting lowland rain forests in Gunung Mulu National Park, Sarawak. III: decomposition processes and nutrient losses from leaf litter. *Journal of Ecology,* **71**, 503–28.

Anon. (1893). Dye plants. *Agricultural Bulletin of the Malay Peninsula,* **3**, 44–51.

Anon. (1895). Notes on soils and lalang-grass. *Agricultural Bulletin of the Malay Peninsula,* **4**, 73–86.

Anon. (1900). The native rubbers of the Malay peninsula. *Agricultural Bulletin of the Malay Peninsula,* **9**, 239–52.

Anon. (1901). Report on the system of rice cultivation practised in Pahang. *Agricultural Bulletin of the Straits and Federated Malay States,* 2nd Series, **1**, 13–19.

Anon. (1902). Gutta rambong (*Ficus elastica*) in Malacca. *Agricultural Bulletin of the Straits and Federated Malay States,* 2nd Series, **1**, 185–8.

Anon. (1903*a*). The history of the introduction of Para rubber into the Malay peninsula. *Agricultural Bulletin of the Straits and Federated Malay States,* 2nd Series, **2**, 2–4.

Anon. (1903*b*). Corrections and observations on the history of the introduction of Para rubber. *Agricultural Bulletin of the Straits and Federated Malay States,* 2nd Series, **2**, p. 61.

Anon. (1904). Forest administration in the Native States in 1903. *Agricultural Bulletin of the Straits and Federated Malay States,* 2nd Series, **3**, p. 165.

Anon. (1910). Historical notes on the rubber industry. *Agricultural Bulletin of the Straits and Federated Malay States,* 2nd Series, **6**, 200–14.

Anon. (1921). Recent research. *Planter,* **2**, 55–7.

Anon. (1938). The role of cover-plants in rubber growing. *Planter,* Supplement to Volume **19**.

Anon. (1967). Forestry and conservation in the lowlands of West Malaysia. *Malayan Forester,* **30**, 243–5.

Anon. (1968). Prerequisites for forest industries. *Malayan Forester,* **31**, 155–6.

Anon. (1979*a*). Forest resource base, policy and legislation of Sarawak. *Malaysian Forester,* **42**, 311–27.

Anon. (1979*b*). Forest resource base, policy and legislation of Sabah. *Malaysian Forester,* **42**, 286–310.

Anon. (1979*c*). Forest resource base, policy and legislation of Peninsular Malaysia. *Malaysian Forester,* **42**, 328–47.

Anon. (1982). 'Tembeling Declaration'. *Malayan Naturalist,* **August**, p. 2.

Anon. (1983). Population and the size of the domestic market: a Malaysian view. *Population and Development Review,* **9**, 389–91.

Anon. (1986*a*). A road to Gunung Tahan? *Malayan Naturalist,* **August**, 2–9.

Anon. (1986*b*). Bukun dam—a recipe for success or disaster? *Alam Sekitar,* **11** (1), 15–19.

Anon. (1987). Have you seen *Peripatus*? *Malayan Naturalist,* **August**, p. 43.

Anon. (1990). Forest conservation and management practices in Malaysia. Paper presented at the 18th session of the General Assembly of IUCN, Perth, Australia, 28 November–5 December 1990.

Apin, T. (1987). The Sarawak timber blockade. *The Ecologist,* **17**, 186–8.

Arnot, D. B. (1929). Timber extraction in Johore. *Empire Forestry Journal,* **8**, 233–7.

Arnot, D. B. and Smith, J. S. (1937). Shifting cultivation in Brunei and Trengganu. *Malayan Forester,* **6**, 13–17.

Ashton, P. S. (1964). *Ecological studies in the mixed dipterocarp forests of Brunei State.* Oxford Forest Memoir, No. 25. Oxford University Press.

Ashton, P. S. (1969). Speciation among tropical trees: some deductions in the light of recent evidence. *Biological Journal of the Linnean Society,* **1**, 155–96.

Ashton, P. S. (1971). The plants and vegetation of the Bako National Park. *Malayan Nature Journal,* **24**, 151–62.

Ashton, P. S. (1976*a*). Mixed dipterocarp forest and its variation with habitat in the Malayan lowlands: a re-evaluation of Pasoh. *Malaysian Forester,* **39**, 56–72.

Ashton, P. S. (1976*b*). Factors affecting the development and conservation of tree genetic resources in South-east Asia. In *Tropical trees: variation, breeding and conservation* (ed. J. Burley and B. T. Styles), pp. 189–98. Academic Press, London.

Ashton, P. S. (1977). A contribution of rain forest research to evolutionary theory. *Annals of the Missouri Botanical Garden,* **64**, 694–705.

Ashton, P. S. (1982). *Dipterocarpaceae,* Flora Malesiana, Series I—Spermatophyta (Flowering Plants), Vol. 9, Part 2. Martinus Nijhoff, The Hague.

Audley-Charles, M. G. (1981). Geological history of the region of Wallace's line. In *Wallace's line and plate tectonics* (ed. T. C. Whitmore), pp. 25–35. Clarendon Press, Oxford.

Audley-Charles, M. G. (1987). Dispersal of Gondwanaland: relevance to evolution of the angiosperms. In *Biogeographical evolution of the Malay archipelago* (ed. T. C. Whitmore), pp. 5–25. Clarendon Press, Oxford.

Audley-Charles, M. G., Carter, D. J., and Milsom, J. S. (1972). Tectonic development of eastern Indonesia in relation to Gondwanaland dispersal. *Nature*, **239**, 18 September, 35–9.

AUSTEC (1974). *Pahang river basin study*. Vol. 3. *Basin hydrology and river behaviour*. Australian Engineering Consultants, Cooma North, New South Wales, Australia.

Austin, M. P., Greig-Smith, P., and Ashton, P. S. (1972). The application of quantitative methods to vegetation survey. III: a re-examination of rain forest data from Brunei. *Journal of Ecology*, **60**, 305–24.

Ayensu, E. S. (ed.) (1980). *Jungles*. Crown Publishers, New York.

Baharuddin Kasran (1988). Effect of logging on sediment yield in a hill dipterocarp forest in Peninsular Malaysia. *Journal of Tropical Forest Science*, **1**, 56–66.

Bahrin, Tunku Shamsul (1967). The pattern of Indonesian migration and settlement in Malaya. *Asian Studies*, **5**, 233–57.

Bahrin, Tunku Shamsul (1988). Land settlement in Malaysia: a case study of the Federal Land Development Authority projects. In *Land settlement policies and population redistribution in developing countries: achievements, problems and prospects* (ed. A. S. Oberai), pp. 89–119. Praeger, New York.

Bahrin, Tunku Shamsul and Perera, P. D. A. (1977). *FELDA 21 years of land development*. Federal Land Development Authority, Kuala Lumpur.

Baillie, I. C. (1976). Further studies on drought in Sarawak, East Malaysia. *Journal of Tropical Geography*, **43**, 20–9.

Baillie, I. C., Ashton, P. S., Court, M. N., Anderson, J. A. R., Fitzpatrick, E. A., and Tinsley, J. (1987). Site characteristics and the distribution of tree species in mixed dipterocarp forest on Tertiary sediments in central Sarawak, Malaysia. *Journal of Tropical Ecology*, **3**, 201–20.

Balestier, J. (1848). View of the state of agriculture in the British possessions in the Straits of Malacca. *Journal of the Indian Archipelago and Eastern Asia*, **2**, 139–50.

Balgooy, M. M. J. van (1976). Phytogeography. In *New Guinea vegetation* (ed. K. Paijmans), pp. 1–22. Australian National University Press, Canberra.

Barlow, B. A. (1981). The Australian flora: its origin and evolution. In *Flora of Australia*. Vol. 1 *Introduction*, pp. 25–75. Australian Government Publishing Service, Canberra.

Barlow. H. S. (1985). The Malayan Nature Society/Star Endau–Rompin expedition. *Journal of the Malaysian Branch of the Royal Asiatic Society*, **58**, 135–42.

Barlow, H. S. (1988). Forest Lepidoptera. In *Malaysia* (ed. Earl of Cranbrook), pp. 212–24. Pergamon Press, Oxford.

Barney, G. O. (ed.) (1980). *The global 2000 report to the President: entering the twenty-first century*; Vol. 2: Technical report. US Government Printing Office, Washington, DC.

Barraclough, S. (1984). Political participation and its regulation in Malaysia: opposition to the Societies (Amendment) Act 1981. *Pacific Affairs*, **57**, 450–61.

Bartelmus, P. (1986). *Environment and development*. Allen and Unwin, Boston, Massachusetts.

Baur, G. N. (1968). *The ecological basis of rainforest management*. Government Printer, Sydney, New South Wales, Australia.

Beaman, R. S., Beaman, J. H., Marsh, C. W., and Woods, P. V. (1985). Drought and forest fires in Sabah in 1983. *Sabah Society Journal*, **8**, 10–30.

Beard, J. S. (1944). Climax vegetation in tropical America. *Ecology*, **25**, 127–58.

Beard, J. S. (1955). The classification of tropical American vegetation-types. *Ecology*, **36**, 89–100.

Begbie, P. J. (1834/1967). *The Malayan peninsula*. Vepery Mission Press, Madras. Reprinted 1967, Oxford University Press, Kuala Lumpur.

Bellwood, P. (1978). *Man's conquest of the Pacific: the prehistory of Southeast Asia and Oceania*. Collins, Auckland, New Zealand.

Bellwood, P. (1985). *Prehistory of the Indo-Malaysian archipelago*. Academic Press Australia, North Ryde, Australia.

Benjamin, G. (1985). In the long term: three themes in Malayan cultural ecology. In *Cultural values and human ecology in Southeast Asia* (ed. K. L. Hutterer, A. T. Rambo, and G. Lovelace), pp. 219–78. Michigan Papers on South and Southeast Asia, No. 27. Center for South and Southeast Asian Studies, University of Michigan, Ann Arbor, Michigan.

Bennett, E. L. and Caldecott, J. O. (1986). Letter from

Sarawak. *Far Eastern Economic Review*, **17 July**, 80–1.

Bennett, J. W. (1976). *The ecological transition: cultural anthropology and human adaptation*. Pergamon Press, New York.

Berenger, L. U. (1922). Cover-crops vs clean weeding. *Planter*, **2**, 214–17.

Berry, A. J. (1972). The natural history of the West Malaysian mangrove fauna. *Malayan Nature Journal*, **25**, 135–62.

Beven, S., Connor, E. F., and Beven, K. (1984). Avian biogeography in the Amazon basin and the biological model of diversification. *Journal of Biogeography*, **11**, 383–99.

Bigarella, J. J. and Ferreira, A. M. M. (1985). Amazonian geology and the Pleistocene and the Cenozoic environments and paleoclimates. In *Amazonia* (ed. G. T. Prance and T. E. Lovejoy), pp. 49–71. Pergamon Press, Oxford.

Biswas, B. (1973). Quaternary changes in sea-level in the South China Sea. *Bulletin of the Geological Society of Malaysia*, **6**, 229–56.

Black, I. (1983). *A gambling style of government: the establishment of the Chartered Company's rule in Sabah, 1878–1915*. Oxford University Press, Kuala Lumpur.

Blaikie, P. (1986). Natural resource use in developing countries. In *A world in crisis? geographical perspectives* (ed. R. J. Johnston and P. J. Taylor), pp. 107–26. Blackwell, Oxford.

Blaikie, P. and Brookfield, H. (1987). *Land degradation and society*. Methuen, London.

Blower, J. (1984). National parks for developing countries. In *National parks, conservation, and development: the role of protected areas in sustaining society* (ed. J. A. McNeely and K. R. Miller), pp. 722–7. Smithsonian Institution Press, Washington, DC.

Bolin, B., Döös, B. R., Jäger, J., and Warrick, R. A. (1986). *The greenhouse effect, climatic change, and ecosystems*. Wiley, Chichester, UK.

Borneo Post (Sibu), 20 June 1986; 11 February, 5 September 1987; 29 May, 15 June, 17 July, 20 August 1990.

Boserup, E. (1965). *The conditions of agricultural growth: the economics of agrarian change under population pressure*. Aldine, Chicago.

Bourlière, F. (1983). Animal species diversity in tropical forests. In *Tropical rain forest ecosystems: structure and function* (ed. F. B. Golley), pp. 77–92. Elsevier, Amsterdam.

Bowie, A. (1988). Redistribution with growth? the dilemmas of state-sponsored economic development in Malaysia. In *State and development* (ed. C. Clark and J. Lemco), pp. 52–66. Brill, Leiden.

Brandani, A., Hartshorn, G. S., and Orians, G. H. (1988). Internal heterogeneity of gaps and tropical tree species richness. *Journal of Tropical Ecology*, **4**, 99–119.

Briggs, J. G. (1985). The current Nepenthes situation in Borneo. *Malayan Naturalist*, **February**, 46–8.

Brockway, L. H. (1979). *Science and colonial expansion: the role of the British Royal Botanic Gardens*. Academic Press, New York.

Brokaw, N. V. L. (1985). Treefalls, regrowth, and community structure in tropical forests. In *The ecology of natural disturbance and patch dynamics* (ed. S. T. A. Pickett and P. S. White), pp. 53–69. Academic Press, Orlando, Florida.

Bronson, B. (1978). Angkor, Anuradhapura, Prambanan, Tikal: Maya subsistence in an Asian perspective. In *Pre-Hispanic Maya agriculture* (ed. P. D. Harrison and B. L. Turner II), pp. 255–300. University of New Mexico Press, Albuquerque, New Mexico.

Brookfield, H. C. (1972). Intensification and disintensification in Pacific agriculture: a theoretical approach. *Pacific Viewpoint*, **13**, 30–48.

Brookfield, H. [C.] (1988). The new great age of clearance and beyond. In *People of the tropical rain forest* (ed. J. S. Denslow and C. Padoch), pp. 209–24. University of California Press, Berkeley.

Brown, G. S. (1955). Timber extraction methods in N. Borneo. *Malayan Forester*, **18**, 121–32.

Browne, F. G. (1949). Storm forest in Kelantan. *Malayan Forester*, **12**, 28–33.

Browne, F. G. (1952). The kerangas lands of Sarawak. *Malayan Forester*, **15**, 61–73.

Bruijnzeel, L. A. (1990). *Hydrology of moist tropical forests and effects of conversion: a state of knowledge review*. Unesco International Hydrological Programme, Humid Tropics Programme Publication, Paris.

Brünig, E. F. (1964). A study of damage attributed to lightning in two areas of *Shorea albida* forest in Sarawak. *Empire Forestry Review*, **43**, 134–44.

Brünig, E. F. (1969a). The classification of forest types in Sarawak. *Malayan Forester*, **32**, 143–79.

Brünig, E. F. (1969b). On the seasonality of drought in the lowlands of Sarawak (Borneo). *Erdkunde*, **23**, 127–33.

Brünig, E. F. (1970). Stand structure, physiognomy and environmental factors in some lowland forests in Sarawak. *Tropical Ecology*, **11**, 26–43.

Brünig, E. F. (1971). On the ecological significance of drought in the equatorial wet evergreen (rain) forest of Sarawak (Borneo). In *The water relations of Malesian forests*, University of Hull, Department of Geography Miscellaneous Series No. 11 (ed. J. R. Flenley), pp. 66–88. University of Hull, Hull, UK.

Brünig, E. F. (1973). Species richness and stand diversity in relation to site and succession of forests in Sarawak and Brunei (Borneo). *Amazoniana*, **4**, 293–320.

Brünig, E. F. (1977). The tropical rain forest—a wasted asset or an essential biospheric resource? *Ambio*, **6**, 187–91.

Brünig, E. F. (1985). Deforestation and its ecological implications for the rain forests in South East Asia. *The Environmentalist*, **5** (Supplement No. 10), 17–26.

Brünig, E. F. (1987). The forest ecosystem: tropical and boreal. *Ambio*, **16**, 68–79.

Burbridge, N. T. (1960). The phytogeography of the Australian region. *Australian Journal of Botany*, **8**, 75–212.

Burgess, P. F. (1961). Wild life conservation in North Borneo. In *Nature conservation in western Malaysia, 1961* (ed. J. Wyatt-Smith and P. R. Wycherley), pp. 143–51. Malayan Nature Society, Kuala Lumpur.

Burgess, P. F. (1968). An ecological study of the hill forests of the Malay peninsula. *Malayan Forester*, **31**, 314–25.

Burgess, P. F. (1971). Effect of logging on hill dipterocarp forest. *Malayan Nature Journal*, **24**, 231–7.

Burgess, P. F. (1973). The impact of commercial forestry on the hill forests of the Malay peninsula. In *Proceedings of the symposium on biological resources and national development* (ed. E. Soepadmo and K. G. Singh), pp. 131–6. Malayan Nature Society, Kuala Lumpur.

Burgess, P. F. (1975). *Silviculture in the hill forests of the Malay peninsula*. Forest Research Institute, Research Pamphlet No. 66. Kepong, Malaysia.

Burgess, P. F. and Tang, H. T. (1972). Prospects for the natural regeneration of the hill forests of the Malay peninsula. In *Proceedings of the Fourth Forestry Conference*, Forestry Department, Kuala Lumpur, 6 pp.

Burkill, H. M. (1971). A plea for the inviolacy of Taman Negara. *Malayan Nature Journal*, **24**, 206–9.

Burkill, I. H. (1943). The biogeographic division of the Indo-Australian archipelago. 2: a history of the divisions which have been proposed. *Proceedings of the Linnean Society of London*, **154**, 127–38.

Burnham, C. P. (1984). The forest environment: soils. In *Tropical rain forests of the Far East* (2nd edn) (T. C. Whitmore), Chapter 11. Clarendon Press, Oxford.

Burtt-Davy, J. (1938). *The classification of tropical woody vegetation*, Imperial Forestry Institute, Paper No. 13. University of Oxford.

Caldecott, J. (1985). Letter from Long Lellang. *Far Eastern Economic Review*, **26 September**, p. 134.

Caldecott, J. (1987). *Hunting and wildlife management in Sarawak*. (2nd impression). World Wildlife Fund Malaysia and National Parks and Wildlife Office, Sarawak Forest Department, Kuching.

Carey, I. (1976). *The Orang Asli: the aboriginal tribes of Peninsular Malaysia*. Oxford University Press, Kuala Lumpur.

Carruthers, J. B. (1908a). Weeding in Para rubber cultivation. *Agricultural Bulletin of the Straits and Federated Malay States*, 2nd Series, **7**, 383–6.

Carruthers, J. B. (1908b). Report of the Director of Agriculture F. M. S. for 1907. *Agricultural Bulletin of the Straits and Federated Malay States*, 2nd series, **8**, 523–48.

Carson, G. L. (1968). Conservation in Sabah, Malaysia. In *Conservation in tropical South East Asia*, IUCN Publications, New Series, No. 10 (ed. L. M. Talbot and M. H. Talbot), pp. 492–7. IUCN, Morges, Switzerland.

Cassells, D. S., Bonell, M., Hamilton, L. S., and Gilmour, D. A. (1987). The protective role of tropical forests: a state-of-knowledge review. In *Agroforestry in the humid tropics: its protective and ameliorative roles to enhance productivity and sustainability* (ed. N. T. Vergara and N. D. Briones), pp. 31–58. Environment and Policy Institute, East-West Center, Honolulu, Hawaii, USA and Southeast Asian Regional Center for Graduate Study and Research in Agriculture, Los Baños, Laguna, Philippines.

Caufield, C. (1983). Dam the Amazon, full steam ahead. *Natural History*, **92**, 60–7.

Chadwick, A. C. and Sutton, S. L. (eds.) (1984). *Tropical rain-forest: the Leeds symposium*. Leeds Philosophical and Literary Society, Leeds, UK.

Chai, D. N. P. and Udarbe, M. P. (1977). The effectiveness of current silvicultural practice in Sabah. *Malaysian Forester*, **40**, 27–31.

Chai, Hon-Chan (1964). *The development of British Malaya 1896–1909*. Oxford University Press, Kuala Lumpur.

Chai, P. P. K. (1975). Mangrove forests in Sarawak. *Malaysian Forester*, **38**, 108–32.

Champion, H. G. (1936). A preliminary survey of the forest types of India and Burma. *Indian Forest Records*, New Series, **1**, 1–286.

Chia, L. S. (1977). Seasonal rainfall distribution over West Malaysia. *Malayan Nature Journal*, **31**, 11–40.

Chin, S. C. (1977). The limestone hill flora of Malaya. I. *Gardens' Bulletin Singapore*, **30**, 165–219.

Chin, S. C. (1979). The limestone hill flora of Malaya. II. *Gardens' Bulletin Singapore*, **32**, 64–203.

Chin, S. C. (1985). *Agriculture and resource utilization in a lowland rainforest Kenyah community*. Sarawak Museum Journal Special Monograph No. 4. Kuching.

Chin, S. C. (1987). Deforestation and environmental degradation in Sarawak. *Wallaceana*, **48** and **49**, 6–8.

Chin, S. C. (1988). Looking after our forests. *Star* (Kuala Lumpur), 28 December.

Chin, S. C. (1989). Managing Malaysia's forests for sustained production. *Wallaceana*, **55** and **56**, 1–11.

Chivers, D. J. (1978). The gibbons of Peninsular Malaysia. *Malayan Nature Journal*, **30**, 565–91.

Clark, W. C. and Munn, R. E. (eds.) (1986). *Sustainable development of the biosphere*. Cambridge University Press.

Coats, A. M. (1970). *The plant hunters: being a history of the horticultural pioneers, their quests and their discoveries from the Renaissance to the twentieth century*. McGraw-Hill, New York.

Cockburn, P. F. (1974). The origin of Sook Plain, Sabah. *Malaysian Forester*, **37**, 61–3.

Colchester, M. (1989). *Pirates, squatters and poachers: the political ecology of dispossession of the native peoples of Sarawak*. Survival International and INSAN, London and Petaling Jaya.

Cole, R. (1959a). Temiar Senoi agriculture: a note on aboriginal shifting cultivation in Ulu Kelantan, Malaya. Part I. *Malayan Forester*, **22**, 191–207.

Cole, R. (1959b). Temiar Senoi agriculture: a note on aboriginal shifting cultivation in Ulu Kelantan, Malaya. Part II. *Malayan Forester*, **22**, 260–71.

Colinvaux, P. A. (1989). The past and future Amazon. *Scientific American*, **260** (5), 102–8.

Collins, M. and Wells, S. (1983). Invertebrates—who needs them? *New Scientist*, **98** (1358), 441–4.

Comyn-Platt, Sir T. (1937). A report on fauna preservation in Malaya. *Journal of the Society for the Preservation of the Fauna of the Empire*, New Series, Part **XXX**, 45–52.

Conklin, H. C. (1957). *Hanunóo agriculture: a report on an integral system of shifting cultivation in the Philippines*, FAO Forestry Development Paper No. 12. Food and Agriculture Organization of the United Nations, Rome.

Conklin, H. C. (1961). The study of shifting cultivation. *Current Anthropology*, **2**, 27–61.

Connell, J. H. (1978). Diversity in tropical rain forests and coral reefs. *Science*, **199** (4335), 1302–10.

Connor, E. F. (1986). The role of Pleistocene forest refugia in the evolution and biogeography of tropical biotas. *Trends in Ecology and Evolution*, **1**, 165–9.

Corbet, A. S. and Pendlebury, H. M. (1978). *The butterflies of the Malay Peninsula* (3rd edn). Malayan Nature Society, Kuala Lumpur.

Cousens, J. E. (1958). A study of 155 acres of tropical rain forest by complete enumeration of all large trees. *Malayan Forester*, **21**, 155–64.

Cox, E. H. M. (1961). *Plant hunting in China: a history of botanical exploration in China and the Tibetan marches*. Oldbourne, London.

Cramb, R. A. (1985). The importance of secondary crops in Iban hill rice farming. *Sarawak Museum Journal*, New Series, **34**, 37–45.

Cramb, R. A. (1988). The commercialization of Iban agriculture. In *Development in Sarawak: historical and contemporary perspectives* (ed. R. A. Cramb and R. H. W. Reece), pp. 105–34. Monash Paper on Southeast Asia No. 17, Centre of Southeast Asian Studies, Monash University, Australia.

Cranbrook, Earl of (1981). The vertebrate faunas. In *Wallace's line and plate tectonics* (ed. T. C. Whitmore), pp. 57–69. Clarendon Press, Oxford.

Cranbrook, Earl of (1983). *The wild mammals of Malaya (Peninsular Malaysia) and Singapore* (3rd edn). Oxford University Press, Kuala Lumpur.

Cranbrook, Earl of (1984). New and interesting records of mammals from Sarawak. *Sarawak Museum Journal*, New Series, **54**, 137–44.

Cranbook, Earl of (1988). Mammals: distribution and ecology. In *Malaysia* (ed. Earl of Cranbrook), pp. 146–66. Pergamon Press, Oxford.

Crisswell, C. N. (1978). *Rajah Charles Brooke: monarch of all he surveyed*. Oxford University Press, Kuala Lumpur.

Crosby, A. W. (1986). *Ecological imperialism: the biological expansion of Europe, 900–1900*. Cambridge University Press.

Cross, M. (1990). Logging agreement fails to protect Sarawak. *New Scientist*, **128** (1745), p. 7.

Crow, B. and Thomas, A. (1983). *Third World atlas*. Open University Press, Milton Keynes, UK.

Cruz, R. E. de la and Vergara, N. T. (1987). Protective and ameliorative roles of agroforestry: an overview. In *Agroforestry in the humid tropics: its protective and ameliorative roles to enhance productivity and sustainability* (ed. N. T. Vergara and N. D. Briones), pp. 3–30. Environment and Policy Institute, East–West Center, Honolulu, Hawaii, USA and Southeast Asian Regional Center for Graduate Study and Research in Agriculture, Los Baños, Laguna, Philippines.

Cubitt, G. E. S. (1920). *Forestry in the Malay peninsula: a statement prepared for the British Empire forestry conference, London, 1920*. Federated Malay States Printing Office, Kuala Lumpur.

Curtin, P. D. (1984). *Cross-cultural trade in world history*. Cambridge University Press.

Dabral, B. G., Premnath, and Ramswarup (1963). Some preliminary investigations on the rainfall interception by leaf litter. *Indian Forester*, **89**, 112–16.

Dakeyne, N. H. (1929). Reconditioning of old hill rubber and soil conservation generally. *Planter*, **9**, 233–6.

Dale, W. L. (1959). The rainfall of Malaya, Part I. *Journal of Tropical Geography*, **13**, 23–37.

Darby, H. C. (1956). The clearing of the woodland in Europe. In *Man's role in changing the face of the earth* (ed. W. L. Thomas, Jun.), pp. 183–216. University of Chicago Press, Chicago.

Darlington, P. J. (1957). *Zoogeograpy: the geographical distribution of animals*. Wiley, New York.

Das, K. (1982a). Societies in shock. *Far Eastern Economic Review*, **26 November**, 12–13.

Das, K. (1982b). Withdrawn from debate. *Far Eastern Economic Review*, **10 December**, p. 12.

Dasmann, R. F. (1984). The relationship between protected areas and indigenous peoples. In *National parks, conservation, and development: the role of protected areas in sustaining society* (ed. J. A. McNeely and K. R. Miller), pp. 667–71. Smithsonian Institution Press, Washington, DC.

Davies, G. and Payne, J. (n.d.). *A faunal survey of Sabah*. World Wildlife Fund, Kuala Lumpur.

Davis, S. D. *et al.* (1986). *Plants in danger: what do we know?* International Union for Conservation of Nature and Natural Resources, Gland, Switzerland.

Davison, G. W. H. (1987). The birds of Ulu Endau, Johore, Malaysia, with special reference to birds of heath and fan palm forests. *Malayan Nature Journal*, **41**, 425–34.

Davison, G. W. H. and Kiew, B. H. (1987). Mammals of Ulu Endau, Johore, Malaysia. *Malayan Nature Journal*, **41**, 435–40.

Day, M. J. (1980). Landslides in the Gunung Mulu National Park. *Geographical Journal*, **146**, 7–13.

Delcourt, H. R. (1987). The impact of prehistoric agriculture and land occupation on natural vegetation. *Ecology*, **34**, 341–6.

Denevan, W. M. (1980). Latin America. In *World systems of traditional resource management* (ed. G. A. Klee), pp. 217–44. Halsted Press, New York.

Denevan, W. M. (1983). Adaptation, variation, and cultural geography. *Professional Geographer*, **35**, 399–407.

Denslow, J. S. (1988). The tropical rain-forest setting. In *People of the tropical rain forest* (ed. J. S. Denslow and C. Padoch), pp. 25–36. University of California Press, Berkeley.

Dentan, R. K. (1968). *The Semai: a nonviolent people of Malaya*. Holt, Rinehart, and Winston, New York.

Desch, H. E. (1938). The forests of the Malay peninsula and their exploitation. *Malayan Forester*, **7**, 169–80.

Deshmukh, I. (1986). *Ecology and tropical biology*. Blackwell, Oxford.

Devall, B. and Sessions, G. (1985). *Deep ecology: living as if nature mattered*. Peregrine Smith Books, Layton, Utah, USA.

Diamond, J. M. (1980). Patchy distributions of tropical birds. In *Conservation biology: an evolutionary-ecological perspective* (ed. M. E. Soulé and B. A. Wilcox), pp. 57–74. Sinauer Associates, Sunderland, Massachusetts.

Dickinson III, J. C. (1972). Alternatives to monoculture in the humid tropics of Latin America. *Professional Geographer*, **24**, 217–22.

Dickinson, R. E. (1981). Effects of tropical deforestation on climate. In *Blowing in the wind: deforestation and long-range implications*, Studies in Third World Societies, No. 14, pp. 411–41. Department of Anthropology, College of William and Mary, Williamsburg, Virginia.

Dittus, W. P. J. (1985). The influence of cyclones on the dry evergreen forest of Sri Lanka. *Biotropica*, **17**, 1–14.

Douglas, I. (1972). *The environment game* (inaugural lecture). University of New England, Armidale, New South Wales, Australia.

Dover, M. J. and Talbot, L. M. (1987). *To feed the earth: agro-ecology for sustainable development*. World Resources Institute, Washington, DC.

Doyle, P. (1879). *Tin mining in Larut*. Spon, London.

Drabble, J. H. (1973). *Rubber in Malaya 1876–1922: the genesis of the industry*. Oxford University Press, Kuala Lumpur.

Drainage and Irrigation Department (1936). *Annual report for the Federated Malay States and Straits Settlements*. Government Press, Kuala Lumpur.

Drainage and Irrigation Department (1937). *Annual report for the Federated Malay States and Straits Settlements*. Government Press, Kuala Lumpur.

Dransfield, J. and Johnson, D. (1989). The conservation status of palms in Sabah. *Malayan Naturalist*, **November**, 16–19.

Duckett, J. E. (1976). Plantations as a habitat for wild-life in Peninsular Malaysia with particular reference to the oil palm (*Elaesis guineensis*). *Malayan Nature Journal*, **29**, 176–82.

Dunn, F. L. (1975). *Rain-forest collectors and traders: a study of resource utilization in modern and ancient Malaya*. Monographs of the Malaysian Branch of the Royal Asiatic Society, No. 5. Council of the Malaysian Branch of the Royal Asiatic Society, Kuala Lumpur.

Dunster, J. A. (1990). Forest conservation strategies in Canada: a challenge for the nineties. *Alternatives*, **16** (4) and **17** (1), 44–51.

Durant, C. L. (1936). Erosion. *Malayan Forester*, **5**, 109–11.

Eaton, B. J. (1935). Mr. Ridley and rubber in Malaya. *Gardens' Bulletin of the Straits Settlements*, **9**, 39–41.

Eckholm, E. P. (1982). *Down to earth: environment and human needs*. Pluto Press, London.

Edgar, A. T. (1958). *Manual of rubber planting (Malaya)*. Incorporated Society of Planters, Kuala Lumpur.

Edgar, A. T. and Shebbeare, E. O. (1961). The Malayan Nature Society 1940–1961. In *Nature conservation in western Malaysia, 1961* (ed. J. Wyatt-Smith and P. R. Wycherley), pp. 11–13. Malayan Nature Society, Kuala Lumpur.

Edwards, J. P. (1930). Growth of Malayan forest trees as shown by sample plot records, 1915–1928. *Malayan Forest Records*, **9**. Forestry Department, Federated Malay States, Kuala Lumpur.

Edwards, P. J. (1982). Studies of mineral cycling in a montane rain forest in New Guinea. V: rates of cycling in throughfall and litter fall. *Journal of Ecology*, **70**, 807–27.

Ehrenfeld, D. W. (1976). The conservation of non-resources. *American Scientist*, **64**, 648–56.

Ehrenfeld, D. (1978). *The arrogance of humanism*. Oxford University Press, New York.

Ehrlich, P. R. (1980). The strategy of conservation, 1980–2000. In *Conservation biology: an evolutionary–ecological perspective* (ed. M. E. Soulé and B. A. Wilcox), pp. 329–44. Sinauer Associates, Sunderland, Massachusetts.

Ehrlich, P. [R.] and Ehrlich, A. [H.] (1983). *Extinction: the causes and consequences of the disappearance of species*. Ballantine Books, New York.

Ehrlich, P. R. and Ehrlich, A. H. (1986). World population crisis. *Bulletin of the Atomic Scientists*, **42**, 13–17.

Ehrlich, P. R. and Mooney, H. A. (1983). Extinction, substitution, and ecosystem services. *BioScience*, **33**, 248–54.

Emerson, R. (1937). *Malaysia: a study in direct and indirect rule*. Macmillan, New York.

Endicott, K. (1979). The impact of economic modernisation on the *Orang Asli* (aborigines) of northern Peninsular Malaysia. In *Issues in Malaysian development* (ed. J. C. Jackson and M. Rudner), pp. 167–204. Heinemann, Singapore.

Endler, J. A. (1982). Pleistocene forest refuges: fact or fancy? In *Biological diversification in the tropics* (ed. G. T. Prance), pp. 641–57. Columbia University Press, New York.

Environmental Protection Society Malaysia (1981). The Fourth Malaysia Plan—so disappointing on the environment. *Alam Sekitar*, **6** (2), 7–8.

Evernden, N. (1985). *The natural alien: humankind and environment*. University of Toronto Press, Toronto.

Ewel, J. and Conde, L. F. (1980). *Potential ecological impact of increased intensity of tropical forest utilization*. BIOTROP, SEAMEO Regional Center for Tropical Biology, Bogor, Indonesia.

Eyles, R. J. (1967). Laterite at Kerdau, Pahang, Malaya. *Journal of Tropical Geography*, **25**, 18–23.

FAO/UNEP, Food and Agriculture Organization of the United Nations/United Nations Environment Programme (1981*a*). *Los recursos forestales de la America tropical*, Technical Report No. 1. FAO, Rome.

FAO/UNEP, Food and Agriculture Organization of the United Nations/United Nations Environment Programme (1981b). *Forest resources of tropical Africa*. Technical Report No. 2. FAO, Rome.

FAO/UNEP, Food and Agriculture Organization of the United Nations/United Nations Environment Programme (1981c). *Forest resources of tropical Asia*. Technical Report No. 3. FAO, Rome.

Fearnside, P. M. (1979). The development of the Amazon rain forest: priority problems for the formulation of guidelines. *Interciencia*, **4**, 338–43.

Fearnside, P. M. (1982). Deforestation in the Brazilian Amazon: how fast is it occurring? *Interciencia*, **7**, 82–8.

Fearnside, P. M. (1983). Development alternatives in the Brazilian Amazon: an ecological evaluation. *Interciencia*, **8**, 65–78.

Fearnside, P. M. (1986). Spatial concentration of deforestation in the Brazilian Amazon. *Ambio*, **15**, 74–81.

Fearnside, P. M. (1987a). Rethinking continuous cultivation in Amazonia. *BioScience*, **37**, 209–14.

Fearnside, P. M. (1987b). Deforestation and international economic development projects in Brazilian Amazonia. *Conservation Biology*, **1**, 214–21.

Federov, A. (1966). The structure of tropical rain forest and speciation in the humid tropics. *Journal of Ecology*, **54**, 1–11.

Fermor, L. L. (1939). *Report upon the mining industry of Malaya*. Government Press, Kuala Lumpur.

Fernandez, J. (1985). Letter from Bakun. *Far Eastern Economic Review*, **26 December**, p. 78.

Fernie, J. and Pitkethly, A. S. (1985). *Resources: environment and policy*. Harper and Row, London.

Fisher, C. A. (1964). *South-east Asia: a social, economic and political geography*. Methuen, London.

Fitch, F. H. (1952). *The geology and mineral resources of the neighbourhood of Kuantan, Pahang*, Geological Survey Memoir, New Series, No. 6. Kuala Lumpur.

Fitch, T. J. (1970). Earthquake mechanisms and island arc tectonics in the Indonesian–Philippine region. *Bulletin of the Geological Society of America*, **60**, 565–91.

Fitter, R. (1986). *Wildlife for man: how and why we should conserve our species*. Collins, London.

Fitzgerald, S. (1986). The US debate over the environmental performance of four multilateral development banks. *Ambio*, **15**, 291–5.

Flenley, J. (1979). *The equatorial rain forest: a geological history*. Butterworths, London.

Flynn, R. W. (1980). *Endau–Rompin national park management plan (preliminary draft)*. Department of Wildlife and National Parks, Kuala Lumpur.

Flynn, R. W. and Mohd. Tajuddin Abdullah (1984). Distribution and status of the Sumatran rhinoceros in Peninsular Malaysia. *Biological Conservation*, **28**, 253–73.

Foenander, E. C. (1952). *Big game of Malaya: their types, distribution and habits*. Batchworth Press, London.

Foster, R. B. (1980). Heterogeneity and disturbance in tropical vegetation. In *Conservation biology: an evolutionary–ecological perspective* (ed. M. E. Soulé and B. A. Wilcox), pp. 75–92. Sinauer Associates, Sunderland, Massachusetts.

Fox, J. E. D. (1968). Defect, damage and wastage. *Malayan Forester*, **31**, 157–64.

Fox, J. E. D. (1969). The soil damage factor in present day logging in Sabah. *Sabah Society Journal*, **5**, 43–52.

Fox, J. E. D. (1978). The natural vegetation of Sabah, Malaysia 1: the physical environment and classification. *Tropical Ecology*, **19**, 218–39.

Fox, J. E. D. (1983). The natural vegetation of Sabah, Malaysia 2: the *Parashorea* forests of the lowlands. *Tropical Ecology*, **24**, 94–112.

Frankel, O. H. and Soulé, M. E. (1981). *Conservation and evolution*. Cambridge University Press.

Franklin, I. R. (1980). Evolutionary change in small populations. In *Conservation biology: an evolutionary–ecological perspective* (ed. M. E. Soulé and B. A. Wilcox), pp. 135–49. Sinauer Associates, Sunderland, Massachusetts.

Fryer, D. W. (1979). *Emerging Southeast Asia* (2nd edn). Philip, London.

Furnivall, H. (1903). Mangrove swamps in the Federated Malay States. *Agricultural Bulletin of the Straits and Federated Malay States*, 2nd Series, **3**, 3–4.

Furtado, J. I. (1978). Endau–Rompin national park (a scientific case). In *Malaysian Scientific Association Annual Report 1976–77*, pp. 33–6. Kuala Lumpur.

Fyfe, A. J. (1964). Forestry in Sabah. *Malayan Forester*, **27**, 82–95.

Gardner, J. and Roseland, M. (1989a). Thinking globally: the role of social equity in sustainable development. *Alternatives*, **16** (3), 26–34.

Gardner, J. and Roseland, M. (1989b). Acting locally:

community strategies for equitable sustainable development. *Alternatives*, **16** (3), 36–48.

Garwood, N. C., Janos, D. P., and Brokaw, N. (1979). Earthquake-caused landslides: a major disturbance to tropical forests. *Science*, **205** (4410), 997–9.

Gautam-Basak, M. and Proctor, J. (1983). Micronutrients, aluminium, silicon and ash in leaf litterfall from forests in Gunung Mulu National Park, Sarawak. *Malaysian Forester*, **46**, 224–32.

Geddes, W. R. (1954). *The Land Dayaks of Sarawak*. HMSO, London.

Geertz, C. (1963). *Agricultural involution: the process of ecological change in Indonesia*. University of California Press, Berkeley.

Gentry, A. H. (1986). Endemism in tropical versus temperate plant communities. In *Conservation biology: the science of scarcity and diversity* (ed. M. E. Soulé), pp. 153–81. Sinauer Associates, Sunderland, Massachusetts.

Gentry, A. H. and Lopez-Parodi, J. (1980). Deforestation and increased flooding of the upper Amazon. *Science*, **210** (4476), 1354–6.

Gilbert, L. E. (1980). Food web organization and the conservation of neotropical diversity. In *Conservation biology: an evolutionary–ecological perspective* (ed. M. E. Soulé and B. A. Wilcox), pp. 11–33. Sinauer Associates, Sunderland, Massachusetts.

Gill, E. D. (1975). Evolution of Australia's unique flora and fauna in relation to the plate tectonics theory. *Proceedings of the Royal Society of Victoria*, **87**, 215–34.

Gillis, M. (1988). Malaysia: public policies and the tropical forest. In *Public policies and the misuse of forest resources* (ed. R. Repetto and M. Gillis), pp. 115–64. Cambridge University Press.

Glacken, C. J. (1967). *Traces on the Rhodian shore: nature and culture in western thought from ancient times to the end of the eighteenth century*. University of California Press, Berkeley.

Gliessman, S. R., Garcia, E. R., and Amador, A. M. (1981). The ecological basis for the application of traditional agricultural technology in the management of tropical agro-ecosystems. *Agro-ecosystems*, 7, 173–85.

Global Tomorrow Coalition (1986). Sustainable development and how to achieve it. A paper prepared by nongovernmental organizations for submission to the World Commission on Environment and Development. Global Tomorrow Coalition and Natural Resources Defense Council, Washington, DC.

Godfrey-Smith, W. (1979). The value of wilderness. *Environmental Ethics*, **1**, 309–19.

Goh, K. C. (1971). A comparative study of the rainfall–runoff characteristics of a developed and a forested catchment. Unpublished MA thesis, University of Malaya.

Goldman, C. R. (1979). Ecological aspects of water impoundment in the tropics. *Unasylva*, **31** (123), 2–11.

Gomes, A. G. (1988). The Semai: the making of an ethnic group in Malaysia. In *Ethnic diversity and the control of natural resources in Southeast Asia* (ed. A. T. Rambo, K. Gillogly, and K. L. Hutterer), pp. 99–117. Michigan Papers on South and Southeast Asia, No. 32. Center for South and Southeast Asian Studies, University of Michigan, Ann Arbor, Michigan.

Gómez-Pompa, A. (1987). On Maya silviculture. *Mexican Studies*, **3**, 1–17.

Goreau, T. J. and de Mello, W. Z. (1988). Tropical deforestation: some effects on atmospheric chemistry. *Ambio*, **17**, 275–81.

Goudie, A. (1986). *The human impact on the natural environment* (2nd edn). Blackwell, Oxford.

Gradwohl, J. and Greenberg, R. (1988). *Saving the tropical forests*. Earthscan, London.

Graham, E. (1984). *The modern plantation in the Third World*, E. Graham with I. Floering (ed. D. Fieldhouse). Croom Helm, London.

Grainger, A. (1983). Improving the monitoring of deforestation in the humid tropics. In *Tropical rain forest: ecology and management* (ed. S. L. Sutton, T. C. Whitmore, and A. C. Chadwick), pp. 387–95. Blackwell, Oxford.

Grainger, A. (1984). Quantifying changes in forest cover in the humid tropics: overcoming current limitations. *Journal of World Forest Resource Management*, **1**, 3–63.

Grainger, A. (1988). Estimating areas of degraded tropical lands requiring replenishment of forest cover. *International Tree Crops Journal*, **5**, 31–62.

Grandison, A. G. C. (1978). Snakes of West Malaysia and Singapore. *Annalen des Naturhistorischen Museums*, **81**, 282–303.

Green, K. M. (1983). Using Landsat to monitor tropical forest ecosystems: realistic expectations of digital processing technology. In *Tropical rain forest: ecology and management* (ed. S. L. Sutton, T. C.

Whitmore, and A. C. Chadwick), pp. 397–409. Blackwell, Oxford.

Greenland, D. J. (1975). Bringing the green revolution to the shifting cultivator. *Science*, **190** (4217), 841–4.

Greenwood, S. R. (1987). The role of insects in tropical forest food webs. *Ambio*, **16**, 267–71.

Grist, D. H. (1922). *Wet padi planting in Negri Sembilan*, Department of Agriculture, Federated Malay States Bulletin No. 33, Kuala Lumpur.

Gullick, J. M. (1958). *Indigenous political systems of western Malaya*. The Athlone Press, London.

Gullick, J. M. (1975). Selangor, 1876–1882: the Bloomfield Douglas diary. *Journal of the Malaysian Branch of the Royal Asiatic Society*, **48** (2), 1–51.

Gullick, J. [M.] (1981). *Malaysia: economic expansion and national unity*. Benn, London.

Gullick, J. M. (1983). *The story of Kuala Lumpur (1857–1939)*. Eastern Universities Press, Singapore.

Guppy, N. (1984). Tropical deforestation: a global view. *Foreign Affairs*, **62**, 928–65.

Gurmit Singh, K. S. (1984). *Malaysian societies: friendly or political?* Environmental Protection Society Malaysia and Selangor Graduates Society, Petaling Jaya, Malaysia.

Gurmit Singh, K. S. (1986). Environmental and technical dimensions of Bakun controversy. Paper presented at the Forum on the Bakun Hydro Electric Power Project, 22 February, Kuching [mimeo].

Hadley, M. and Lanly, J.-P. (1983). Tropical forest eco-systems: identifying differences, seeking similarities. *Nature and Resources*, **19**, 2–18.

Haile, N. S. (1958). The snakes of Borneo, with a key to the species. *Sarawak Museum Journal*, **8**, 743–71.

Haines, W. B. (1929). Some considerations on silt-pittings. *Quarterly Journal of the Rubber Research Institute of Malaya*, **1**, 20–8.

Hallé, F., Oldeman, R. A. A., and Tomlinson, P. B. (1978). *Tropical trees and forests: an architectural analysis*. Springer-Verlag, Berlin.

Hames, R. B. and Vickers, W. T. (eds.) (1983). *Adaptive responses of native Amazonians*. Academic Press, New York.

Hamilton, A. C. (1982). *Environmental history of East Africa: a study of the Quaternary*. Academic Press, London.

Hamilton, L. S. (with King, P. N.) (1983). *Tropical forested watersheds: hydrologic and soils response to major uses or conversions*. Westview Press, Boulder, Colorado.

Hamilton, L. S. (1985a). Overcoming myths about soil and water impacts of tropical forest land uses. In *Soil erosion and conservation* (ed. S. A. El-Swaify, W. C. Moldenhauer, and A. Lo), pp. 680–90. Soil Conservation Society of America, Ankeny, Iowa.

Hamilton, L. S. (1985b). Some water/soil consequences of modifying tropical rain forests. *The Environmentalist*, **5** (Supplement No. 10), 69–80.

Hamilton, L. S. (1987a). Semantics, definitions and deforestation. *IUCN Special Report Bulletin*, **18** (4–6), 8–9.

Hamilton, L. S. (1987b). Tropical watershed forestry—aiming for greater accuracy. *Ambio*, **16**, 372–3.

Hamilton, W. (1973). Tectonics of the Indonesian region. *Bulletin of the Geological Society of Malaysia*, **6**, 3–10.

Hamilton, W. (1979). *Tectonics of the Indonesian region*, United States Geological Survey Professional Paper, No. 1078. United States Government Printing Service, Washington, DC.

Hanbury-Tenison, R. (1980). *Mulu: the rain forest*. Weidenfeld and Nicolson, London.

Hanbury-Tenison, R. (1990). No surrender in Sarawak. *New Scientist*, **128** (1745), 12–13.

Han, Wai Toon (1985). Notes on Bornean camphor imported into China. *Brunei Museum Journal*, **6**, 1–31.

Hardesty, D. L. (1986). Rethinking cultural adaptation. *Professional Geographer*, **38**, 11–18.

Hare, F. K. (1980). The planetary environment: fragile or sturdy? *Geographical Journal*, **146**, 379–95.

Harrison, P. D. (1978). So the seeds shall grow: some introductory comments. In *Pre-Hispanic Maya agriculture* (ed. P. D. Harrison and B. L. Turner II), pp. 1–11. University of New Mexico Press, Albuquerque, New Mexico.

Harrison, P. D. and Turner II, B. L. (eds.) (1978). *Pre-Hispanic Maya agriculture*. University of New Mexico Press, Albuquerque, New Mexico.

Hartley, C. W. S. (1949). Soil erosion in Malaya. *Corona*, **1**, 25–7.

Hatch, T. (1982). *Shifting cultivation in Sarawak—a review*. Technical Paper No. 8, Soils Division Research Branch, Department of Agriculture, Kuching.

Hecht, S. B. (1981). Deforestation in the Amazon basin: magnitude, dynamics and soil resource effects. In *Where have all the flowers gone?: deforestation in the Third World*, Studies in Third World Societies, No. 13, pp. 61–108. Department of Anthropology, College of William and Mary, Williamsburg, Virginia.

Hecht, S. [B.] (1983). Cattle ranching in the eastern Amazon: environmental and social implications. In *The dilemma of Amazonian development* (ed. E. F. Moran), pp. 158–88. Westview Press, Boulder, Colorado.

Hecht, S. [B.] (1985). Environment, development and politics: capital accumulation and the livestock sector in eastern Amazonia. *World Development*, **13**, 663–84.

Helvey, J. D. and Patric, J. M. (1965). Canopy and litter interception of rainfall by hardwoods of eastern United States. *Water Resources Research*, **1**, 193–206.

Higgins, B. (1982). Development planning. In *The political economy of Malaysia* (ed. E. K. Fisk and H. Osman-Rani), pp. 148–83. Oxford University Press, Kuala Lumpur.

Hill, K. and Hurtado, A. M. (1989). Hunter–gatherers of the New World. *American Scientist*, **77**, 436–43.

Hill, R. D. (1977). *Rice in Malaya: a study in historical geography*. Oxford University Press, Kuala Lumpur.

Hing, Ai Yun (1984). Women and work in West Malaysia. *Journal of Contemporary Asia*, **14**, 204–18.

Hirschman, C. (1986). The recent rise in Malay fertility: a new trend or a temporary lull in a fertility transition? *Demography*, **23**, 161–84.

Ho, C. C., Newbery, D. McC., and Poore, M. E. D. (1987). Forest composition and inferred dynamics in Jengka Forest Reserve, Malaysia. *Journal of Tropical Ecology*, **3**, 25–56.

Hodgkin, E. P. (1956). *The transmission of malaria in Malaya*. Studies from the Institute for Medical Research, Federation of Malaya, No. 27. Kuala Lumpur.

Hoffman, C. (1986). *The Punan: hunters and gatherers of Borneo*. UMI Research Press, Ann Arbor, Michigan.

Hong, E. (1987). *Natives of Sarawak: survival in Borneo's vanishing forests*. Institute Masyarakat, Pulau Pinang, Malaysia.

Hooijer, D. A. (1975). Quaternary animals west and east of Wallace's line. *Netherlands Journal of Zoology*, **25**, 46–56.

Hose, C. (1893). *Mammals of Borneo*. Abbott, Diss, Norfolk, UK.

Hose, C. (1929). *The field-book of a jungle-wallah*. Witherby, London.

Hubback, T. R. (1905). *Elephant and seladang hunting in the Federated Malay States*. Rowland Ward, London.

Hubback, T. R. (1923). Game in Malaya. *Journal of the Society for the Preservation of the Fauna of the Empire*, New Series, Part **III**, 20–6.

Hubback, T. R. (1924). Letter from Theodore R. Hubback to the Chairman of the Fauna Society on the preservation of the fauna of Malaya. *Journal of the Society for the Preservation of the Fauna of the Empire*, New Series, Part **IV**, 60–3.

Hubback, T. R. (1926). Conservation of Malayan fauna. *Journal of the Society for the Preservation of the Fauna of the Empire*, New Series, Part **VI**, 35–41.

Hubback, T. [R.] (1938*a*). Malayan gaur or seladang. *Journal of the Bombay Natural History Society*, **40**, 8–19.

Hubback, T. [R.] (1938*b*). Principles of wild life conservation. *Journal of the Bombay Natural History Society*, **40**, 100–11.

Hubbell, S. P. (1979). Tree dispersion, abundance, and diversity in a tropical dry forest. *Science*, **203** (4387), 1299–309.

Hubbell, S. P. and Foster, R. B. (1983). Diversity of canopy trees in a neotropical forest and implications for conservation. In *Tropical rain forest: ecology and management* (ed. S. L. Sutton, T. C. Whitmore, and A. C. Chadwick), pp. 25–41. Blackwell, Oxford.

Hubbell, S. P. and Foster, R. B. (1986). Commonness and rarity in a neotropical forest: implications for tropical tree conservation. In *Conservation biology: the science of scarcity and diversity* (ed. M. E. Soulé), pp. 205–31. Sinauer Associates, Sunderland, Massachusetts.

Humphrey, J. W. (1982). The urbanization of Singapore's rural landscape. In *Too rapid rural development: perceptions and perspectives from Southeast Asia* (ed. C. MacAndrews and Chia, Lin Sien), pp. 334–66. Ohio University Press, Athens, Ohio.

Humphrey, S. R. (1985). How species become vulnerable to extinction and how we can meet the crises. In *Animal extinctions: what everyone should know* (ed. R. J. Hoage), pp. 9–29. Smithsonian Institution Press, Washington, DC.

Hunter, R. P. (1928). Problems of wash in old rubber on hill lands. *Planter*, **9**, 56–8.

Hunting Technical Services. (1971). *Johor Tenggara regional master plan: the master plan*. Hunting Technical Services, Boreham Wood, England, UK.

Hurst, P. (1987). Forest destruction in South East Asia. *The Ecologist*, **17**, 170–4.

Hutterer, K. L. (1982). Early Southeast Asia: old wine in new skins?—a review article. *Journal of Asian Studies*, **41**, 559–70.

Hutterer, K. L. (1983). The natural and cultural history of Southeast Asian agriculture. *Anthropos*, **78**, 169–212.

Hutterer, K. L. (1988). The prehistory of the Asian rain forests. In *People of the tropical rain forest* (ed. J. S. Denslow and C. Padoch), pp. 63–72. University of California Press, Berkeley.

Huxley, A. (1984). *Green inheritance: the World Wildlife Fund book of plants*. Collins, London.

ICIHI, Independent Commission on International Humanitarian Issues (1986). *The vanishing forest*. Zed Books, London.

ICIHI, Independent Commission on International Humanitarian Issues (1987). *Indigenous peoples: a global quest for justice*. Zed Books, London.

Indonesia, Department of Information (n.d.). *Indonesia 1988: an official handbook*. Jakarta, Indonesia.

Inger, R. F. (1966). The systematics and zoogeography of the amphibia of Borneo. *Fieldiana Zoology*, **52**, 1–402.

Inger, R. F. (1979). Abundance of amphibians and reptiles in tropical forests of South-east Asia. In *The abundance of animals in Malesian rain forests*, University of Hull, Department of Geography Miscellaneous Series No. 22 (ed. A. G. Marshall), pp. 93–112. University of Hull, Hull, UK.

Inskipp. T. and Wells, S. (1979). *International trade in wildlife*. International Institute for Environment and Development, London.

Irion, G. (1978). Soil infertility in the Amazonian rain forest. *Naturwissenschaften*, **65**, 515–19.

Ismail bin Haji Ali (1966). A critical review of Malayan silviculture in the light of changing demand and form of timber utilisation. *Malayan Forester*, **29**, 228–33.

ITTO, International Tropical Timber Organization, ITTO Mission (1990). The promotion of sustainable forest management: a case study in Sarawak, Malaysia. Report submitted to the international Tropical Timber Council, 8th session, Denpasar, Bali, Indonesia, 16–23 May 1990.

IUCN/UNEP/WWF, International Union for Conservation of Nature and Natural Resources/United Nations Environment Programme/World Wildlife Fund (1980). *World conservation strategy: living resource conservation for sustainable development*. IUCN, Gland, Switzerland.

Jackson, J. C. (1967). Oil palm: Malaya's post-independence boom crop. *Geography*, **52**, 319–21.

Jackson, J. C. (1968a). *Sarawak: a geographical survey of a developing state*. University of London Press.

Jackson, J. C. (1968b). *Planters and speculators: Chinese and European agricultural enterprise in Malaya 1786–1921*. University of Malaya Press, Kuala Lumpur.

Janzen, D. H. (1973). Tropical agroecosystems. *Science*, **182** (4118), 1212–19.

Jayl Langub (1988). The Penan strategy. In *People of the rain forest* (ed. J. S. Denslow and C. Padoch), pp. 207–8. University of California Press, Berkeley.

Jeyakumar Devaraj (1989). Logging accidents in Sarawak. In *Logging against the natives of Sarawak*, pp. 31–54. INSAN (Institute of Social Analysis), Petaling Jaya, Malaysia.

John, D. W. (1974). The timber industry and forest administration in Sabah under Chartered Company rule. *Journal of Southeast Asian Studies*, **5**, 55–81.

Johns, A. D. (1981). The effects of selective logging on the social structure of resident primates. *Malaysian Applied Biology*, **10**, 221–6.

Johns, A. [D.] (1983a). Wildlife can live with logging. *New Scientist*, **99** (1367), 206–9.

Johns, A. D. (1983b). Tropical forest primates and logging—can they co-exist? *Oryx*, **17**, 114–18.

Johns, A. D. (1985). Selective logging and wildlife conservation in tropical rain-forest: problems and recommendations. *Biological Conservation*, **31**, 355–75.

Johns, A. D. (1986). Effects of selective logging on the behavioral ecology of West Malaysian primates. *Ecology*, **67**, 684–94.

Johns, A. D. (1987). The use of primary and selectively logged rain-forest by Malaysian hornbills (Bucerotidae) and implications for their conservation. *Biological Conservation*, **40**, 179–90.

Johns, A. D. (1988a). Economic development and wildlife conservation in Brazilian Amazonia. *Ambio*, **17**, 302–6.

Johns, A. D. (1988b). Effects of 'selective' timber extraction on rain forest structure and composition and some consequences for frugivores and folivores. *Biotropica*, **20**, 31–7.

Johns, R. J. (1982). Plant migration. In *Biogeography and ecology of New Guinea* (ed. J. L. Gressitt), pp. 309–30. Dr. W. Junk, The Hague.

Johns, R. J. (1986). The instability of the tropical ecosystem in New Guinea. *Blumea*, **31**, 341–71.

Jomo, K. S. (1989). The pillage of Sarawak's forests. In *Logging against the natives of Sarawak*, pp. v–xii. INSAN (Institute of Social Analysis), Petaling Jaya, Malaysia.

Jones, A. (1968). The Orang Asli: an outline of their progress in modern Malaya. *Journal of Southeast Asian History*, **9**, 286–305.

Jones, D. T. (1984). Portraits of threatened plants. 5. *Melicope suberosa* Stone. *Malayan Naturalist*, **May**, 5–6.

Jones, G. W. and Tan, P. C. (1985). Recent and prospective population trends in Malaysia. *Journal of Southeast Asian Studies*, **16**, 262–80.

Jones, L. W. (1966). *The population of Borneo: a study of the people of Sarawak, Sabah and Brunei*. University of London, Athlone Press.

Jordan, C. F. (1982). Amazon rain forests. *American Scientist*, **70**, 394–401.

Jordan, C. F. (1985). *Nutrient cycling in tropical forest ecosystems: principles and their application in management and conservation*. Wiley, Chichester, UK.

Jordan, C. F. (1986). Local effects of tropical deforestation. In *Conservation biology: the science of scarcity and diversity* (ed. M. E. Soulé), pp. 410–26. Sinauer Associates, Sunderland, Massachusetts.

Jordan, C. F. (1987a). Conclusion: comparison and evaluation of case studies. In *Amazonian rain forests: ecosystem disturbance and recovery: case studies of ecosystem dynamics under a spectrum of land use-intensities* (ed. C. F. Jordan), pp. 100–21. Springer-Verlag, New York.

Jordan, C. F. (ed.) (1987b). *Amazonian rain forests: ecosystem disturbance and recovery: case studies of ecosystem dynamics under a spectrum of land use-intensities*. Springer-Verlag, New York.

Kamarrudin Sharif (1982). The demographic situation. In *The political economy of Malaysia* (ed. E. K. Fisk and H. Osman-Rani), pp. 66–87. Oxford University Press, Kuala Lumpur.

Kartawinata, K. (1981). The environmental consequences of tree removal from the forest in Indonesia. In *Where have all the flowers gone? Deforestation in the Third World*, Studies in Third World Societies Publication, No. 13, pp. 191–214. Department of Anthropology, College of William and Mary, Williamsburg, Virginia.

Kartawinata, K., Abdulhadi, R., and Partomihardjo, T. (1981a). Composition and structure of a lowland dipterocarp forest at Wanariset, East Kalimantan. *Malaysian Forester*, **44**, 397–406.

Kartawinata, K., Soenartono Adisoemarto, Soedarsono Riswan, and A. P. Vayda (1981b). The impact of man on a tropical forest in Indonesia. *Ambio*, **10**, 115–19.

Keast, J. A. (1983). In the steps of Alfred Russel Wallace: biogeography of the Asian–Australian interchange zone. In *Evolution, time and space: the emergence of the biosphere*, Systematics Association Special Volume No. 25 (ed. R. W. Sims, J. H. Price, and P. E. S. Whalley), pp. 367–407. Academic Press, London.

Kemper, C. (1988). The mammals of Pasoh Forest Reserve, Peninsular Malaysia. *Malayan Nature Journal*, **41**, 435–40.

Kenworthy, J. B. (1971). Water and nutrient cycling in a tropical forest. In *The water relations of Malesian forests*, University of Hull, Department of Geography Miscellaneous Series, No. 11 (ed. J. R. Flenley), pp. 49–58. University of Hull, Hull, UK.

Kershaw, A. P. (1978). Record of last interglacial-glacial cycle from northeastern Queensland. *Nature*, **272** (9 March), 159–61.

Kershaw, A. P. (1981). Quaternary vegetation and environments. In *Ecological biogeography of Australia* (ed. A. Keast), pp. 82–101. Dr. W. Junk, The Hague.

Khoo, Kay Kim (1972). *The western Malay states 1850–1873: the effects of commercial development on Malay politics*. Oxford University Press, Kuala Lumpur.

Kiew, B. H. (1982a). Conservation status of the Malaysian fauna. I. Mammalia. *Malayan Naturalist*, **May**, 3–19.

Kiew, B. H. (1982b). Conservation in the country: Fourth Malaysia Plan. *Malayan Naturalist*, **January**, 3–6.

Kiew, B. H. (1986a). Progress report on Endau-Rompin expedition for the period June–November 1985. *Malayan Naturalist*, **May**, 3–7.

Kiew, B. H. (1986b). Assessment of the Endau-Rompin expedition. *Malayan Naturalist*, **November**, 5–9.

Kiew, B. H. (1987). The Malaysian Heritage and Scientific Expedition—Endau-Rompin: a Third World experience. *Wallaceana*, **47**, 3–5.

Kiew, B. H. (1988). Cameron Highlands–Fraser's Hill–Genting Highlands road. *Malayan Naturalist*, **August**, 36–7.

Kiew, B. H. and Davison, G. (1982). Conservation status of the Malaysian fauna. II. Birds. *Malayan Naturalist*, **November**, 2–34.

Kiew, B. H., Kiew, R., Chin, S. C., Davison, G., and Ng, F. S. P. (1985). Malaysia's 10 most endangered animals, plants, and areas. *Malayan Naturalist*, **May**, 3–4.

Kiew, B. H., Davison, G. W. H., and Kiew, R. (1987). The Malaysian Heritage and Scientific Expedition: Endau–Rompin, 1985–1986. *Malayan Nature Journal*, **41**, 83–92.

Kiew, R. (1983a). Portraits of threatened plants. *Malayan Naturalist*, **November**, 6–7.

Kiew, R. (1983b). Conservation of Malaysian plant species. *Malayan Naturalist*, **August**, 2–5.

Kiew, R. (1988a). Portraits of threatened plants. 16. *Phoenix paludosa* Roxb. *Malayan Naturalist*, **August**, p. 16.

Kiew, R. (1988b). Herbaceous flowering plants. In *Malaysia* (ed. Earl of Cranbrook), pp. 56–76. Pergamon Press, Oxford.

Kiew, R. (1989a). Utilization of palms in Peninsular Malaysia. *Malayan Naturalist*, **November**, 43–67.

Kiew, R. (1989b). Conservation status of palms in Peninsular Malaysia. *Malayan Naturalist*, **November**, 3–15.

Kiew, R. and Dransfield, J. (1987). The conservation of palms in Malaysia. *Malayan Naturalist*, **August**, 24–31.

King, K. F. S. (1979). Agroforestry and the utilisation of fragile ecosystems. *Forest Ecology and Management*, **2**, 161–8.

King, V. T. (1986). Land settlement schemes and the alleviation of rural poverty in Sarawak, East Malaysia: a critical commentary. *Southeast Asian Journal of Social Science*, **14**, 71–99.

King, V. T. (1988). Models and realities: Malaysian national planning and East Malaysian development problems. *Modern Asian Studies*, **22**, 263–98.

Kira, T. (1978). Primary productivity of Pasoh Forest—a synthesis. *Malayan Nature Journal*, **30**, 292–7.

Kitchener, H. J. (1961). The bleak future for the seladang or Malayan gaur. In *Nature conservation in western Malaysia, 1961* (ed. J. Wyatt-Smith and P. R. Wycherley), pp. 197–201. Malayan Nature Society, Kuala Lumpur.

Klee, G. A. (ed.) (1980a). *World systems of traditional resource management*. Halsted Press, New York.

Klee, G. A. (1980b). Traditional wisdom and the modern resource manager. In *World systems of traditional resource management* (ed. G. A. Klee), pp. 283–5. Halsted Press, New York.

Kochummen, K. M. (1978). Natural plant succession after farming at Kepong. *Malaysian Forester*, **28**, 146–51.

Koopmans, B. N. and Stauffer, P. H. (1966). Glacial phenomena on Mt. Kinabalu. *Geological Survey of Malaysia(Borneo) Bulletin*, **8**, 25–35.

Koteswaram, P. (1974). Climate and meteorology of humid tropical Asia. In *Natural resources of humid tropical Asia*, pp. 27–86. Unesco, Paris.

Kumar, R. (1986). *The forest resources of Malaysia: their economics and development*. Oxford University Press, Singapore.

Kunstadter, P. (1978). Alternatives for the development of upland areas. In *Farmers in the forest: economic development and marginal agriculture in northern Thailand* (ed. P. Kunstadter, E. C. Chapman, and Sanja Sabhasri), pp. 289–308. University Press of Hawaii, Honolulu, Hawaii.

Laarman, J. G. (1988). Export of tropical hardwoods in the twentieth century. In *World deforestation in the twentieth century* (ed. J. F. Richards and R. P. Tucker), pp. 147–63. Duke University Press, Durham, North Carolina.

Lall Singh Gill and Dato Haji Wan Hassan bin Abdul. Halim (1968). Some problems of natural regeneration establishment in the hill dipterocarp forests of Perak. *Malayan Forester*, **31**, 297–304.

Lam, H. J. (1934). Materials towards a study of the flora of the island of New Guinea. *Blumea*, **1**, 113–59.

Lanly, J.-P. (1982). *Tropical forest resources*, FAO Forestry Paper 30. Food and Agriculture Organization of the United Nations, Rome.

Lanly, J.-P. (1983). Present situation and evaluation of tropical forest resources. *Mazingira*, **7**, 2–15.

Lanly, J.-P. and Clement, J. (1979). *Present and future forest and plantation areas in the tropics*. Food and Agriculture Organization of the United Nations, Rome.

Lau, Buong Tiing (1979). The effects of shifting cultivation on sustained yield management for Sarawak national forests. *Malaysian Forester*, **42**, 418–22.

Lee, H. S. (1982). The development of silvicultural systems in the hill forests of Malaysia. *Malaysian Forester*, **45**, 1–9.

Lee, P. C. (1968). The implications of the land capability classification on forest reservation in West Malaysia. *Malayan Forester*, **31**, 74–7.

Lee, R. B. and DeVore, I. (eds.) (1968). *Man the hunter*. Aldine, Chicago.

Lee, Weng Chung (1990). Loggers unite to fight greens. *Far Eastern Economic Review*, **18 January**, 45–6.

Lee, Y. L. (1961). Some aspects of shifting cultivation in British Borneo. *Malayan Forester*, **24**, 102–9.

Lee, Y. L. (1965). *North Borneo (Sabah): a study in settlement geography*. Eastern Universities Press, Singapore.

Lee, Y. L. (1968). Land use in Sarawak. *Sarawak Museum Journal, New Series*, **16**, 282–308.

Leigh, C. H. (1973). Land development and soil erosion in West Malaysia. *Area*, **5**, 213–17.

Leigh, C. H. (1978a). Slope hydrology and denudation in the Pasoh Forest Reserve. I: surface wash: experimental techniques and some preliminary results. *Malayan Nature Journal*, **30**, 179–97.

Leigh, C. H. (1978b). Slope hydrology and denudation in the Pasoh Forest Reserve. II: throughflow: experimental techniques and some preliminary results. *Malayan Nature Journal*, **30**, 199–210.

Leigh, C. H. (1982a). Sediment transport by surface wash and throughflow at the Pasoh Forest Reserve, Negri Sembilan, Peninsular Malaysia. *Geografiska Annaler*, **64A**, 171–80.

Leigh, C. H. (1982b). Urban development and soil erosion in Kuala Lumpur, Malaysia. *Journal of Environmental Management*, **15**, 35–45.

Leigh, C. H. and Low, K. S. (1978). The flood hazard in Peninsular Malaysia: government policies and actions. *Pacific Viewpoint*, **19**, 47–64.

Lewis, K. V., Cassell, P. A., and Fricke, T.J. (1975). *Urban design standards and procedures for Peninsular Malaysia*. Ministry of Agriculture and Rural Development, Kuala Lumpur.

Lian, F. J. (1988). The economics and ecology of the production of the tropical rainforest resources by tribal groups of Sarawak. In *Changing tropical forests: historical perspectives on today's challenges in Asia, Australasia and Oceania* (ed. J. Dargavel, K. Dixon, and N. Semple), pp. 113–25. Centre for Resource and Environmental Studies, Australian National University, Canberra.

Liew, That Chim (1974). A note on soil erosion study at Tawau Hills Forest Reserve. *Malayan Nature Journal*, **27**, 20–6.

Lim, B. H. (1908). The Chinese method of rotation of crops and reclamation of lalang land. *Agricultural Bulletin of the Straits and Federated Malay States*, 2nd Series, **7**, 450–2.

Lim, B. L., Illar Muul, and Chai, K. S. (1977). Zoonotic studies of small animals in the canopy transect of Bukit Lanjan Forest Reserve, Selangor, Malaysia. *Malayan Nature Journal*, **31**, 127–40.

Lim, D. (1982/83). Malaysian development planning. *Pacific Affairs*, **55**, 613–39.

Lim, D. (1986). East Malaysia in Malaysian development planning. *Journal of Southeast Asian Studies*, **17**, 156–70.

Lim, M. T. (1978). Litterfall and mineral nutrient content of litter in Pasoh forest. *Malayan Nature Journal*, **30**, 375–80.

Lim, Teck Ghee (1977). *Peasants and their agricultural economy in colonial Malaya 1874–1941*. Oxford University Press, Kuala Lumpur.

Locke, A. (1954). *The tigers of Trengganu*. Museum Press, London.

Logan, J. R. (1851). Notes at Pinang, Kidah, etc. *Journal of the Indian Archipelago and Eastern Asia*, **5**, 53–65.

Longman, K. A. and Jeník, J. (1974). *Tropical forest and its environment*. Longman, London.

Lovejoy, T. E., Bierregaard, R. O., Rankin, J. M., and Schubart, H. O. R. (1983). Ecological dynamics of tropical forest fragments. In *Tropical rain forest: ecology and management* (ed. S. L. Sutton, T. C. Whitmore, and A. C. Chadwick), pp. 377–84. Blackwell, Oxford.

Lovejoy, T. E. *et al.* (1986). Edge and other effects of isolation on Amazon forest fragments. In *Conservation biology: the science of scarcity and diversity* (ed. M. E. Soulé), pp. 257–85. Sinauer Associates, Sunderland, Massachusetts.

Low, J. (1836/1972). *A dissertation on the soil and agriculture of the British settlement of Penang or Prince of Wales Island*. Singapore. Reprinted 1972 as *The British Settlement of Penang*, Oxford University Press, Singapore.

Low, K. S. (1972). Interception loss in the humid forested areas. *Malayan Nature Journal*, **25**, 104–11.

Lowry, K. and Carpenter, R. A. (1985). Institutionalizing sustainable development: experiences in five countries. *Environmental Impact Assessment Review*, **5**, 239–54.

Lugo, A. E. (1988). The future of the forests. *Environment*, **30**, 17–20, 41–5.

Lugo, A. E. and Brown, S. (1982). Deforestation in the Brazilian Amazon. *Interciencia*, **7**, 361–2.

Lugo, A. E., Applefield, M., Pool, D. J., and McDonald, R. B. (1983). The impact of hurricane David on the forests of Dominica. *Canadian Journal of Forest Research*, **13**, 201–11.

Mabberley, D. J. (1983). *Tropical rain forest ecology*. Blackie, Glasgow.

MacArthur, R. H. and Wilson, E. O. (1967). *The theory of island biogeography*. Princeton University Press, Princeton, New Jersey.

Mackie, C. (1984). The lessons behind East Kalimantan's forest fires. *Borneo Research Bulletin*, **16** (7), 63–74.

MacKinnon, J. R. and MacKinnon, K. S. (1980). Niche differentiation in a primate community. In *Malayan forest primates: ten years' study in a tropical rain forest* (ed. D. J. Chivers), pp. 167–90. Plenum Press, London.

MacKinnon, K. S. (1978). Stratification and feeding differences among Malayan squirrels. *Malayan Nature Journal*, **30**, 593–608.

MacNeill, J. (1989). Strategies for sustainable economic development. *Scientific American*, **261** (3), 154–65.

Mahrus Ibrahim (1986). Hidden value of mangrove forests. *New Straits Times* (Kuala Lumpur), 3 July.

Majid, S. and Majid, A. (1983). Public sector land settlement: rural development in West Malaysia. In *Rural development and the state: contradictions and dilemmas in developing countries* (ed. D. A. M. Lea and D. P. Chaudhri), pp. 66–99. Methuen, London.

Malaya, Government of Malaya (1956). *First Malaya Plan 1956–1960*. Government Printer, Kuala Lumpur.

Malaya, Government of Malaya (1961). *Second Malaya Plan 1961–1965*. Government Printer, Kuala Lumpur.

Malaya, Rubber Research Institute of Malaya (1928). *Annual Report*. Government Printer, Kuala Lumpur.

Malayan Nature Society (1971). The grounds for conservation. *Malayan Nature Journal*, **24**, 193–4.

Malayan Nature Society (1974). A blueprint for conservation in Peninsular Malaysia. *Malayan Nature Journal*, **27**, 1–16.

Malayan Nature Society (1975). The tiger. *Malayan Naturalist*, **2** (1), 5–7.

Malaysia, Government of Malaysia (1966). *First Malaysia Plan 1966–1970*. Government Printer, Kuala Lumpur.

Malaysia, Government of Malaysia (1969). *Mid-term review of the First Malaysia Plan 1966–1970*. Government Printer, Kuala Lumpur.

Malaysia, Government of Malaysia (1971). *Second Malaysia Plan 1971–1975*. Government Printer, Kuala Lumpur.

Malaysia, Government of Malaysia (1973). *Mid-term review of the Second Malaysia Plan 1971–1975*. Government Printer, Kuala Lumpur.

Malaysia, Government of Malaysia (1976). *Third Malaysia Plan 1976–1980*. Government Press, Kuala Lumpur.

Malaysia, Government of Malaysia (1981). *Fourth Malaysia Plan 1981–1985*. National Printing Department, Kuala Lumpur.

Malaysia, Government of Malaysia (1984). *Mid-term review of the Fourth Malaysia Plan 1981–1985*. Government Printer, Kuala Lumpur.

Malaysia, Government of Malaysia (1986). *Fifth Malaysia Plan 1986–1990*. National Printing Department, Kuala Lumpur.

Malaysia, Rubber Research Institute of Malaysia (1974). Management of soils under hevea in Peninsular Malaysia. *Planters' Bulletin of the Rubber Research Institute of Malaysia*, **134**, 147–54.

Malingreau, J.-P. and Tucker, C. J. (1988). Large-scale deforestation in the southeastern Amazon basin of Brazil. *Ambio*, **17**, 49–55.

Malingreau, J.-P., Stephens, G., and Fellows, L. (1985). Remote sensing of forest fires: Kalimantan and North Borneo in 1982–83. *Ambio*, **14**, 314–21.

Maloney, B. K. (1985). Man's impact on the rainforests of West Malesia: the palynological record. *Journal of Biogeography*, **12**, 537–58.

Manning, E. W. (1990). Conservation strategies: providing the vision for sustainable development. *Alternatives*, **16** (4) and **17** (1), 24–9.

Manokaran, N. (1977). Nutrients in various phases of water movement in a lowland tropical rain forest in Peninsular Malaysia. Unpublished MSc thesis, University of Malaya.

Manokaran, N. (1979). Stemflow, throughflow and rainfall interception in a lowland tropical rainforest in Peninsular Malaysia. *Malaysian Forester*, **42**, 174–201.

Manokaran, N. (1980). The nutrient contents of precipitation, throughfall, and stemflow in a lowland tropical rain forest in Peninsular Malaysia. *Malaysian Forester*, **43**, 266–89.

Marsh, G. P. (1864/1965). *Man and nature; or, physical geography as modified by human action*. Scribner, New York. Reprinted 1965, Belknap Press, Cambridge, Massachusetts.

Martyn, H. S. (1966). Wood harvesting, logging and transport in Sabah. *Malayan Forester*, **29**, 296–301.

Masing, J. (1988). The role of resettlement in rural development. In *Development in Sarawak: historical and contemporary perspectives* (ed. R. A. Cramb and R. H. W. Reece), pp. 57–68. Monash Paper on Southeast Asia No. 17, Centre of Southeast Asian Studies, Monash University, Australia.

Maskayu [Monthly Timber Bulletin of the Malaysian Timber Industry Board].

Maxwell, G. (1907/1960). *In Malay forests*. Blackwood, Edinburgh. Reprinted 1960, Eastern Universities Press, Singapore.

Mayr, E. (1944). Wallace's line in the light of recent zoogeographic studies. *Quarterly Review of Biology*, **19**, 1–14.

McClure, H. E. and Hussein b. Othman, (1965). Avian bionomics of Malaya 2: the effect of forest destruction upon a local population. *Bird Banding*, **36**, 242–69.

McDowell, M. A. (1989). Development and the environment in ASEAN. *Pacific Affairs*, **62**, 307–29.

McNamara, R. S. (1984). Time bomb or myth: the population problem. *Foreign Affairs*, **62**, 1107–31.

McNeely, J. A. and Pitt, D. (eds.) (1985). *Culture and conservation: the human dimension in environmental planning*. Croom Helm, London.

Mead, J. P. (1925). Forestry in Sarawak. *Empire Forestry Journal*, **4**, 92–9.

Meade, M. [S.] (1976). Land development and human health in West Malaysia. *Annals of the Association of American Geographers*, **66**, 428–39.

Meade, M. S. (1977). Medical geography as human ecology: the dimensions of population movement. *Geographical Review*, **67**, 379–93.

Meade, M. S. (1978). Community health and changing hazards in a voluntary agricultural resettlement. *Social Science and Medicine*, **12** (D), 95–102.

Meade, M. S., Florin, J. W., and Gesler, W. M. (1988). *Medical geography*.The Guilford Press, New York.

Means, G. P. (1985/86). The Orang Asli: aboriginal policies in Malaysia. *Pacific Affairs*, **58**, 637–52.

Medway, Lord (1972). The Quaternary mammals of Malesia: a review. In *The Quaternary era in Malesia*, University of Hull, Department of Geography Miscellaneous Series, No. 13 (ed. P. Ashton and M. Ashton), pp. 63–83. University of Hull, Hull, UK.

Medway, Lord (1977). *Mammals of Borneo: field keys and annotated checklist*, Malaysian Branch of the Royal Asiatic Society Monograph, No. 7. Kuala Lumpur.

Medway, Lord and Wells, D. R. (1971). Diversity and density of birds and mammals at Kuala Lompat, Pahang. *Malayan Nature Journal*, **24**, 238–47.

Medway, Lord and Wells, D. R. (eds.) (1976). *The birds of the Malay Peninsula*, Vol. V. *Conclusion, and survey of every species*. Witherby, London.

Mehmet, O. (1986). *Development in Malaysia: poverty, wealth and trusteeship*. Croom Helm, London.

Meiggs, R. (1982). *Trees and timber in the ancient Mediterranean world*. Oxford University Press.

Meijer, W. (1971). Plant life in Kinabalu National Park. *Malayan Nature Journal*, **24**, 184–9.

Melillo, J. M., Palm, C. A., Houghton, R. A., Woodwell, G. M., and Myers, N. (1985). A comparison of two recent estimates of disturbance in tropical forests. *Environmental Conservation*, **12**, 37–40.

Menon, K. D. (1969). A brief history of forest research in Malaya. *Malayan Forester*, **32**, 3–29.

Metcalfe, G. T. C. (1961). Rhinoceros in Malaya and their future. In *Nature conservation in western Malaysia, 1961* (ed. J. Wyatt-Smith and P. R. Wycherley), pp. 183–91. Malayan Nature Society, Kuala Lumpur.

Mikesell, M. W. (1969). The deforestation of Mount Lebanon. *Geographical Review*, **59**, 1–28.

Miller, K. R. (1982). Parks and protected areas: considerations for the future. *Ambio*, **11**, 315–17.

Milne, R. S. and Mauzy, D. K. (1986). *Malaysia: tradition, modernity, and Islam*. Westview Press, Boulder, Colorado.

Mitchell, B. A. (1957). Malayan tin tailings—prospects of rehabilitation. *Malayan Forester*, **20**, 181–6.

Moeller, B. B. (1984). Is the Brazilian Amazon being destroyed? *Journal of Forestry*, **82**, 472–5.

Mohamed Suffian bin Hashim (1976). *An introduction to the constitution of Malaysia*. Government Printer, Kuala Lumpur.

Mohd. Khan bin Momin Khan (1977). Reproduction, productivity and mortality of the Malayan elephant. *Malayan Nature Journal*, **30**, 25–30.

Mohd. Khan bin Momin Khan (1988). Animal conservation strategies. In *Malaysia* (ed. Earl of Cranbrook), pp. 251–72. Pergamon Press, Oxford.

Mohd. Khan bin Momin Khan, Sivananthan T. Elagupillay, and Zolkifli bin Zainal (1983). Species conservation priorities in the tropical rain forests of Peninsular Malaysia. *Malayan Naturalist*, **May**, 2–8.

Mohun, J. and Omar Sattaur (1987). The drowning of a culture. *New Scientist*, **113** (1543), 37–42.

Monosowski, E. (1983). The Tecuruí experience. *International Water Power and Dam Construction*, **35**, 11–14.

Moran, E. F. (1986). Anthropological approaches to the study of human impacts. In *Natural resources and people: conceptual issues in interdisciplinary research* (ed. K. A. Dahlberg and J. W. Bennett), pp. 107–27. Westview Press, Boulder, Colorado.

Moran, E. F. (1988). Following the Amazonian highways. In *People of the rain forest* (ed. J. S. Denslow and C. Padoch), pp. 155–62. University of California Press, Berkeley.

Morgan, J. R. (1984). Southeast Asian place names: some lessons learned. *American Cartographer*, **11**, 5–13.

Morgan, R. P. C. (1974). Estimating regional variations in soil erosion hazard in Peninsular Malaysia. *Malayan Nature Journal*, **28**, 94–106.

Morley, R. J. and Flenley, J. R. (1987). Late Cainozoic vegetational and environmental changes in the Malay archipelago. In *Biogeographical evolution of the Malay archipelago* (ed. T. C. Whitmore), pp. 50–9. Clarendon Press, Oxford.

Morrison, A. (1988). Development in Sarawak in the colonial period: a personal memoir. In *Development in Sarawak: historical and contemporary perspectives* (ed. R. A. Cramb and R. H. W. Reece), pp. 35–47. Monash Paper on Southeast Asia No. 17, Centre of Southeast Asian Studies, Monash University, Australia.

Muhammad Jabil (1980). The national forest policy. *Malaysian Forester*, **43**, p. 1.

Muller, J. (1972). Palynological evidence for change in geomorphology, climate and vegetation in the Mio-Pliocene of Malesia. In *The Quaternary era in Malesia*, University of Hull, Department of Geography Miscellaneous Series, No. 13 (ed. P. Ashton and M. Ashton), pp. 6–16. University of Hull, Hull, UK.

Murphey, R. (1951). The decline of North Africa since the Roman occupation: climatic or human? *Annals of the Association of American Geographers*, **41**, 116–32.

Musa bin Nordin (1983). Management of wildlife reserves in Peninsular Malaysia. *Journal of Wildlife and Parks*, **2**, 103–14.

Myers, N. (1979). *The sinking ark: a new look at the problem of disappearing species*. Pergamon Press, Oxford.

Myers, N. (1980). *Conversion of tropical moist forests*. National Academy of Sciences, Washington, DC.

Myers, N. (1981). The hamburger connection: how Central America's forests become North America's hamburgers. *Ambio*, **10**, 3–8.

Myers, N. (1983a). Conservation of rain forests for scientific research, for wildlife conservation, and for recreation and tourism. In *Tropical rain forest ecosystems: structure and function* (ed. F. B. Golley), pp. 325–34. Elsevier, Amsterdam.

Myers, N. (1983b). *A wealth of wild species: storehouse for human welfare*. Westview Press, Boulder, Colorado.

Myers, N. (1984). *The primary source: tropical forests and our future*. Norton, New York.

Myers, N. (1985a). Tropical deforestation and species extinctions: the latest news. *Futures*, **17**, 451–63.

Myers, N. (1985b). A look at the present extinction spasm and what it means for the future evolution of species. In *Animal extinctions: what everyone should know* (ed. R. J. Hoage), pp. 47–57. Smithsonian Institution Press, Washington, DC.

Myers, N. (1986a). Environmental repercussions of deforestation in the Himalayas. *Journal of World Forest Resource Management*, **2**, 63–72.

Myers, N. (1986b). Tropical deforestation and a mega-extinction spasm. In *Conservation biology: the science of scarcity and diversity* (ed. M. E. Soulé), pp. 394–409. Sinauer Associates, Sunderland, Massachusetts.

Myers, N. (1986c). Tree-crop based agroecosystems in Java. *Forest Ecology and Management*, **17**, 1–11.

Myers, N. (1989). *Deforestation rates in tropical forests and their climatic implications*. Friends of the Earth, London.

Myers, N. and Ayensu, E. S. (1983). Reduction of biological diversity and species loss. *Ambio*, **12**, 72–4.

Myers, N. and Myers, D. (1983). How the global community can respond to international environmental problems. *Ambio*, **12**, 20–6.

Myers, N. and Tucker, R. (1987). Deforestation in Central America: Spanish legacy and North American consumers. *Environmental Review*, **11**, 55–71.

Naess, A. (1986). Intrinsic value: will the defenders of nature please rise? In *Conservation biology: the science of scarcity and diversity* (ed. M. E. Soulé), pp. 504–15. Sinauer Associates, Sunderland, Massachusetts.

National Research Council (1982). *Ecological aspects of development in the humid tropics.* National Academy Press, Washington, DC.

Nations, J. D. and Komer, D. I. (1983). Central America's tropical rainforests: positive steps for survival. *Ambio*, **12**, 232–8.

Nectoux, F. and Kuroda, Y. (1989). *Timber from the South Seas.* World Wildlife Fund International, Geneva, Switzerland.

Nelson, J. G. and Eidsvick, H. E. (1990). Sustainable development, conservation strategies and heritage: three basic tools for influencing the global future. *Alternatives*, **16** (4) and **17** (1), 62–71.

Netting, R. McC. (1971). *The ecological approach in cultural study*, a McCaleb Module in Anthropology, Module 6. Cummings Publishing Company, Menlo Park, California.

Netting, R. McC. (1974). Agrarian ecology. *Annual Review of Anthropology*, **3**, 21–56.

Newberry, D. Mc. and Proctor, J. (1984). Ecological studies in four contrasting lowland rain forests in Gunung Mulu National Park, Sarawak. IV: associations between tree distributions and soil factors. *Journal of Ecology*, **72**, 475–93.

New Straits Times (Kuala Lumpur), 7 May 1977; 21 June, 3 July 1984; 5 March, 19 March, 20 March, 4 April, 5 April, 21 August, 9 October, 24 December 1986; 22 March, 11 June, 13 June, 16 June, 22 June, 17 July, 18 July 1987; 13 June, 12 October, 1 November, 3 November, 9 November 1988; 14 June, 16 June, 19 June, 7 July 1990.

New Sunday Times (Kuala Lumpur), 22 February, 25 October 1987.

Ng, F. S. P. (1983). Ecological principles of tropical lowland rain forest conservation. In *Tropical rain forest: ecology and management* (ed. S. L. Sutton, T. C. Whitmore, and A. C. Chadwick), pp. 359–75. Blackwell, Oxford.

Ng, F. S. P. and Low, C. M. (1982). *Check list of endemic trees of the Malay peninsula*, Forest Research Institute, Research Pamphlet, No. 88. Kepong, Malaysia.

Ngau, H., Jalong Apoi, T., and Chee, Yoke Ling (1987). Malaysian timber: exploitation for whom? *The Ecologist*, **17**, 175–9.

Nicholaides III, J. J. *et al.* (1985). Agricultural alternatives for the Amazon basin. *BioScience*, **35**, 279–85.

Nicholas, C. (1986). Damning the people: the social implications of the Bukun dam project. Paper presented at the Forum on the Bakun Hydro Electric Power Project, 22 February, Kuching [mimeo].

Nicholson, D. I. (1958). An analysis of logging damage in tropical rain forest, North Borneo. *Malayan Forester*, **21**, 235–45.

Nicholson, D. I. (1965). A review of natural regeneration in the dipterocarp forests of Sabah. *Malayan Forester*, **28**, 4–25.

Nisbet, I. T. C. (1968). The utilization of mangrove by Malayan birds. *Ibis*, **110**, 348–52.

Oberai, A. S. (1988). An overview of settlement policies and programs. In *Land settlement policies and population redistribution in developing countries: achievements, problems and prospects* (ed. A. S. Oberai), pp. 7–47. Praeger, New York.

Office of Technology Assessment (1987). *Technologies to maintain biological diversity* (summary). Congress of the United States, Washington, DC.

Oldeman, R. A. A. (1983). Tropical rain forest, architecture, silvigenesis and diversity. In *Tropical rain forest: ecology and management* (ed. S. L. Sutton, T. C. Whitmore, and A. C. Chadwick), pp. 139–50. Blackwell, Oxford.

Oldfield, M. L. (1981). Tropical deforestation and genetic resources conservation. In *Blowing in the wind: deforestation and long-range implications*, Studies in Third World Societies, No. 14, pp. 277–345. Department of Anthropology, College of William and Mary, Williamsburg, Virginia.

Oldfield, M. L. and Alcorn, J. B. (1987). Conservation of traditional agroecosystems. *BioScience*, **37**, 199–208.

Oliphant, J. N. (1932). The constitution of reserved forests in the Malay peninsula. *Empire Forestry Journal*, **11**, 239–43.

Oliphant, L. (1860). *Narrative of the Earl of Elgin's mission to China and Japan in the years 1857, '58, '59,* 2 vols. Blackwood, Edinburgh.

Ong, A. S. H., Maheswaran, A., and Ma, Ah Ngan (1987). Malaysia. In *Environmental management in Southeast Asia: directions and current status* (ed. Chia, Lin Sien), pp. 14–76. Faculty of Science, National University of Singapore, Singapore.

Ong, Jin Eong (1982). Mangroves and aquaculture in Malaysia. *Ambio*, **11**, 252–7.

Ooi, Jin-Bee (1955). Mining landscapes of Kinta. *Malayan Journal of Tropical Geography*, **4**, 1–58.

Ooi, Jin-Bee (1961). The rubber industry of the Federation of Malaya. *Journal of Tropical Geography*, **15**, 46–65.

Ooi, Jin-Bee (1976). *Peninsular Malaysia*. Longman, London.

Orians, O. H. (1982). The influence of tree-falls in tropical forests on tree species richness. *Tropical Ecology*, **23**, 255–79.

O'Riordan, T. (1988). The politics of sustainability. In *Sustainable environmental management: principles and practice* (ed. R. K. Turner), pp. 29–49. Belhaven Press, London.

Osborne, M. (1979). *Southeast Asia: an introductory history*. George Allen and Unwin, Sydney.

Padoch, C. (1982). *Migration and its alternatives among the Iban of Sarawak*. Martinus Nijhoff, The Hague.

Padoch, C. (1988). People of the floodplain and forest. In *People of the tropical rain forest* (ed. J. S. Denslow and C. Padoch), pp. 127–41. University of California Press, Berkeley.

Padoch, C. and Vayda, A. P. (1983). Patterns of resource use and human settlement in tropical forests. In *Tropical rain forest ecosystems: structure and function* (ed. F. B. Golley), pp. 301–13. Elsevier, Amsterdam.

Pádua, M. T. J. and Quintão, A. T. B. (1982). Parks and biological reserves in the Brazilian Amazon. *Ambio*, **11**, 309–14.

Page, R. (1934). Wild life in Malaya. *Journal of the Society for the Preservation of the Fauna of the Empire*, New Series, Part **XXIII**, 34–42.

Paijmans, K. (1975). *Explanatory notes to the vegetation map of Papua New Guinea*, Land Research Series, No. 35. Commonwealth Scientific and Industrial Organization, Melbourne.

Paijmans, K. (1976). Vegetation. In *New Guinea vegetation* (ed. K. Paijmans), pp. 23–105. Commonwealth Scientific and Industrial Organization, Canberra.

Passmore, J. (1980). *Man's responsibility for nature* (2nd edn). Duckworth, London.

Payne, J. B. (1979). Abundance of diurnal squirrels at the Kuala Lompat Post of the Krau Game Reserve, Peninsular Malaysia. In *The abundance of animals in Malesian rain forests*, University of Hull, Department of Geography Miscellaneous Series, No. 22 (ed. A. G. Marshall), pp. 37–51. University of Hull, Hull, UK.

Payne, J., Francis, C. M., and Phillipps, K. (1985). *A field guide to the mammals of Borneo*. Sabah Society, Kota Kinabalu, Sabah.

Pearce, F. (1990). Hit and run in Sarawak. *New Scientist*, **126** (1716), 24–7.

Pearce, K. G. (1989a). Utilization of palms in Sarawak. *Malayan Naturalist*, **November**, 68–91.

Pearce, K. G. (1989b). Conservation status of palms in Sarawak. *Malayan Naturalist*, **November**, 20–36.

Peet, R. (1986). The destruction of regional cultures. In *A world in crisis? geographical perspectives* (ed. R. J. Johnston and P. J. Taylor), pp. 150–72. Blackwell, Oxford.

Peh, C. H. (1976). Rates of sediment transport by surface wash in three forested areas of Peninsular Malaysia. Unpublished MA Thesis, University of Malaya.

Pelzer, K. J. (1978). Swidden cultivation in Southeast Asia: historical, ecological, and economic perspectives. In *Farmers in the forest: economic development and marginal agriculture in northern Thailand* (ed. P. Kúnstadter, E. C. Chapman, and Sanga Sabhasri), pp. 271–86. University Press of Hawaii, Honolulu, Hawaii.

Persson, R. (1974). *World forest resources*. Royal College of Forestry, Stockholm.

Petr, T. (1978). Tropical man-made lakes—their ecological impact. *Archiv fuer Hydrobiologie*, **81**, 368–85.

Plumwood, V. and Routley, R. (1982). World rainforest destruction—the social factors. *The Ecologist*, **12**, 4–22.

Poore, M. E. D. (1963). Problems in the classification of tropical rain forest. *Journal of Tropical Geography*, **17**, 12–19.

Poore, M. E. D. (1968). Studies in Malaysian rain forest. 1: the forest on Triassic sediments in Jengka Forest Reserve. *Journal of Ecology*, **56**, 143–96.

Poore, D. [M. E. D.] (1976). *Ecological guidelines for development in tropical rain forests*. IUCN Books, Morges, Switzerland.

Poore, D. [M. E. D.] (1983a). Deforestation and the population factor. *Parks*, **8** (2), 11–12.

Poore, D. [M. E. D.] (1983b). Why replenish?—world forests, past, present and future. *Commonwealth Forestry Review*, **62**, 163–8.

Population Reference Bureau (1989). *1989 world population data sheet*. Population Reference Bureau, Washington, DC.

Posey, D. A. (1985). Indigenous management of tropical forest ecosystems: the case of the Kayapô indians of the Brazilian Amazon. *Agroforestry Systems*, **3**, 139–58.

Postel, S. and Heise, L. (1988). *Reforesting the earth*,

Worldwatch Paper 83. Worldwatch Institute, Washington, DC.

Prance, G. T. (ed.) (1982). *Biological diversification in the tropics*. Columbia University Press, New York.

Prance, G. T. (1985). The changing forests. In *Amazonia* (ed. G. T. Prance and T. E. Lovejoy), pp. 146–65. Pergamon Press, Oxford.

Pringle, R. (1970). *Rajahs and rebels: the Ibans of Sarawak under Brooke rule, 1841–1941*. Macmillan, London.

Proctor, J. (1983). Mineral nutrients in tropical forests. *Progress in Physical Geography*, **7**, 422–31.

Proctor, J. (1985). Tropical rain forest: ecology and physiology. *Progress in Physical Geography*, **9**, 402–13.

Proctor, J. (1986). Tropical rain forest: structure and function. *Progress in Physical Geography*, **10**, 383–400.

Proctor, J. (1987). Tropical rain forest. *Progress in Physical Geography*, **11**, 406–18.

Proctor, J., Anderson, J. M., Chai, P., and Vallack, H. W. (1983*a*). Ecological studies in four contrasting lowland rain forests in Gunung Mulu National Park, Sarawak. I: forest environment, structure and floristics. *Journal of Ecology*, **71**, 237–60.

Proctor, J., Anderson, J. M., and Vallack, H. W. (1983*b*). Comparative studies on soils and litterfall in forests at a range of altitudes on Gunung Mulu, Sarawak. *Malaysian Forester*, **46**, 60–76.

Putz, F. E. and Chai, P. (1987). Ecological studies of lianas in Lambir National Park, Sarawak, Malaysia. *Journal of Ecology*, **75**, 523–31.

Rachagan, S. S. (1983). Malaysia abandons Tembeling. *International Water Power and Dam Construction*, **35**, 43–7.

Rachagan, S. S. and Bahrin, Tunku Shamsul (1983). Development without destruction: the need for a pragmatic forest policy in Malaysia. *Planter*, **58**, 491–506.

Raemaekers, J. J. (1978). The sharing of food sources between two gibbon species in the wild. *Malayan Nature Journal*, **31**, 181–8.

Rahman-Ali, A. (1968). Forest conservation in Malaya. In *Conservation in tropical South East Asia*, IUCN Publications, New Series, No. 10 (ed. L. M. Talbot and M. H. Talbot), pp. 115–24. IUCN, Morges, Switzerland.

Rahman-Ali, A. and Wong, Yew Kwan (1968). The virgin jungle reserve project of the Malayan forest department. In *Conservation in tropical South East*

Asia, IUCN Publications, New Series, No. 10 (ed. L. M. Talbot and M. H. Talbot), pp. 364–5. IUCN, Morges, Switzerland.

Rainforest Information Centre (1989*a*). Penan action alert! increasing, intensive logging in Sarawak. *World Rainforest Report*, No. 13 (July), 3–6.

Rainforest Information Centre (1989*b*). Sarawak update 3 Oct 1989. *World Rainforest Report*, No. 14 (November), 3–5.

Rainforest Information Centre (1990). Stop press: Penan update. *World Rainforest Report*, No. 15 (February), 1–10.

Rambo, A. T. (1978). Bows, blowpipes and blunderbusses: ecological implications of weapons change among the Malaysian Negritos. *Malayan Nature Journal*, **32**, 209–16.

Rambo, A. T. (1979*a*). Primitive man's impact on the genetic resources of the Malaysian tropical rain forest. *Malaysian Applied Biology*, **8**, 59–65.

Rambo, A. T. (1979*b*). Human ecology of the Orang Asli: a review of research on the environmental relations of the aborigines of Peninsular Malaysia. *Federation Museums Journal*, New Series, **24**, 41–71.

Rambo, A. T. (1980*a*). Fire and the energy efficiency of swidden agriculture. *Asian Perspectives*, **23**, 309–16.

Rambo, A. T. (1980*b*). Of stones and stars: Malaysian Orang Asli environmental knowledge in relation to their adaptation to the tropical rain forest ecosystem. *Federation Museums Journal*, New Series, **25**, 77–88.

Rambo, A. T. (1982). Orang Asli adaptive strategies: implications for Malaysian natural resource development planning. In *Too rapid rural development: perceptions and perspectives from Southeast Asia* (ed. C. MacAndrews and Chia, Lin Sien), pp. 251–99. Ohio University Press, Athens, Ohio.

Rambo, A. T. (1988*a*). People of the forest. In *Malaysia* (ed. Earl of Cranbrook), pp. 273–88. Pergamon Press, London.

Rambo, A. T. (1988*b*). Why are the Semang? ecology and ethnogenesis of aboriginal groups in Peninsular Malaysia. In *Ethnic diversity and the control of natural resources in Southeast Asia* (ed. A. T. Rambo, K. Gillogly, and K. L. Hutterer), pp. 19–35. Michigan Papers on South and Southeast Asia, No. 32. Center for South and Southeast Asian Studies, University of Michigan, Ann Arbor, Michigan.

Ranee of Sarawak (1913). *My life in Sarawak.* Methuen, London.

Raven, P. H. (1976). Ethics and attitudes. In *Conservation of threatened plants* (ed. J. B. Simmons, R. I. Beyer, P. E. Brandham, G. Gl. Lucas, and V. T. H. Parry), pp. 155–79. Plenum Press, New York.

Raven, P. H. (1979). Plate tectonics and southern hemisphere biogeography. In *Tropical botany* (ed. K. Larsen and L. B. Holm-Nelson), pp. 3–24. Academic Press, London.

Raven, P. H. (1984). Third World in the global future. *Bulletin of the Atomic Scientists*, **40**, 17–20.

Raven, P. H. (1988). The cause and impact of deforestation. In *Earth '88: changing geographic perspectives* (ed. H. J. de Blij), pp. 212–29. National Geographic Society, Washington, DC.

Raven, P. H. and Axelrod, D. I. (1972). Plate tectonics and Australian paleobiogeography. *Science*, **176** (4042), 1379–86.

Redclift, M. (1987). *Sustainable development: exploring the contradictions.* Methuen, London.

Reece, R. H. W. (1988). Economic development under the Brookes. In *Development in Sarawak: historical and contemporary perspectives* (ed. R. A. Cramb and R. H. W. Reece), pp. 21–34. Monash Paper on Southeast Asia No. 17, Centre of Southeast Asian Studies, Monash University, Australia.

Reed, R. R. (1979). The colonial genesis of hill stations: the Genting exception. *Geographical Review*, **69**, 463–8.

Repetto, R. (1986). Soil loss and population pressure on Java. *Ambio*, **15**, 14–18.

Repetto, R. (1987). Creating incentives for sustainable forest development. *Ambio*, **16**, 94–9.

Repetto, R. (1988a). *The forest for the trees? government policies and the misuse of forest resources.* World Resources Institute, Washington, DC.

Repetto, R. (1988b). Overview. In *Public policies and the misuse of forest resources* (ed. R. Repetto and M. Gillis), pp. 1–41. Cambridge University Press.

Repetto, R. and Gillis, M. (eds.) (1988). *Public policies and the misuse of forest resources.* Cambridge University Press.

Repetto, R. and Holmes, T. (1983). The role of population in resource depletion in developing countries. *Population and Development Review*, **9**, 609–32.

Rich, B. M. (1985). Multi-lateral development banks: their role in destroying the global environment. *The Ecologist*, **15**, 56–68.

Richards, J. F. (1986). World environmental history

and economic development. In *Sustainable development of the biosphere* (ed. W. C. Clark and R. E. Munn), pp. 53–71. Cambridge University Press.

Richards, J. F. and Tucker, R. P. (eds.) (1988). *World deforestation in the twentieth century.* Duke University Press, Durham, North Carolina.

Richards, P. W. (1936a). Ecological observations on the rain forest of Mount Dulit, Sarawak. Part I. *Journal of Ecology*, **24**, 1–37.

Richards, P. W. (1936b). Ecological observations on the rain forest of Mount Dulit, Sarawak. Part II. *Journal of Ecology*, **24**, 340–60.

Richards, P. W. (1952). *The tropical rain forest.* Cambridge University Press.

Richards, P. W. (1969). Speciation in the tropical rain forest and the concept of the niche. *Biological Journal of the Linnean Society*, **1**, 149–53.

Richards, P. W. (1973). The tropical rain forest. *Scientific American*, **229** (6), 58–67.

Ricklefs, R. E. (1973). *Ecology.* Chiron Press, Newton, Massachusetts.

Ridley, H. N. (1892). Gambier. *Agricultural Bulletin of the Malay Peninsula*, **2**, 20–41.

Ridley, H. N. (1903). Reclaiming abandoned mining land. *Agricultural Bulletin of the Straits and Federated Malay States*, 2nd Series, **2**, 63–4.

Ridley, H. N. (1905). The history and development of agriculture in the Malay peninsula. *Agricultural Bulletin of the Straits and Federated Malay States*, 2nd series, **4**, 292–317.

Ridley, H. N. (1910). Tillage of the soil. *Agricultural Bulletin of the Straits and Federated Malay States*, 2nd Series, **9**, 80–4.

Robbins, R. G. and Wyatt-Smith, J. (1964). Dry land forest formations and forest types in the Malayan peninsula. *Malayan Forester*, **27**, 188–216.

Rocheleau, D. E. and Raintree, J. B. (1986). Agroforestry and the future of food production in developing countries. *Impact of Science on Society*, No. **142**, 127–41.

Rolston III, H. (1985). Duties to endangered species. *BioScience*, **35**, 718–26.

Rolston III, H. (1988). *Environmental ethics: duties to and values in the natural world.* Temple University Press, Philadelphia.

Rosen, B. R. (1981). The tropical high diversity enigma—the corals'-eye view. In *The evolving biosphere* (ed. P. L. Forey), pp. 103–29. Cambridge University Press.

Rubeli, K. (1976). The Tembeling hydro-electric project from the Taman Negara viewpoint. *Malayan Nature Journal*, **29**, 307–14.

Rubeli, K. (1986*a*). Endau–Rompin: a refuge for Malaysia's rainforest. *Habitat*, **14** (4), 19–22.

Rubeli, K. (1986*b*). Taman Negara park 'progress': wise and otherwise. *Habitat*, **14** (4), p. 23.

Rubeli, K. (1989). Pride and protest in Malaysia. *New Scientist*, **124** (1687), 27–30.

Ruthenberg, H. (1976). *Farming systems in the tropics* (2nd edn). Clarendon Press, Oxford.

Rutter, O. (1922). *British North Borneo: an account of its history, resources and native tribes*. Constable, London.

Sabah (n.d.). *Sabah. Annual Report 76–77*. Government of Sabah, Kota Kinabalu.

Sabah, Forest Department (1967). *Annual report of the Forest Department for the year 1966*. Government Printing Office, Jesselton, Sabah.

Sadka, E. (1968). *The protected Malay states, 1874–1895*. University of Malaya Press, Kuala Lumpur.

Sahabat Alam Malaysia (1989). *The battle for Sarawak's forests*. World Rainforest Movement and Sahabat Alam Malaysia, Penang.

Sahlins, M. (1974). *Stone age economics*. Tavistock Publications, London.

Salas, R. (1987). Population and sustainable development. In *Conservation with equity: strategies for sustainable development* (ed. P. Jacobs and D. A. Munro), pp. 55–61. International Union for Conservation of Nature and Natural Resources, Gland, Switzerland.

Salati, E. and Vose, P. B. (1983). Depletion of tropical rain forests. *Ambio*, **12**, 67–71.

Salati, E. and Vose, P. B. (1984). Amazon basin: a system in equilibrium. *Science*, **225** (4658), 129–38.

Salleh Mohd. Nor (1982). Developmental research in tropical forestry: a Malaysian experience. Paper presented at the Forestry and Development in Asia Conference, Bangalore, India, 19–23 April [mimeo].

Salleh Mohd. Nor (1983). Forestry in Malaysia. *Journal of Forestry*, **81**, 164–6 and p. 187.

Salleh Mohd. Nor (1988). Forest management. In *Malaysia* (ed. Earl of Cranbrook), pp. 126–37. Pergamon Press, Oxford.

Salo, J. *et al.* (1986). River dynamics and the diversity of Amazon lowland forest. *Nature*, **322** (17 July), 254–8.

Sanchez, P. A. (1976). *Properties and management of soils in the tropics*. Wiley, New York.

Sanchez, P. A. and Buol, S. W. (1975). Soils of the tropics and the world food crisis. *Science*, **188** (4188), 598–603.

Sanchez, P. A., Bandy, D. E., Villachica, J. H., and Nicholaides, J. J. (1982). Amazon basin soils: management for continuous crop production. *Science*, **216** (4548), 821–7.

Sandhu, K. S. (1969). *Indians in Malaya: some aspects of their immigration and settlement (1786–1957)*. Cambridge University Press.

Sanford, Jun., R. L., Saldarriaga, J., Clark, K. E., Uhl, C., and Herrera, R. (1985). Amazon rain-forest fires. *Science*, **227** (4682), 53–5.

Sandin, B. (1967). *The Sea Dayaks of Borneo before white rajah rule*. Macmillan, London.

Sandosham, A. A. (1970). Malaria in rural Malaya. *Medical Journal of Malaya*, **24**, 221–6.

Sarawak (n.d.*a*). *Sarawak national parks*. National Parks and Wildlife Branch, Forest Department, Kuching [brochure].

Sarawak (n.d.*b*). *Sarawak annual report 1962*. Sarawak Government Printing Office, Kuching.

Sarawak Study Group (1989). Logging in Sarawak: the Belaga experience. In *Logging against the natives of Sarawak*, pp. 1–28. INSAN (Institute of Social Analysis), Petaling Jaya, Malaysia.

Sayer, J. A. and Whitmore, T. C. (1991). Tropical moist forests: destruction and species extinction. *Biological Conservation*, **55**, 199–213.

Schimper, A. F. H. (1903). *Plant-geography upon a physiological basis*. Clarendon Press, Oxford.

Schuster, R. M. (1972). Continental movements, 'Wallace's line' and Indomalayan–Australasian dispersal of land plants: some eclectic concepts. *Botanical Review*, **38**, 3–86.

Scott, M. (1988). Loggers and locals fight for the heart of Borneo. *Far Eastern Economic Review*, **28 April**, 44–5 and p. 47.

Scrivenor, J. B. (1928). *A sketch of Malayan mining*. Mining Publications, London.

Seaward, N. (1987*a*). The core of the matter. *Far Eastern Economic Review*, **13 August**, p. 79.

Seaward, N. (1987*b*). At loggerheads with power: Sarawak tribal chiefs protest against logging on native lands. *Far Eastern Economic Review*, **2 July**, p. 32.

Sedjo, R. A. and Clawson, M. (1983). How serious is tropical deforestation? *Journal of Forestry*, **81**, 792–4.

Sedjo, R. A. and Clawson, M. (1984). Global forests. In *The resourceful earth: a response to Global 2000* (ed.

J. L. Simon and H. Kahn), pp. 128–70. Blackwell, Oxford.

Segal, J. (1983). The third time around. *Far Eastern Economic Review*, **31 March**, p. 14.

Semple, E. C. (1919). Climatic and geographic influences on ancient Mediterranean forests and the lumber trade. *Annals of the Association of American Geographers*, **9**, 13–40.

Service, E. R. (1966). *The hunters*. Prentice-Hall, Englewood Cliffs, New Jersey.

Setten, G. G. K. (1962). The need for a forest estate in Malaya. *Malayan Forester*, **25**, 184–98.

SGSNRM, Stockholm Group for Studies on Natural Resources Management (1988). *Perspectives of sustainable development: some critical issues related to the Brundtland report*. SGSNRM, Stockholm.

Shafruddin, B. H. (1987). *The federal factor in the government and politics of Peninsular Malaysia*. Oxford University Press, Singapore.

Shallow, P. G. D. (1956). *Riverflow in the Cameron Highlands*. Hydro-Electricity Memorandum, No. 3. Central Electricity Board, Kuala Lumpur.

Shamsul, A. B. (1983). The politics of poverty eradication: the implementation of development projects in a Malaysian district. *Pacific Affairs*, **56**, 455–76.

Sharp, T. (1983). Malaysia's environment in danger. *Ambio*, **12**, 275–6.

Sharp, T. (1984). ASEAN nations tackle common environmental problems. *Ambio*, **13**, 45–6.

Shelton, N. (1985). Logging versus the natural habitat in the survival of tropical forests. *Ambio*, **14**, 39–41.

Shen, S. (1987). Biological diversity and public policy. *BioScience*, **37**, 709–12.

Shigeo Kurata (1976). *Nepenthes of Mount Kinabalu*, Sabah National Parks Publications, No. 2. Sabah National Parks Trustees, Kota Kinabalu.

Short, D. E. and Jackson, J. C. (1971). The origins of an irrigation policy in Malaya: a review of developments prior to the establishment of the Drainage and Irrigation Department. *Journal of the Malaysian Branch of the Royal Asiatic Society*, **64**, 78–103.

Silva, G. S. de (1968a). Elephants of Sabah. *Sabah Society Journal*, **3**, 169–81.

Silva, G. S. de (1968b). Wildlife conservation in the state of Sabah. In *Conservation in tropical South East Asia*, IUCN Publications, New Series, No. 10 (ed. L. M. Talbot and M. H. Talbot), pp. 144–50. IUCN, Morges, Switzerland.

Sim, Kwang Yang (1986). Damn the dam. Paper presented at the Forum on the Bakun Hydro Electric Power Project, 22 February, Kuching [mimeo].

Simberloff, D. (1986). Are we on the verge of a mass extinction in tropical rain forests? In *Dynamics of extinction* (ed. D. K. Elliott), pp. 165–80. Wiley, New York.

Simberloff, D. and Abele, L. G. (1982). Refuge design and island biogeographic theory: effects of fragmentation. *American Naturalist*, **120**, 41–50.

Simon, J. L. (1984). Bright global future. *Bulletin of the Atomic Scientists*, **40**, 14–17.

Simon, J. L. (1986). Disappearing species, deforestation and data. *New Scientist*, **110** (1508), 60–3.

Simon, J. L. and Kahn, H. (1984). Introduction. In *The resourceful earth: a response to Global 2000* (ed. J. L. Simon and H. Kahn), pp. 1–49. Blackwell, Oxford.

Simon, J. L. and Wildavsky, A. (1984). On species loss, the absence of data, and risks to humans. In *The resourceful earth: a response to Global 2000* (ed. J. L. Simon and H. Kahn), pp. 171–83. Blackwell, Oxford.

Simonett, D. S. (1967). Landslide distribution and earthquakes in the Bewani and Torricelli mountains, New Guinea: statistical analysis. In *Landform studies from Australia and New Guinea* (ed. J. N. Jennings and J. A. Mabbutt), pp. 64–84. Australian National University Press, Canberra.

Simpson, G. G. (1977). Too many lines: the limits of the Oriental and Australian zoogeographical regions. *Proceedings of the American Philosophical Society*, **121**, 107–20.

Skolimowski, H. (1981). *Eco-philosophy: designing new tactics for living*. Marion Boyers, London.

Smiet, F. (1987). Tropical watershed forestry under attack. *Ambio*, **16**, 156–8.

Smil, V. (1983). Deforestation in China. *Ambio*, **12**, 226–31.

Smith, J. M. B. (1982). Origins of the tropialpine flora. In *Biogeography and ecology of New Guinea* (ed. J. L. Gressitt), pp. 287–308. Dr. W. Junk, The Hague.

Smith, N. J. H. (1985). The impact of cultural and ecological change on Amazonian fisheries. *Biological Conservation*, **32**, 355–73.

Smythies, B. E. (1963). History of forestry in Sarawak. *Malayan Forester*, **26**, 232–53.

Smythies, B. E. (1981). *The birds of Borneo*. (3rd edn). Sabah Society and Malayan Nature Society, Kota Kinabalu, Sabah.

Soepadmo, E. (1971). Plants and vegetation along the

paths to Gunong Tahan. *Malayan Nature Journal*, **24**, 118–24.

Sommer, A. (1976). Attempt at an assessment of the world's tropical moist forests. *Unasylva*, **28** (112–13), 5–25.

Soong, Ngin Kwi, Haridas, G., Yeoh, Choon Seng, and Tan, Pen Hua (1980). *Soil erosion and conservation in Peninsular Malaysia*. Rubber Research Institute of Malaysia, Kuala Lumpur.

Soulé, M. E. (1985). What is conservation biology? *Bio-Science*, **35**, 727–34.

Soulé, M. E. and Simberloff, D. (1986). What do genetics and ecology tell us about the design of nature reserves? *Biological Conservation*, **35**, 19–40.

Soulé, M. E. and Wilcox, B. A. (1980). Preface. In *Conservation biology: an evolutionary–ecological perspective* (ed. M. E. Soulé and B. A. Wilcox), pp. xi–xiv. Sinauer Associates, Sunderland, Massachusetts.

Sowunmi, M. A. (1986). Change of vegetation with time. In *Plant ecology in West Africa: systems and processes* (ed. G. W. Lawson), pp. 273–307. Wiley, Chichester, UK.

Spears, J. S. (1979). Can the wet tropical forest survive? *Commonwealth Forestry Review*, **58**, 165–80.

Spears, J. S. (1983). Tropical reforestation: an achievable goal? *Commonwealth Forestry Review*, **62**, 201–17.

Spears, J. (1985). Malaysia agricultural sector assessment mission: forestry subsector discussion paper. World Bank, Washington, DC [mimeo].

Spencer, J. E. (1966). *Shifting cultivation in Southeast Asia*. University of California Publications in Geography, No. 19. University of California Press, Berkeley.

Spoehr, A. (1956). Cultural differences in the interpretation of natural resources. In *Man's role in changing the face of the earth* (ed. W. L. Thomas, Jun.), pp. 93–102. University of Chicago Press, Chicago.

Spurway, B. J. C. (1937). Shifting cultivation in Sarawak. *Malayan Forester*, **6**, 124–8.

Star (Kuala Lumpur), 10 June 1985, 6 July 1988.

Steenis, C. G. G. J. van (1950). The delimitation of Malaysia and its main plant geographical divisions. *Flora Malesiana*, Series I, **1**, lxx–lxxv.

Steenis, C. G. G. J. van (1962). The land bridge theory in botany. *Blumea*, **11**, 235–372.

Steenis, C. G. G. J. van (1972). *The mountain flora of Java*. Brill, Leiden.

Steenis, C. G. G. J. van (1979). Plant-geography of east Malesia. *Botanical Journal of the Linnean Society of London*, **79**, 97–178.

Steenis, C. G. G. J. van (1981). Editorial. *Flora Malesiana Bulletin*, **34**, p. 3541.

Stenson, M. (1980). *Class, race and colonialism in West Malaysia: the Indian case*. University of Queensland Press, St. Lucia, Australia.

Stevens, W. E. (1968). *The conservation of wildlife in West Malaysia*. Office of the Chief Game Warden, Federal Game Department, Ministry of Lands and Mines, Seremban, Negeri Sembilan.

Strong, M. F. (1986). Conservation strategies for a changing world. In *Pieces of the global puzzle: international approaches to environmental concerns* (ed. A. M. Blackburn), pp. 4–18. Fulcrum, Golden, Colorado.

Strugnell, E. J. (1938). Silviculture in Malaya. *Empire Forestry Journal*, **17**, 188–94.

Suhaini Aznam (1987*a*). The unofficial thorns. *Far Eastern Economic Review*, **1 January**, 16–17.

Suhaini Aznam (1987*b*). Alternative societies. *Far Eastern Economic Review*, **15 October**, 33–4 and p. 36.

Suhaini Aznam (1987*c*). Mahathir cracks down. *Far Eastern Economic Review*, **5 November**, p. 14.

Suhaini Aznam (1987*d*). The great crackdown. *Far Eastern Economic Review*, **12 November**, 12–14.

Suhaini Aznam (1987*e*). The Dayak awakening. *Far Eastern Economic Review*, **30 April**, p. 14.

Suhaini Aznam (1990). Volatile mixture. *Far Eastern Economic Review*, **21 June**, 12–13.

Suhaini Aznam (1991). Links under stress. *Far Eastern Economic Review*, **17 January**, 10–11.

Sunday Star (Kuala Lumpur), 24 May 1987.

Sutton, S. L. (1979). The vertical distribution of flying insects in the lowland rain forest of Brunei: preliminary report. *Brunei Museum Journal*, **4**, 161–73.

Sutton, S. L. (1983). The spatial distribution of flying insects in tropical rain forests. In *Tropical rain forest: ecology and management* (ed. S. L. Sutton, T. C. Whitmore, and A. C. Chadwick), pp. 77–92. Blackwell, Oxford.

Sutton, S. L., Whitmore, T. C., and Chadwick, A. C. (eds.) (1983). *Tropical rain forest: ecology and management*. Blackwell, Oxford.

Symington, C. F. (1943/1974). Foresters' manual of dipterocarps. *Malayan Forest Records*, **16**. Forest Department, Kuala Lumpur. Reprinted 1974, University of Malaya Press, Kuala Lumpur.

Tan, K. L. (1965). The oil palm industry in Malaya. Unpublished MA thesis, University of Malaya.

Tang, H. T. (1987). Problems and strategies for regenerating dipterocarp forests in Malaysia. In *Natural management of tropical moist forests: silvicultural and management prospects of sustained utilization* (ed. F. Mergen and J. R. Vincent), pp. 23–45. Yale University, School of Forestry and Environmental Studies, New Haven, Connecticut.

Tangley, L. (1987). Beyond the green revolution. *BioScience*, **37**, 176–80.

Tee, A. C. (1982). Some aspects of the ecology of the mangrove forest at Sungai Buloh, Selangor. II: distribution pattern and population dynamics of tree-dwelling fauna. *Malayan Nature Journal*, **35**, 267–77.

Teoh, Teck Seng (1971). Where does all the rainfall go? *Planters' Bulletin of the Rubber Research Institute of Malaysia*, **115**, 215–19.

Teoh, Teck Seng (1973). Some effects of *Hevea* plantations on rainfall redistribution. In *Proceedings of the symposium on biological resources and national development* (ed. E. Soepadmo and K. G. Singh), pp. 131–6. Malayan Nature Society, Kuala Lumpur.

Teoh, Teck Seng (1977). Throughfall, stemflow and interception studies on *Hevea* stands in Peninsular Malaysia. *Malayan Nature Journal*, **31**, 141–5.

Terborgh, J. (1986). Keystone plant resources in the tropical forest. In *Conservation biology: the science of scarcity and diversity* (ed. M. E. Soulé), pp. 330–44. Sinauer Associates, Sunderland, Massachusetts.

Terborgh, J. and Winter, B. (1980). Some causes of extinction. In *Conservation biology: an evolutionary-ecological perspective* (ed. M. E. Soulé and B. A. Wilcox), pp. 119–33. Sinauer Associates, Sunderland, Massachusetts.

Terborgh, J. and Winter, B. (1983). A method for siting parks and reserves with special reference to Colombia and Ecuador. *Biological Conservation*, **27**, 45–58.

Thang, H. C. (1985). Timber supply and domestic demand in Peninsular Malaysia. *Malaysian Forester*, **48**, 87–100.

Thirgood, J. V. (1981). *Man and the Mediterranean forest: a history of resource depletion*. Academic Press, New York.

Tho, Y. P. and Lim, R. (1985). Unveiling the natural treasures of Endau–Rompin. In *The New Straits Times Annual '86*, pp. 29–37. Berita Publishing, Kuala Lumpur.

Thomson, J. T. (1850). General report on the Residency of Singapore, drawn up principally with a view of illustrating its agricultural statistics. *Journal of the Indian Archipelago and Eastern Asia*, **4**, 134–43.

Thomson, J. T. (1864/1984). *Glimpses into life in Malayan lands*. Richardson, London. Reprinted 1984, Oxford University Press, Singapore.

Toebes, C. and Goh, Kiam Seng (1975). *Notes on some hydrological effects of land use changes in Peninsular Malaysia*. Drainage and Irrigation Department Water Resources Publication, No. 4. Ministry of Agriculture and Rural Development, Kuala Lumpur.

Tomlinson, P. B. (1978). Branching and axis differentiation in tropical trees. In *Tropical trees as living systems* (ed. P. B. Tomlinson and M. H. Zimmermann), pp. 187–208. Cambridge University Press.

Tregonning, K. G. (1965). *A history of modern Sabah (North Borneo 1881–1963)*. University of Malaya Press, Kuala Lumpur.

Tribe, L. H., Schelling, C. S., and Voss, J. (eds.) (1976). *When values conflict: essays on environmental analysis, discourse, and decision*. Ballinger, Cambridge, Massachusetts.

Troup, R. S. (1940). *Colonial forest administration*. Oxford University Press.

Truswell, E. M., Kershaw, A. P., and Sluiter, I. R. (1987). The Australian–southeast Asian connection: evidence from the palaeobotanical record. In *Biogeographical evolution of the Malay archipelago* (ed. T. C. Whitmore), pp. 32–49. Clarendon Press, Oxford.

Tsuruoka, D. (1990). Led from the front. *Far Eastern Economic Review*, **6 September**, p. 23.

Tuan, Y.-F. (1968). Discrepancies between environmental attitude and behaviour: examples from Europe and China. *Canadian Geographer*, **12**, 176–91.

Tuan, Y.-F. (1984). *Dominance and affection: the making of pets*. Yale University Press, New Haven, Connecticut.

Tucker, R. P. and Richards, J. F. (eds.) (1983*a*). *Global deforestation and the nineteenth-century world economy*. Duke University Press, Durham, North Carolina.

Tucker, R. P. and Richards, J. F. (1983*b*). Introduction. In *Global deforestation and the nineteenth-century world economy* (ed. R. P. Tucker and J. F. Richards), pp. xi–xviii. Duke University Press, Durham, North Carolina.

Turnbull, C. M. (1972). *The Straits Settlements 1826–67: Indian presidency to crown colony*. The Athlone Press, London.

Turnbull, C. M. (1981). *A short history of Malaysia, Singapore and Brunei*. Brash, Singapore.

Turner II, B. L. and Brush, S. B. (1987). Purpose, classification, and organization. In *Comparative farming systems* (ed. B. L. Turner II and S. B. Brush), pp. 3–10. Guilford Press, New York.

Turner II, B. L., Hanham, R. Q., and Portararo, A. V. (1977). Population pressure and agricultural intensity. *Annals of the Association of American Geographers*, **67**, 384–96.

Turvey, N. D. (1974). Water in the nutrient cycle of a Papuan rain forest. *Nature*, **251** (4 October), 414–15.

Uhl, C. (1983). You can keep a good forest down. *Natural History*, **92** (4), 70–9.

Uk Tinal and Palenewen, J. L. (1978). Mechanical logging damage after selective cutting in the lowland dipterocarp forest at Beloro, East Kalimantan. In *Proceedings of the symposium on the long-term effects of logging in Southeast Asia*, BIOTROP Special Publication, No. 3 (ed. Rahardjo S. Suparto *et al*.), pp. 91–6. Bogor, Indonesia.

Unesco, United Nations Educational, Scientific and Cultural Organization (1978). *Tropical forest ecosystems: a state-of-knowledge report prepared by Unesco/UNEP/FAO*. Unesco, Paris.

Vayda, A. P. and Ahmad Sahur (1985). Forest clearing and pepper farming by Bugis migrants in East Kalimantan: antecedents and impact. *Indonesia*, **39**, 93–110.

Vayda, A. P., Colfer, C. J. P., and Mohamad Brotokusumo (1980). Interaction between people and forests in East Kalimantan. *Impact of Science on Society*, **30**, 179–90.

Vergara, N. T. (1985). Agroforestry systems: a primer. *Unasylva*, **37** (No. 147), 22–8.

Vergara, N. T. and Briones, N. D. (eds.) (1987). *Agroforestry in the humid tropics: its protective and ameliorative roles to enhance productivity and sustainability*. Environment and Policy Institute, East–West Center, Honolulu, Hawaii, USA and Southeast Asian Regional Center for Graduate Study and Research in Agriculture, Los Baños, Laguna, Philippines.

Verstappen, H. Th. (1975). On palaeo climates and landform development in Malesia. In *Modern Quaternary research in Southeast Asia* (ed. G.-J. Bartstra and W. A. Casparie), pp. 3–36. Balkema, Rotterdam.

Vickers, W. T. (1983). Tropical forest mimicry in swiddens: a reassessment of Geertz's model with Amazonian data. *Human Ecology*, **11**, 35–45.

Vincent, J. R. (1988). Optimal tariffs on resource-based intermediate and final goods. Unpublished PhD thesis, Yale University.

Voon, Phin Keong (1981). The rural development programme in Sabah, Malaysia, with reference to the 1970s. *Malaysian Journal of Tropical Geography*, **3**, 53–67.

Wain, B. (1978). Fight over a forest. *Asian Wall Street Journal*, **22 February**, p. 1 and p. 7.

Walker, B. W. (1987). Strategy needed to establish institutional building blocks. In *Conservation with equity: strategies for sustainable development* (ed. P. Jacobs and D. A. Munro), pp. 109–18. International Union for Conservation of Nature and Natural Resources, Gland, Switzerland.

Walker, D. (1982). Speculations on the origin and evolution of Sunda-Sahul rain forests. In *Biological diversification in the tropics* (ed. G. T. Prance), pp. 554–75. Columbia University Press, New York.

Walker, F. S. (1950). Three-axle 'weapon carriers' in south Perak. *Malayan Forester*, **13**, 143–7.

Wallace, A. R. (1859). On the zoological geography of the Malay archipelago. *Linnean Society of London Journal of Zoology*, **4**, 172–84.

Wallace, A. R. (1869). *The Malay archipelago*. Macmillan, London.

Wang Gungwu (1958). The Nanhai trade: a study of the early history of Chinese trade in the South China Sea. *Journal of the Malayan Branch of the Royal Asiatic Society*, **31**, 1–135.

Warnford-Lock, C. G. (1907). *Mining in Malaya for gold and tin*. Crowther and Goodman, London.

Water Resources Committee (1971). *Report on the flood problems in West Malaysia*. Drainage and Irrigation Division, Ministry of Agriculture and Lands, Kuala Lumpur.

Watson, J. G. (1928). Mangrove swamps of the Malay Peninsula. *Malayan Forest Records*, **6**. Forest Department, Kuala Lumpur.

Watson, J. G. (1934). Forest research in Malaya. *Empire Forestry Journal*, **13**, 223–31.

Watson, J. G. (1950). Some materials for a forest history of Malaya. *Malayan Forester*, **13**, 63–72.

Watts, I. E. M. (1954). Line squalls of Malaya. *Malayan Journal of Tropical Geography*, **3**, 1–14.

WCED, World Commission on Environment and Development (1987). *Our common future*. Oxford University Press.

Webb, B. E. (1950). Logging by tractors. *Malayan Forester*, **13**, 148–50.

Webb, L. G. (1982). Ecological values of the tropical rain forest resource. *Proceedings of the Linnean Society of New South Wales*, **106**, 263–74.

Webb, L. J. and Tracey, J. G. (1984). A floristic framework of Australian rainforests. *Australian Journal of Ecology*, **9**, 169–98.

Weber, B. E. (1972). A parks system for West Malaysia. *Oryx*, **11**, 461–9.

Wee, Y. C. (1970). The development of pineapple cultivation in West Malaysia. *Journal of Tropical Geography*, **30**, 68–75.

Wegener, A. (1924). *The origin of continents and oceans*. Methuen, London.

Wells, D. R. (1971). Survival of the Malaysian bird fauna. *Malayan Nature Journal*, **24**, 248–56.

Wells, D. R. (1976). Resident birds. In *The birds of the Malay peninsula*. Vol. V. *Conclusion, and survey of every species* (ed. Lord Medway and D. R. Wells), pp. 1–34. Witherby, London.

Wells, D. R. (1988). Birds. In *Malaysia* (ed. Earl of Cranbrook), pp. 167–95. Pergamon Press, Oxford.

Westman, W. E. (1977). How much are nature's services worth? *Science*, **197** (4307), 960–4.

Wetterberg, G. B., Prance, G. T., and Lovejoy, T. E. (1981). Conservation progress in Amazonia: a structural review. *Parks*, **6** (2), 5–10.

Wheatley, P. (1959). Geographical notes on some commodities involved in Sung maritime trade. *Journal of the Malayan Branch of the Royal Asiatic Society*, **32**, 5–140.

Wheatley, P. (1961). *The golden khersonese*. University of Malaya Press, Kuala Lumpur.

White, P. S. and Bratton, S. P. (1980). After preservation: philosophical and practical problems of change. *Biological Conservation*, **18**, 241–55.

Whitmore, T. C. (1971). Wild fruit trees and some trees of pharmacological potential in the rain forest of Ulu Kelantan. *Malayan Nature Journal*, **24**, 222–4.

Whitmore, T. C. (1973). Plate tectonics and some aspects of Pacific plant geography. *New Phytologist*, **72**, 1185–90.

Whitmore, T. C. (1978). Gaps in the forest canopy. In *Tropical trees as living systems* (ed. P. B. Tomlinson and M. H. Zimmermann), pp. 639–55. Cambridge University Press.

Whitmore, T. C. (1980). The conservation of tropical rain forest. In *Conservation biology: an evolutionary-ecological perspective* (ed. M. E. Soulé and B. A. Wilcox), pp. 303–18. Sinauer Associates, Sunderland, Massachusetts.

Whitmore, T. C. (1981). Wallace's line and some other plants. In *Wallace's line and plate tectonics* (ed. T. C. Whitmore), pp. 70–80. Clarendon Press, Oxford.

Whitmore, T. C. (1982). On pattern and process in forests. In *The plant community as a working mechanism* (ed. E. I. Newman), pp. 45–59. Blackwell, Oxford.

Whitmore, T. C. (1983). Secondary succession from seed in tropical rain forests. *Forestry Abstracts*, **44**, 767–79.

Whitmore, T. C. (1984*a*). *Tropical rain forests of the Far East* (2nd edn). Clarendon Press, Oxford.

Whitmore, T. C. (1984*b*). A vegetation map of Malesia at scale 1 : 5 million. *Journal of Biogeography*, **11**, 461–71.

Whitmore, T. C. (1988). Forest types and forest zonation. In *Malaysia* (ed. Earl of Cranbrook), pp. 20–30. Pergamon Press, Oxford.

Whitmore, T. C. (1990). *An introduction to tropical rain forests*. Clarendon Press, Oxford.

Whitmore, T. C. and Burnham, C. P. (1969). The altitudinal sequence of forests and soils on granite near Kuala Lumpur. *Malayan Nature Journal*, **22**, 99–118.

Whitmore, T. C. and Ho, S. Y. (1968). Logging by elephant in north Kelantan. *Malayan Forester*, **31**, 222–4.

Whitmore, T. C., Peralta, R., and Brown, K. (1985). Total species count in a Costa Rican tropical rain forest. *Journal of Tropical Ecology*, **1**, 375–8.

Wild Life Commission (1933). Wild Life Commission of Malaya: précis of report of. *Journal of the Society for the Preservation of the Fauna of the Empire*, New Series, Part **XIX**, 37–90.

Wild Life Commission of Malaya (1932). *Report of the Wild Life Commission*, 3 vols. Government Printer, Singapore.

Williams, M. (1989). Deforestation: past and present. *Progress in Human Geography*, **13**, 176–208.

Wilson, C. C. and Wilson, W. L. (1975). The influence of selective logging on primates and some other animals in East Kalimantan. *Folia Primatologia*, **23**, 245–74.

Wilson, E. O. (1984). *Biophilia*. Harvard University Press, Cambridge, Massachusetts.

Wilson, E. O. (1985). The biological diversity crisis. *BioScience*, **35**, 700–6.

Wilson, E. O. (1989). Threats to biodiversity. *Scientific American*, **261** (3), 108–16.

Wilson, W. L. and Johns, A. D. (1982). Diversity and abundance of selected animal species in undisturbed forest, selectively logged forest and plantations in East Kalimantan, Indonesia. *Biological Conservation*, **24**, 205–18.

Winstedt, R. O. (1927). The great flood, 1926. *Journal of the Malayan Branch of the Royal Asiatic Society*, **5**, 295–309.

Winstedt, R. O. (1961). *The Malays: a cultural history* (6th edn). Routledge and Kegan Paul, London.

Winterbottom, R. and Hazlewood, P. T. (1987). Agroforestry and sustainable development: making the connection. *Ambio*, **16**, 100–10.

Winzeler, R. L. (1988). Ethnic groups and the control of natural resources in Kelantan, Malaysia. In *Ethnic diversity and the control of natural resources in Southeast Asia* (ed. A. T. Rambo, K. Gillogly, and K. L. Hutterer), pp. 83–97. Michigan Papers on South and Southeast Asia, No. 32. Center for South and Southeast Asian Studies, University of Michigan, Ann Arbor, Michigan.

Wolf, E. C. (1986). *Beyond the green revolution: new approaches for Third World agriculture*, Worldwatch Paper 73. Worldwatch Institute, Washington, DC.

Wong, C. S. (1978). Atmospheric input of carbon dioxide from burning wood. *Science*, **200** (4338). 197–9.

Wong, F. O. (1973). A study of the growth of the main commercial species in the Segaliud–Lokan F. R. Sandakan, Sabah. *Malaysian Forester*, **36**, 95–112.

Wong, Lin Ken (1965). *The Malayan tin industry to 1914.* Association for Asian Studies Monographs and Papers, No. 14. University of Arizona Press, Tucson, Arizona.

Wong, Y. K. and Whitmore, T. C. (1970). On the influence of soil properties on species distribution in a Malayan lowland dipterocarp rain forest. *Malayan Forester*, **33**, 42–54.

Woods, P. (1989). Effects of logging, drought, and fire on structure and composition of tropical forests in Sabah, Malaysia. *Biotropica*, **21**, 290–8.

Woodwell, G. M. (1978). The carbon dioxide question. *Scientific American*, **238** (1), 34–43.

Woon, W. C. and Lim, H. F. (1990). The non-government organizations and government policy on environmental issues in Malaysia. *Wallaceana*, **59** and **60**, 10–15.

WWFM, World Wildlife Fund Malaysia (1985). Proposals for a conservation strategy for Sarawak. Kuala Lumpur.

WWFNM, World Wide Fund for Nature Malaysia (1990). Conservation strategy Malaysia: briefing paper. Kuala Lumpur.

Wray, L. (1894). *Alluvial tin prospecting.* Government Printing Office, Taiping.

Wright, A. and Reid, T. H. (1912). *The Malay peninsula: a record of British progress in the Middle East.* Unwin, London.

Wyatt-Smith, J. (1949). A note on tropical lowland evergreen rain-forest in Malaya. *Malayan Forester*, **12**, 58–64.

Wyatt-Smith, J. (1950). Virgin jungle reserves. *Malayan Forester*, **13**, 40–5.

Wyatt-Smith, J. (1954). Storm forest in Kelantan. *Malayan Forester*, **17**, 5–11.

Wyatt-Smith, J. (1958a). Shifting cultivation in Malaya. *Malayan Forester*, **21**, 139–54.

Wyatt-Smith J. (1958b). Report for the Federation of Malaya. In *Study of tropical vegetation: proceedings of the Kandy Symposium* (Kandy, Ceylon, 19–21 March 1956), pp. 40–2. Unesco, Paris.

Wyatt-Smith, J. (1959). Peat swamp forest in Malaya. *Malayan Forester*, **22**, 5–32.

Wyatt-Smith, J. (1961a). A note on the fresh-water swamp, lowland and hill forest types of Malaya. *Malayan Forester*, **24**, 110–21.

Wyatt-Smith, J. (1961b). A review of Malayan silviculture to-day. *Malayan Forester*, **24**, 5–18.

Wyatt-Smith, J. (1961c). The Malayan forest department and conservation. In *Nature conservation in western Malaysia, 1961* (ed. J. Wyatt-Smith and P. R. Wycherley), pp. 37–42. Malayan Nature Society, Kuala Lumpur.

Wyatt-Smith, J. (1963). Manual of Malayan silviculture for inland forests. *Malayan Forest Records*, **23**.

Wyatt-Smith, J. (1964). A preliminary vegetation map of Malaya with descriptions of the vegetation types. *Journal of Tropical Geography*, **18**, 200–13.

Wyatt-Smith, J. (1987). Problems and prospects for natural management of tropical moist forests. In *Natural management of tropical moist forests: silvicultural and management prospects of sustained utilization* (ed. F. Mergen and J. R. Vincent), pp. 5–22. Yale University, School of Forestry and Environmental Studies, New Haven, Connecticut.

Wyatt-Smith, J. and Foenander, E. C. (1962). Damage to regeneration as a result of logging. *Malayan*

Forester, **25**, 40–4.

Wyatt-Smith, J. and Vincent, A. J. (1962). Progressive development in the management of tropical lowland evergreen rain forest and mangrove forest in Malaya. *Malayan Forester*, **25**, 199–223.

Wyatt-Smith, J. and Wycherley, P. R. (eds.) (1961). *Nature conservation in western Malaysia, 1961*. Malayan Nature Society, Kuala Lumpur.

Yip, Yat Hoong (1969). *The development of the tin mining industry of Malaya*. University of Malaya Press, Kuala Lumpur.

Yoda, K. (1978). Organic carbon, nitrogen and mineral nutrient stocks in the soils of Pasoh forest. *Malayan Nature Journal*, **30**, 229–51.

Yoneda, T., Yoda, K., and Kira, T. (1978). Accumulation and decomposition of wood litter in Pasoh forest. *Malayan Nature Journal*, **30**, 381–9.

Zaharah bte. Haji Mahmud (1966). Moving frontiers of agricultural settlement in 18th and 19th century Kedah. *Geographica*, **2**, 33–8.

Zimmermann, E. W. (1933). *World resources and industries: a functional appraisal of the availability of agricultural and industrial resources*. Harper, New York.

Zulkifli Zainal (1983). A preliminary observation on the effects of development and logging on the numbers, distribution and movement of seladang in Ulu Lepar, Pahang. *Journal of Wildlife and Parks*, **2**, 145–52.

AUTHOR INDEX

SUBJECT INDEX